INTRODUCTION TO
SOLAR RADIO ASTRONOMY AND RADIO PHYSICS

GEOPHYSICS AND ASTROPHYSICS MONOGRAPHS

AN INTERNATIONAL SERIES OF FUNDAMENTAL TEXTBOOKS

VOLUME 16

INTRODUCTION TO SOLAR RADIO ASTRONOMY AND RADIO PHYSICS

by

A. KRÜGER

Akademie der Wissenschaften der D.D.R., Zentralinstitut für solar-terrestrische Physik (Heinrich-Hertz-Institut)
1199 Berlin-Adlershof

D. REIDEL PUBLISHING COMPANY

DORDRECHT : HOLLAND / BOSTON : U.SA.

LONDON : ENGLAND

Library of Congress Cataloging in Publication Data

Krüger, Albrecht.
Introduction to solar radio astronononomy and radio physics.

(Geophysics and astrophysics monographs; v. 16)
Bibliography: p.
Includes index.
1. Sun. 2. Radio astronomy. 3. Cosmic physics. I. Title. II. Series.
QB539.R3K78 523.7′2 79–4270
ISBN 90–277–0957–2 (Hardback)
ISBN 90–277–0997–1 (Paperback)

Published by D. Reidel Publishing Company,
P.O. Box 17, Dordrecht, Holland

Sold and distributed in the U.S.A., Canada, and Mexico
by D. Reidel Publishing Company, Inc.
Lincoln Building, 160 Old Derby Street, Hingham,
Mass. 02043, U.S.A.

Printed in The Netherlands

To my family
for neglecting their own wishes
in favor of the present work.

TABLE OF CONTENTS

'The use of books is endless . . .'
Ecclesiastes, 12, 12.

PREFACE

The plan for the present volume resulted from a desire to have a concise up-to-date survey of the whole field of solar radio physics. According to the frame of the series *Geophysics and Astrophysics Monographs* the volume should contain an extract of basic methods, essential results, and most interesting problems in a form which will not rapidly become obsolete. The book is intended to represent the elementary knowledge concerning solar radio astronomy and to cover a wide range of interests for scientists from various fields.

Since the appearance of the famous monographs on Solar Radio Astronomy by Kundu (1964/65) and Zheleznyakov (1964) a new 11 yr solar cycle has passed by and rapid progress is to be noted regarding e.g. radioheliographic observations, the detection of spectral fine-structure effects, various kinds of extraterrestrial observations, the opening of a new stage of plasma physics, and the successive development of global aspects in solar physics. In such a way solar radio physics has grown to be an extended field, making it rather difficult to review by a single person. Beyond this, other rapidly growing disciplines of solar research (Gamma-, X-ray, EUV-, optical, and IR-astronomy) provide a multitude of information demanding a wider scope and stressing the necessity to construct a summary picture, putting together the radio evidence of important physical processes on the Sun (e.g. at the flare-burst complex) with effects from other spectral regions. Thus, solar radio astronomy can be hardly considered as an isolated branch. Moreover, the field is not only coupled with different branches in solar physics but it is also linked with various other disciplines like geophysics, solar-terrestrial physics, plasma physics, radio sciences, etc.

The book is divided into five chapters. Chapter I contains a brief general introduction to the matter consisting of the history of solar radio astronomy and some fundamentals of astronomy and solar physics necessary for an understanding of solar radio physics. Chapter II presents some topics of the instrumental background of solar radio astronomy and Chapter III gives the main results of observations. Chapter IV comprises the elements of a theoretical interpretation of solar radio observations and Chapter V is aimed to outline a synthesis of both observation and theory contributing to a general picture of solar and solar-terrestrial physics.

Considering the above, the present book can be understood mainly as a consideration of *solar physics* from the point of view of the results and activities of radio astronomy. With respect to the size of this task (requiring e.g. the 'digestion' of a huge amount of literature and personal information, a solid knowledge of the language and a sound optimism), it is hoped that the reader will be soon enabled to distinguish between the inevitably subjective form of the representation and the part of objective scientific content behind it.

A. KRÜGER

Berlin, March 1979

ACKNOWLEDGMENTS

The present book would not have been finished without the help of many friends and colleagues. The author is indebted to Drs. A. Boischot, Ø. Elgarøy, V. V. Fomichev, O. Hachenberg, E. I. Mogilevskij, Y. N. Parijskij, K. V. Sheridan, C. Slottje, S. F. Smerd, and J. P. Wild who kindly supplied glossy photographs. Furthermore it is a pleasant duty to acknowledge the kind permission to reproduce figures and tables of a number of authors, particularly of Drs. H. Alvarez, E. J. Blum, F. W. Crawford, N. Dunckel, J. Eddy, F. Haddock, M. A. Heald, J. Hildebrandt, V. N. Tsytovich, A. Kai, H. Künzel, M. R. Kundu, D. J. McLean, G. Newkirk, Jr., A. M. Peterson, R. Ramaty, A. C. Riddle, K. Sakurai, P. H. Scherrer, R. G. Stone, S. I. Syrovatskij, T. Takakura, H. Tanaka, J. van Nieuwkoop, J. P. Wild, and also publishing houses, among them J. Wiley and Sons Inc. and the D. Reidel Publishing Co., who provided the bulk of reproduced figures.

Thanks are also due to the Academy of Sciences of the G.D.R. for providing me with the opportunity to write this volume and to numerous colleagues both inside and outside the Heinrich Hertz Institute, among them Mrs. Dr. A. Böhme, Prof. Dr. F. W. Jäger, Prof. Dr. L. Mollwo, and Prof. Dr. G. Wallis for a number of suggestions for improving the manuscript, as well as to Mrs. Ch. Kühnlein and the author's wife for inestimable practical help in its preparation. Moreover a number of useful critical hints by the author's colleagues J. Hildebrandt, B. Kliem, and Dr. J. Staude when proof reading different parts of the volume are gratefully mentioned.

Finally it is a pleasure to acknowledge the nice cooperation with the Editor of this series who carefully checked the manuscript and, by numerous valuable advices, has a great share especially in improving the English style. Also the fruitful cooperation with the D. Reidel Publishing Company in all stages of the production of the work is highly acknowledged.

CHAPTER I

INTRODUCTION

1.1. Short History of Solar Radio Astronomy

Since its birth in the forties of our century, solar radio astronomy has grown into an extensive scientific branch comprising a number of quite different topics covering technical sciences, astrophysics, plasma physics, solar-terrestrial physics, and other disciplines. Historically, the story of radio astronomy goes back to the times of James Clerk Maxwell, whose well known phenomenological electromagnetic field equations have become the basis of present-time radio physics. As a direct consequence of these equations, Maxwell was able to prognosticate the existence of radio waves which fifteen years later were experimentally detected by the famous work of Heinrich Hertz (1887/88). However, all attempts to detect radio waves from cosmic objects failed until 1932, which was mainly due to the early stage of development of receiving techniques and the as yet missing knowledge of the existence of a screening ionosphere (which was detected in 1925). Therefore, famous inventors like Thomas Edison and A. E. Kennelly, as well as Sir Oliver Lodge, were unsuccessful in receiving any radio emission from the Sun or other extraterrestrial sources.

Another hindering point was that nobody could *a priori* expect that solar radio emission should have something to do with solar activity so that unfortunately by chance some experiments were carried out just at periods of low solar activity. This was also why Karl Guthe Jansky at the birth of radio astronomy detected galactic radio waves but no emission from the Sun.

So it was not before 1942 (February, 26–28) during World War II, that an English radar station received strongly directed noise signals which at first were believed to be a new kind of jamming by enemy transmitters. But soon it turned out that the strange interferences came from the Sun, evidently connected with the appearance of a large spot group which was present on the solar disk at that time. This remarkable detection was published only some years later (Hey, 1946). Nevertheless, the time was ripe for the observation of solar radio waves so that independently at different stations and in different frequency bands radio noise of solar origin could be identified. Still in 1942, G. C. Southworth, working at the Bell Telephone Laboratories, New York, detected the undisturbed solar microwave emission at three widely spaced wavelengths in the region between 1 and 10 cm; the results were published in 1945. Also Grote Reber, the pioneer of galactic radio astronomy, attempted to see the Sun at 187 cm wavelength and was able to report an actual observation of solar radiation in 1944 (after reporting inconclusive results in 1940 and 1942) so that he was not the first to detect solar radio waves but was the first to publish it. It may be that, in the early forties, noise signals from the Sun were also observed at other places, but due to the war publications were prevented and the basic data were lost (Schott, 1947).

In 1945 after the war a rapid development of solar radio astronomy commenced, initiated by the discovery of the main basic components of solar radio emission: The quiet Sun, the slowly varying component and various types of radio bursts including noise storms. The close connection between intense burst radiations and the flare phenomenon was soon established (Appleton and Hey, 1946).

With regard to the great variability of the solar radio emission the need for a continuous patrol and systematic investigation was recognized, which started in their early beginnings in 1946. At that stage the main progress was achieved by a few centers in Australia, Great Britain, and Canada, where the basic techniques of the observations (flux and polarization measurements, interferometric and spectrographic measurements) were successively worked out.

Later on, particularly stimulated by the enterprises of the International Geophysical Year (July 1957–December 1958) and the International Years of the Quiet Sun (IQSY: 1964/1965) more and more countries took part in the work of solar radio astronomy. After these, besides national programs, a number of international projects incorporating the radio exploration of the Sun were initiated by large scientific communities such as the IAU, IUCSTP–SCOSTP, URSI, COSPAR, and others. After more than one and a half decades of extensive work in the field, the first monographs on solar radio astronomy were published in 1964, independently, by Kundu and Zheleznyakov, providing excellent reviews stressing the observational and theoretical points of view, respectively. Meanwhile after another decade, new investigations e.g. by heliographic and satellite measurements, observations of enhanced time and spectral resolution etc. were accumulated, supplemented by the development of a new generation of plasma theories (Kaplan and Tsytovich, 1972).

1.2. General Views of the Sun

1.2.1. The Sun as a Source of Radio Waves

Due to its proximity to the Earth the Sun plays a very special role among all other cosmic radio sources. But solar radio emission cannot be considered as isolated from general solar physics, which results from the totality of the available radiations covering all spectral ranges. In this way, the solar radio physicist has to be aware of the achievements of e.g., optical solar astronomy which, due to its long history, the easy access by ground-based observations, the facilities of obtaining angular resolution, and the high content of information by line emissions makes a large contribution to solar physics. Moreover, solar radio and optical emissions are nicely complementing each other.

The greatest part of the optical radiation comes from the photosphere, the roughly 350 km thick layer between the Sun's interior and the outer atmosphere, whereas the radio emission originates in the plasma of the outer atmospheric layers, the chromosphere and the corona. The propagation characteristics of the radio waves depend basically on the electron density in these layers: Each value of electron density is related to a certain critical frequency of the radio waves below which propagation is impossible. Therefore, and also because of the emission properties of the radio waves, the frequency spectrum can be roughly related to a height scale in the solar

TABLE I.1
General properties of the Sun

Radius (photosphere)	$6.959\,97 \times 10^{10}$ cm
Mass	1.9892×10^{33} g
Spectral type	G 2 V
Gravitational acceleration (photosphere)	$2.739\,84 \times 10^{4}$ cm s^{-2}
Centrifugal acceleration (photosphere-equator)	-0.587 cm s^{-2}
Rotational energy	2.4×10^{42} erg
Effective temperature (photosphere)	5770 K
Temperature at the center of the Sun	$(13.6 \pm 1.2) \times 10^{6}$ K
Age	4.5×10^{9} yr
Synodic rotation	$13.45° - 3.0° \sin^2 \phi$ per day
Sidereal rotation	$14.44° - 3.0° \sin^2 \phi$ per day (sunspot zone)
Distance to the Earth	1.4710×10^{13} cm (perihelion) 1.5210×10^{13} cm (aphelion) $1.495\,979\,1 \times 10^{13}$ cm (mean) (1 AU)
Solar constant	1.360×10^{6} erg cm^{-2} s^{-1}

atmosphere if the density distribution is known. Since the density decreases with increasing height in the solar atmosphere one can see with higher frequencies into lower levels, i.e., closer to the photosphere. However, before discussing the radio features in more detail, some general characteristics of the solar atmosphere will be recalled to mind, being necessary for a better understanding of solar radio physics. Beyond that, some numerical data characterizing the Sun are compiled in Table I.1.

1.2.2. The Solar Atmosphere

a. *Photosphere*

As it is well known, the photosphere is the surface of the Sun seen in white light. At present the photosphere can be regarded as the best investigated part of the Sun. Physically the photosphere is characterized by a steep gradient of particle density and a minimum of the temperature. The high opacity of the photosphere prevents viewing into the subphotospheric layers. From the photosphere the energetically largest part of the solar radiation emerges represented by the optical continuum which is superimposed by tens of thousands of Fraunhofer absorption lines. As a consequence of the opacity and the number of degrees of freedom radiative equilibrium predominates in the photosphere.

The most noticeable type of structure of the photosphere consists of a cellular pattern visible as the granulation which is intimately connected with the *convective zone* lying beneath the photosphere. The single granules representing individual convective cells have a size of the order of 1000 km and lifetimes of about 5 min. The competition between radiative and convective equilibrium is controlled by the Schwarzschild criterion, according to which convection operates if

$$(\mathrm{d}T/\mathrm{d}h)_{\mathrm{rad}} > (\mathrm{d}T/\mathrm{d}h)_{\mathrm{ad}} \tag{I.1}$$

i.e., if the vertical radiative temperature gradient exceeds the adiabatic one, otherwise the energy transport is entirely determined by radiation.

However, it should be mentioned that, indicated by the existence of a hierarchy of granulation structures (subgranules – granules – supergranules – giant granules – supergiant structures), the real conditions are certainly much more complicated than those described in criterion (I.1).

There are several models describing the undisturbed photosphere and some parts of the chromosphere. Here we note the Harvard-Smithsonian-Reference-Atmosphere which replaced the older Bilderberg-Continuum-Atmosphere (Gingerich *et al.*, 1971) and the improvements attained by the model of Vernazza *et al.* (1973, 1976).

b. *Chromosphere*

The chromosphere can be regarded as a transition zone of some thousands km thickness between the cooler, almost neutral photosphere and the hot, thin plasma of the corona. It is of direct interest for radio waves in the mm and cm region. Due to the increase of temperature and decrease of density with height the ionization of the particles becomes noticeable according to Boltzmann's formula

$$(N_{r,s}/N_r) = (g_{r,s}/g_r)\exp(-\chi_{r,s}/KT) \tag{I.2}$$

and the Saha equation

$$(N_{r+1}/N_r)\,P_e = 2(u_{r+1}/u_r)(2\pi m)^{3/2}\,h^{-3}(KT)^{5/2}\exp(-\chi_r/KT), \tag{I.3}$$

where

$$u_r(T) = \sum_{s=0}^{\infty} g_{r,s}\exp(-\chi_{r,s}/KT)$$

is the partition function of an r-times ionized atom, $N_{r,s}$ – number per cm^3 of r-times ionized atoms of the sth quantum state; $P_e = N_e KT$, $g_{r,s}$ – statistical weight (number of quantum states); and $\chi_{r,s}$ – ionization energy. However, these formulas are valid for (local) thermal equilibrium only. With increasing height above the photosphere, deviations from this condition become remarkable.

The chromosphere is far from being homogeneous and exhibits a multitude of structural elements. A striking feature is the inhomogeneous boundary towards the corona, known as spicular structure or fine mottling. The mottles or spicules represent relatively cold and dense elements embedded in coronal matter. Their characteristics are: Temperature – about (1–4) $\times 10^4$ K; diameter – 500–1000 km; lifetime – several minutes; height range – about 5000–10000 km above the photosphere. Though the spicules occupy only about 1 % of the solar surface, they show a remarkable influence, e.g., on radio observations of the quiet Sun in the microwave region. Special chromospheric models are based on UV, optical, and radio measurements taking into account the existence of hot and cold elements also for the lower chromosphere.

Other chromospheric structures are represented by the coarse mottling (dark elements of 2000–6000 km diameter seen e.g. in the K-line of Ca^+) or flocculi (bright patches in Ca^+, which in active regions are passing over into plages) and the coarse network which corresponds to the supergranulation.

Wave motions play an important role in the energy transport of the chromosphere. Evidence is given e.g. by the occurrence of the 5 min oscillations in the photosphere and chromosphere.

Different aspects of the physics of the solar atmosphere and particularly the chromosphere are considered in detail in several textbooks (cf. e.g. Zirin, 1966; Gibson, 1973; Athay, 1976). Besides this, there are special symposia devoted to chromospheric features (e.g. Kiepenheuer, 1966; Athay and Newkirk, 1969; Athay, 1974).

c. *Corona*

In striking contrast to the relatively thin layers of the photosphere and chromosphere, the corona is much more extended and does not show a fixed, well-defined boundary towards the interplanetary space. The most spectacular feature is the steep gradient of the temperature at the bottom of the corona caused by a drastic dissipation of mechanical shock-wave disturbances traversing the chromosphere and being generated in the convection zone below the photosphere. Hence the transition region between the chromosphere and corona is of particular interest. At heights of about 2000 to 5000 km above the photosphere the temperature rises from 10 000 to more than 1 000 000 degrees and the density drops simultaneously by two orders of magnitude. Though the interesting mechanism of coronal heating is not yet fully clear in all details, it is reasonable to assume that a strong dissipation of wave energy takes place, if, favored by the density profile of the traversed medium, the Mach number increases leading to shock waves. Then the dissipation occurs at relaxation paths of the order of the scale height

$$h = (\mathrm{d}\ln N/\mathrm{d}s)^{-1}. \qquad (1.4)$$

Moreover, in order to get heating it must be required that the amount of dissipated energy overcomes possible energy losses, which may be caused by radiation, heat conduction or convection. The realization of these conditions will account for the steep gradient of the coronal temperature.

The dimensions, shape, and structure of the corona as observed in visible light during solar eclipses depend strongly on the solar cycle, thus indicating variations of density, temperature, and other plasma parameters with solar activity.

From optical observations three components of coronal emission can be distinguished:

(1) The L-corona (line emissions from highly ionized atoms);

(2) The K-corona (partially polarized continuous emission consisting of photospheric light scattered at free electrons); and

(3) The F-corona (presenting absorption lines of the photospheric Fraunhofer spectrum caused by diffraction from interplanetary dust).

The emission of the L-corona yields only 1% of the emission of both the other components forming the 'white-light corona', which in turn delivers only about 10^{-6} of the Sun's total optical brightness.

From photometric measurements, especially of the K-corona electron densities can be estimated. Thus the resulting density profiles are represented on the average by the Baumbach–Allen formula

$$N_e(R) = 10^8[0.036(R/R_\odot)^{-1.5} + 1.55(R/R_\odot)^{-6} + \\ + 2.99(R/R_\odot)^{-16}] \qquad [\mathrm{cm}^{-3}]. \tag{I.5}$$

Though the corona does not exhibit as many different structural elements as the underlying chromosphere, great inhomogeneities and departures from the spherical symmetry postulated in Equation (I.5) do occur. As a matter of fact, it is nearly impossible to separate the structure of the corona from the appearance of solar activity. At high latitudes only, polar plumes and brushes indicate a quiet solar dipole field of the order of about 1 to 2 G, which very probably changes its polarity with the solar cycle.

Further typical structures are connected with the appearance of coronal streamers and rays extending up to several solar radii outwards, which evidently mark magnetic fields and paths of particles leaving the Sun. Recently, three-dimensional maps of the coronal electron density distribution have been computed from K-coronameter data tracing the structure of a streamer as a function of height and determining its nonradial orientation (Perry and Altschuler, 1973).

Besides the optical investigations, the exploration of the solar corona is most effective using the X-ray emission and the radio emission as well as the solar wind. In particular, photographs of the Sun in the soft X-ray range delivered by the Skylab mission have evidenced the existence of coronal holes and bright points.

1.2.3. Solar Activity

a. *General Phenomena*

The solar atmosphere is permanently influenced by a multitude of processes caused by solar activity. Even in periods of minimum activity the undisturbed 'quiet Sun' appears only as an idealized limiting case. Solar activity is rooted in the occurrence of more or less clearly localized active regions. An *active region* consists of the totality of numerous quite different phenomena and processes exhibiting different scales in time and space. As a rough outline some main phenomena of the quiet and active Sun are listed together in Table I.2. A major feature of solar activity is its periodicity in the 11 (22) yr solar cycle.

Solar activity strongly affects the whole picture of the Sun in different spectral regions revealing a wealth of special physical effects influencing the interplanetary medium and the Earth. In particular most of the features of solar radio emission are directly related to solar activity. In the following some aspects of solar activity will be discussed.

b. *Sunspots*

Sunspots are the most popular phenomenon of solar activity and have been observed on a regular basis since the middle of the 18th century, but naked-eye observations can be dated back to antiquity. The outstanding physical property of sunspots

TABLE I.2
Main phenomena of the quiet and active Sun

Phenomenon Layer	Stationary (quiet Sun)	Quasi-stationary (slowly varying active Sun)	Instationary (fast changing active Sun)
Photosphere	Granules (also sub-, super-, and super-giant granulation)	Spots, plages (faculae), Emerging flux regions (EFR)	Bright points, white light flares
Chromosphere	Spicules, flocculi, fine and coarse mottling, coarse network	Plages, evolving magnetic regions (EMR) X- and radio-'plages'	Moustaches, flares in opt. and UV, umbral flashes, X- and γ-ray umbral flashes, bursts,
Corona	Bushels,	'plages' (S-component),	
	(plumes), streamers,	coronal condensations,	eruptive prominences, sprays, surges, disparison brusque,
	quiet-Sun electromagnetic radiation, coronal holes,	quiescent and spot prominences (filaments),	
	solar wind	magnetic arch systems, helmets, R-rays, enhanced solar wind	various types of radio bursts, shock fronts, particle clouds, coronal transients, cosmic and subcosmic radiation

is their magnetic field which ranges in the central parts of the umbrae between about 1000 and 4000 G. The magnetic fields can be regarded as a huge accumulation of energy (up to 10^{34} erg for big spot groups) controlling physical processes inside active regions not only in the photosphere, where the spots are visible, but also to greater heights of the solar atmosphere.

Models of active regions have to involve sunspots as a central part. At photospheric level, sunspots are about 2000 K cooler than the surrounding matter, the extrapolation into higher (where an inversion of the horizontal temperature distribution takes place) and lower levels, however, is still somewhat uncertain. Typical dimensions of sunspot umbrae and penumbrae are about 2000 to 20 000 km and 5000 to 50 000 km, respectively, but the E–W extension of complex spot groups shows maximum values of about 300000 km, i.e., about 25 Earth diameters.

Special features of sunspots are:

The Wilson effect (depression of the solar surface in the spot region); and

The Evershed effect (inward and outward motion in the chromospheric and photospheric penumbral region, respectively).

According to their appearance spots or groups of unipolar, bipolar, and complex form can be distinguished. A classification scheme regarding the magnetic (Mt. Wilson) types and the evolutional (Zürich) types is shown in Figure I.1. The distribution of sunspots over the solar disk is described statistically by Spoerer's law (butterfly diagram, latitude drift of the spot positions towards the equator during the solar cycle). Also the magnetic polarity of the sunspots is closely related to the solar cycle: The opposite polarities of the leading and following spots depending on their

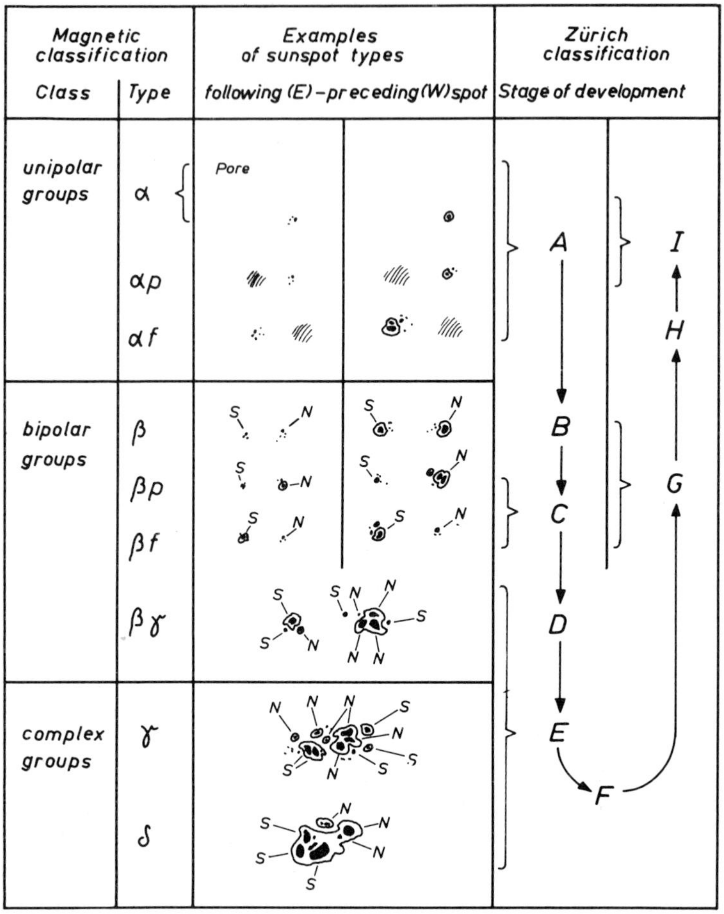

Magnetic classification: Class	Magnetic classification: Type	Examples of sunspot types: following (E) – preceding (W) spot	Zürich classification: Stage of development
unipolar groups	α	Pore	A
	αp		
	αf		
bipolar groups	β	S N	B
	βp	S N	C
	βf	S N	
	βγ	S N S N S N N	D
complex groups	γ	N N N S N S S N S	E
	δ	N S N S N S	

Zürich stages of development: A → B → C → D → E → F → G → H → I

Fig. I.1. Scheme of classification of sunspot groups. W is right on the spot pictures. 'N' and 'S' indicate north and south polarities, respectively. The shaded areas represent K faculae. (Courtesy of H. Künzel).

position on the northern or southern solar hemisphere are reversed after each 11 yr cycle. The magnetic features have strong consequences on the polarization characteristics of the solar radio waves (cf. Table I.3).

For comparison with the radio emission as well as other phenomena integrated sunspot areas or sunspot numbers have been used as a measure of solar activity. The sunspot number (Wolf number or sunspot relative number) is defined by

$$R = k(10g + f), \tag{I.6}$$

where f is the total number of spots on the visible disk (irrespective of their size); g the number of spot groups; and k an individual reduction coefficient depending on the observer and observing instrument. Yearly averages of sunspot numbers exhibiting the periodic nature of solar activity and the so-called Maunder-minimum during the 17th century are reproduced in Figure I.2.

TABLE I.3
Sunspot polarities and orientation of circular polarization of radio waves

	Solar cycle number				Source position on solar hemisphere
	odd (e.g. No. 19 1954–1964)		even (e.g. No. 20 1965–1976)		
Leading spot	+ (north)		− (south)		northern
polarity	− (south)		+ (north)		southern
Sense of polarization	right	left	right	left	—
Corresponding cold plasma wave mode	e	o	o	e	northern
	o	e	e	o	southern

e—extraordinary
o—ordinary.

c. *Plages*

Plages or faculae are bright, structured regions which are detectable in monochromatic light (Hα, Lα, K-line of Ca, HeI, HeII, ...) at chromospheric heights. In integrated light they can be detected close to the solar limb also at the photospheric level. Plages represent hotter parts in the solar atmosphere surrounding sunspots but having weaker magnetic fields ($\lesssim 10$ G) and greater lifetimes than spots. Like sunspots the plages are well correlated with slowly varying radio and X-ray emissions indicating an extension of the plage regions into greater heights. For that reason sometimes also the names 'radio plages' and 'X-ray plages' are used in an attempt to generalize the plage phenomenon. The regions above the optical plages, which are characterized by higher densities and temperatures in comparison to the surrounding atmosphere and being detectable in radio (S-component), soft X-ray, and optical line emissions, are sometimes called coronal condensations.

Plages provide the most extensive (in time and area) phenomenon of solar activity in the lower solar atmosphere. Like sunspots (and all other phenomena of solar activity) plages are situated in two zones parallel to the solar equator at latitudes smaller than 45°, but apart from these zones also quiet-Sun polar faculae occur at latitudes higher than 70° which are a little fainter and shorter in duration.

d. *Flares*

Flares are the most active manifestation of solar activity and most important for solar radio astronomy. In optical light the flares are observed as sudden and short-lived brightenings of plage regions in the neighborhood of sunspots ('chromospheric flares'). They are displayed in most of the Fraunhofer lines, especially at Hα and the H and K lines of Ca. Only in rare cases do very strong flare events also show an emission in the integrated light ('white-light flare' or 'integral eruption'). The

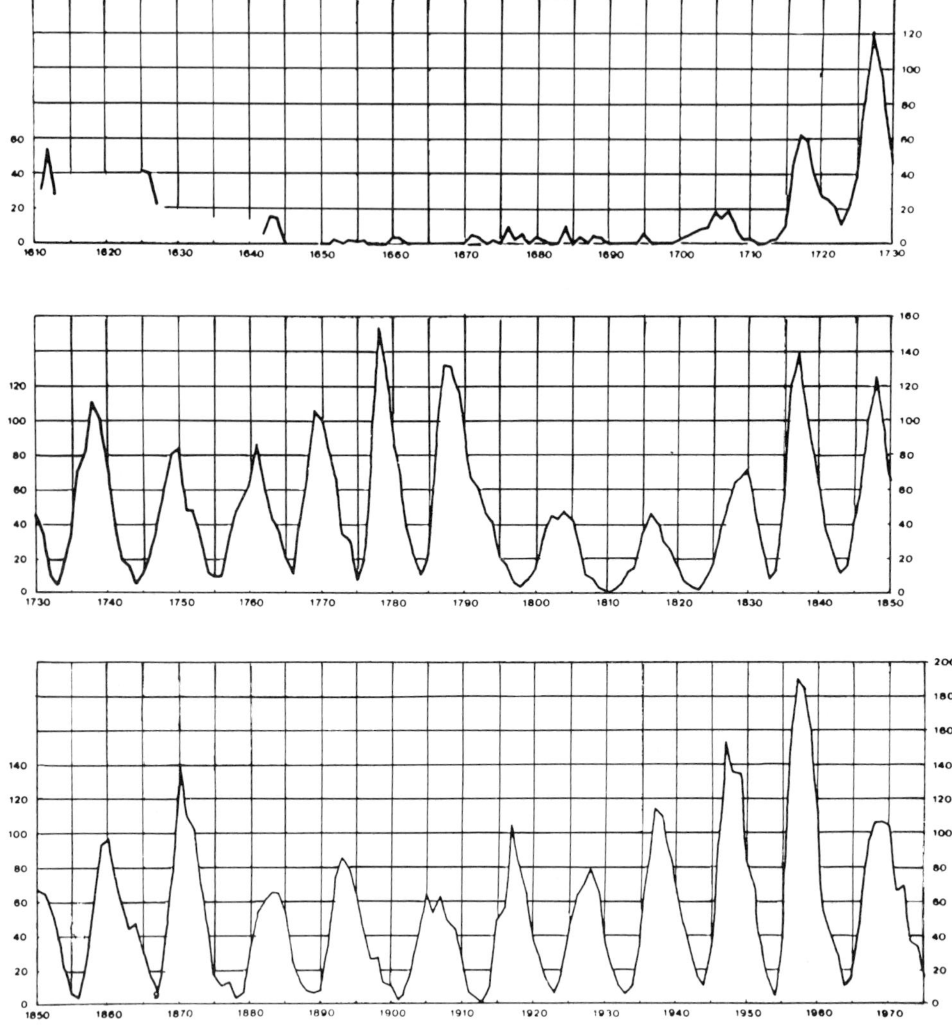

Fig. I.2 Yearly sunspot numbers since 1610 (after Eddy, 1976b).

first reported flare was observed in 1859 in white light by Carrington. Literature on flares has been condensed in special monographs and symposia (Smith and Smith, 1963; Hess, 1964; De Jager and Švestka, 1969; Švestka, 1976).

In periods of high activity solar flares are relatively frequent (more than 1 to 2 h^{-1} at maximum). Obviously flares form a necessary and very typical phase in the development of an active region. Solar flares are classified by different importance groups according to their area and brightness (e.g., intensity in the line center of H-α as reproduced in Table I.4.

As will be considered in more detail in Section 5.2, flares constitute a very complex phenomenon. In the optical region the H-line emissions correspond to temperatures

TABLE I.4
Dual flare-importance classification

Corrected flare area [square degrees]	Importance Faint	Normal	Brilliant*
2.0	s f	s n	s b**
2.1 ... 5.1	1 f	1 n	1 b
5.2 ... 12.4	2 f	2 n	2 b
12.5 ... 24.7	3 f	3 n	3 b
24.7	4 f	4 n	4 b

* Relative intensity evaluation
** 'Subflares' (formerly '1 –')

lower than the ionization temperature of H, i.e., $\lesssim 10^4$ K. In spite of this relatively low temperature the efficiency of the energy exchange in the source region is comparatively high and reaches an appreciable fraction of the total energy emitted by flares through line emissions in the optical and UV region. Concerning the flare heights in the solar atmosphere a scatter ranging from levels near the photosphere (white-light flares) up to the lower corona ('coronal flares', cf. Slonim, 1973) is to be noted. Beside the optical flare emissions, the multitude of radio and X-ray components, as well as the particle emissions accompanying flare events indicate that, in sufficiently strong events, all levels of the solar atmosphere out to interplanetary space are violently disturbed. Therefore, in a wider sense, the term 'flare phenomenon' (or also 'flare-burst phenomenon') is used to describe the totality of all components occurring on the Sun in connection with solar flares.

The main problem connected with the investigation of solar flares at present (the 'flare problem') consists of finding out the physical mechanism; i.e., in building up a consistent model describing the main flare processes. The flare problem is still one of the great puzzles of cosmic physics at our time. The main difficulties of the interpretation of solar flares arise especially from the magnitude (up to 10^{33} erg) and suddenness (~ 100 s) of the energy exchange (cf. Section 5.6.1).

It is to be mentioned that the existence of flare activity is also known to exist on other stars (*flare stars*). From simultaneous observations in the optical and radio ranges it may be concluded that the flare phenomenon is universal in nature. A monograph on flare stars was written e.g. by Gurzadyan (1968) (cf. for reference: Gurzadyan, 1971, 1972).

e. *Prominences*

Beside sunspots, plages, and flares, the filaments or prominences belong to the most conspicuous local phenomena of the visible Sun. They appear either as bright structures ('prominences') beyond the solar limb (usually below 10 000 to 30 000 km height above the photosphere) or as dark filaments on the disk representing matter of about 10^4 K condensed in the surrounding hotter corona. The observation of such prominences provides one of the rare opportunities of getting information of mass motions in the corona which is also of great interest for radio physics.

There are rather numerous types of prominences and different classifications have been worked out. According to Zirin (1966) the following main types may be distinguished:

(1) Short-lived, active prominences (surges, sprays, loops, LPS – loop-prominence systems;

(2) Intermediate prominences (sunspots filaments, AFS – arch-filament systems, eruptive prominences);

(3) Long-lived, quiescent prominences (normal and polar prominences outside active regions).

Characterizing these different types briefly, it may be noted that *surges* are flare-associated eruptions of matter into a restricted solid angle sometimes returning to the direction of the solar surface. *Sprays* are irregular flare-associated eruptions into a larger solid angle, perhaps they may be considered as a visible expression of the 'explosive' flare phase. The arch-filament systems mark closed magnetic field patterns of young, growing spot groups. They exhibit mass motions downward at both sides from the top. The flare-initiated loop-prominence systems show, in some respect, similar behavior, but they are quite different in many other respects, e.g. occurrence and development. Eruptive prominences are pre-existing filaments which suddenly erupt outward. Special interesting phenomena besides sudden filament activations are sudden disappearances of filaments (disparition brusque) which often follow an active stage.

Quiescent prominences are out of the direct scope of our present considerations so far as they are not immediately related to solar activity. A textbook on solar prominences was published by Tandberg-Hanssen (1974).

f. *Active Longitudes, Sector Structure and Coronal Holes*

Active regions are not uniformly distributed on the solar disk even with respect to longitude. In close relation to the large-scale sector structure of the interplanetary and photospheric magnetic field (cf. Section 3.2.6), a statistical preference of major active regions near the sector boundaries ($\Delta\lambda < 50°$) can be stated. In particular zones of active longitudes separated by about 180° (and also by 60°) appear to be maintained over periods of the order of an 11 yr solar cycle.

Sector boundaries can be considered as a manifestation of large-scale neutral (warped current) sheets in the solar corona. The open–closed magnetic field configuration is not only an indication of the field structure itself but it has also important consequences for the distribution of other physical parameters such as plasma density, pressure, and temperature. A tabulation of extraterrestrially observed, well-defined sector boundaries during several years since 1960 was published by Wilcox (1975).

Coronal holes, i.e. extended regions of reduced temperature and density, are clearly displayed by the Skylab soft X-ray heliograms. They can be regarded as the counterpart to solar activity. Evidently coronal holes are characterized by open magnetic field structures which favor the outflow of the solar wind (cf. Section 5.4.1). Two major types of coronal holes can be distinguished: Low-latitude isolated regions and holes of more global extension expanding in north-south direction from one polar region through a channel of filaments into the opposite hemisphere (Timothy *et. al.*, 1975).

1.3. Some Astronomical Fundamentals

1.3.1. SUN–EARTH DISTANCE

According to Kepler's first law, the Earth's orbit around the Sun is an ellipse with the Sun at one of its foci. Though the eccentricity of the ellipse ($e = 0.01673$) is very small, the difference between the shortest Sun–Earth distance R_p (Earth at perihelion) and the longest one R_a (Earth at aphelion) amounts about 5×10^6 km. The consequence is a seasonal variation of the solar flux (corresponding to an apparent variation of the solar diameter) of about a maximum of $\pm 3\%$. For this reason radio flux measurements are often adjusted to a constant Sun–Earth distance of 1 AU:

$$a = (R_p + R_a)/2 = R_p/(1 - e) = R_a/(1 + e) \approx 149.5 \times 10^6 \text{ km}. \qquad (I.7)$$

Corresponding reduction factors (averages) are compiled in Table I.5. Exact values for every particular year can be found in the Nautical Almanac and other sources.

1.3.2. COORDINATE SYSTEMS

a. *Horizontal System*

Two coordinate systems can be used to determine the position of the Sun or a star on the celestial sphere, the horizontal system and the equatorial system. In the

TABLE I.5

Average reduction factors k compensating the seasonal variation of solar fluxes S/S_{AU} and apparent radius of the solar disk R due to the eccentricity of the Earth orbit (without allowance of possible atmospheric attenuation effects)

Date	k	S/S_{AU}	$R_\odot/R_{\odot,AU}$
Jan 01	0.9669	1.034	0.9833
Feb 01	0.9708	1.030	0.9853
Mar 01	0.9821	1.018	0.9910
Apr 01	0.9990	1.001	0.9995
May 01	1.0157	0.985	1.0078
Jun 01	1.0286	0.972	1.0142
Jul 01	1.0337	0.967	1.0167
Aug 01	1.0300	0.971	1.0149
Sep 01	1.0183	0.982	1.0091
Oct 01	1.0020	0.998	1.0010
Nov 01	0.9847	1.016	0.9923
Dec 01	0.9720	1.029	0.9859

horizontal system spherical coordinates are specified by the zenith distance angle z (or also by its complement, the altitude h) and the azimuth angle α counted from the south direction. A certain disadvantage of this system, however, is its dependence on the particular position of the observer on the Earth and on the observing time.

b. *Equatorial System*

The above mentioned inconvenience is overcome by the equatorial system which has its polar axis parallel to the Earth's rotational axis. The spherical coordinates are specified by the polar distance p (or its complement, the declination δ) and the right ascension RA defined as the angular distance of the hour circle from the vernal equinox measured in westward direction in degrees or hours. (The vernal equinox is the point of intersection of the celestial equator and the ecliptic, i.e. of the basic circle perpendicular to the Earth's rotational axis and to the plane of the Earth's orbit.)

Due to its apparent path along the ecliptic (zodiac), the declination and right ascension of the Sun change slowly during a year. The time between two subsequent meridian passages of the Sun is called solar day. We have

$$\begin{aligned}&365.24220 \text{ mean solar days} = 366.24216 \text{ sidereal days, or}\\&1 \text{ mean solar day} = 24 \text{ h } 3 \text{ m } 56.556 \text{ s sidereal time.}\end{aligned} \tag{I.8}$$

For estimations of the position of the Sun it must be further taken into account, that the real solar time does not run uniformly through a year, i.e., in general the mean solar time is different from the true solar time. This occurs for two reasons; the first is of geometrical nature: As seen from the Earth the true Sun moves along the ecliptic; the mean Sun, however, is moving along its projection on the celestial equator. Secondly, according to Kepler's second law, the movement on the ecliptic is not uniform.

The uniform solar time is defined as:

$$\text{mean time} = \text{true time} + \text{time difference (`equation of time')}. \tag{I.9}$$

Daily values of the equation of time can, like other data, be obtained from annually issued almanacs and astronomical ephemeris catalogs. For any particular observer the difference between the zonal time and the geographic longitude of the observational site must also be considered.

c. *Solar Coordinates*

The determination of a fixed coordinate system at the solar surface meets with two difficulties:

(i) The Sun does not rotate as a rigid body. The photospheric angular velocity ζ decreases with latitude approximately as $\zeta = \zeta_0 + \text{const} \cdot \sin^2 \phi$ where ζ_0 denotes the value of ζ at the equator. To certain extent ζ also increases with height above the photosphere.

(ii) Permanent well marked features (with the exception of the poles of the rotational axis) are missing. Therefore, solar coordinates were defined by convention according to Carrington's proposal – measuring heliographic longitudes from the solar meridian that passed through the ascending node of the solar equator on the ecliptic on January 1, 1854, Greenwich mean noon.

Carrington's elements are:

Longitude of the ascending node:

$$\Omega = 73.6667 + (a - 1850) \times 0.01396\,[^\circ]$$

$$(a = \text{year of observation}); \tag{I.10}$$

Inclination of the Sun's rotation axis (angle between the equatorial plane of the Sun and the ecliptic):

$$i = 7^\circ\,15';$$

Sidereal rotation period:

$$T_{sid} = 25.380 \text{ days; and}$$

Synodic rotation period:

$$T_{syn} = 27.275 \text{ days.}$$

Introduced by Carrington, serial numbers of (synodic) solar rotations are counted commencing with No. 1 for the period beginning with November 9, 1853. For practical use the heliographic latitude B_0 and longitude L_0 of the center of the solar disk, as well as the position of the poles of the solar rotation axis P measured in an eastward direction from the north point, can be obtained from tabulations.

Maps of disk coordinates with meridians calibrated either in degrees or days from the central meridian passage are used in form of transparencies as distributed e.g. in the supplement issues for the explanation of data reports of the NOAA Solar-Geophysical Data (World Data Center A, Boulder) for each season.

CHAPTER II

INSTRUMENTAL BACKGROUND

2.1. Fundamentals of Radio Observations

2.1.1. The Electromagnetic Spectrum and the Atmospheric Radio Window

Like optical astronomy, the facilities of ground-based radio astronomy depend on the transparency of the Earth's atmosphere in special regions of the electromagnetic wave spectrum. The whole spectrum of electromagnetic waves and their propagation characteristics are illustrated by Figure II.1

The *frequency extent* of the cosmic radio waves accessible to ground-based observations is about 15 MHz to 30 GHz covering more than three orders of magnitude or about eleven octaves. Thus, the radio spectrum is much more extensive than the visible spectral range. This relatively large *radio window* is limited at the low-frequency side by the ionosphere and at the high frequency side by the troposphere.

The plasma frequency of the ionosphere prevents the propagation of electromagnetic waves below a certain frequency limit: For a stratified medium with positive local gradients of the electron density a successive refraction (frequency-dependent) of the radio waves takes place, which at sufficiently high densities leads to a reversal of the original ray direction appearing as a 'reflection'. In the general case of oblique incidence this reversal occurs well before reaching the level of the local plasma frequency cutoff (cf. Section 4.2.4).

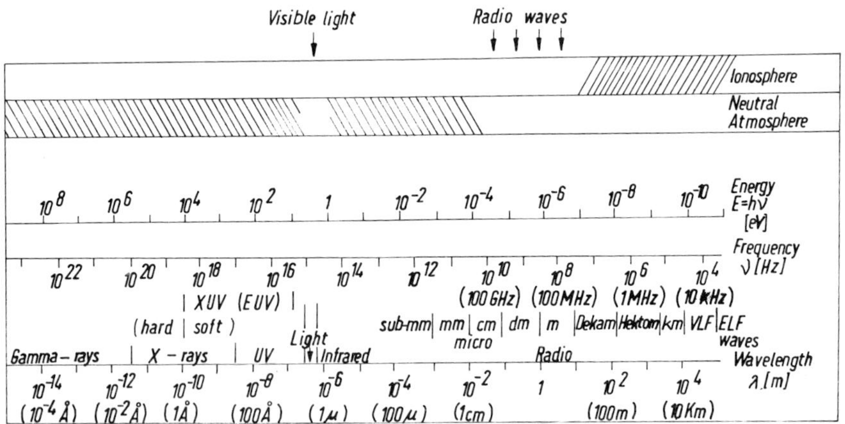

Fig. II.1. The electromagnetic wave spectrum and transparency properties of the Earth's atmosphere (shaded areas refer to opacity).

The ionosphere consists of several layers. The main influence is due to the *F*2 layer – its maximum plasma frequency varies roughly between 6 and 15 MHz, corresponding to 20 to 50 m wavelength, depending on the day time and season (elevation angle of the Sun above the horizon), as well as slightly on solar activity. Besides that, ionospheric absorption effects on radio waves are important, which are most striking in the case of SCNA (sudden cosmic noise absorption) effects due to an excessive *D* layer ionization caused by the incidence of flare-produced radiations in the X-ray spectral region.

The high-frequency limit of the radio window is determined by the influence of the neutral part of the terrestrial atmosphere. Here the main role is played by the troposphere which comprises about 80% of the total mass content of the Earth's atmosphere. Before becoming totally opaque, there is a number of distinct absorption bands caused mainly by the atmospheric water and oxygen content starting at

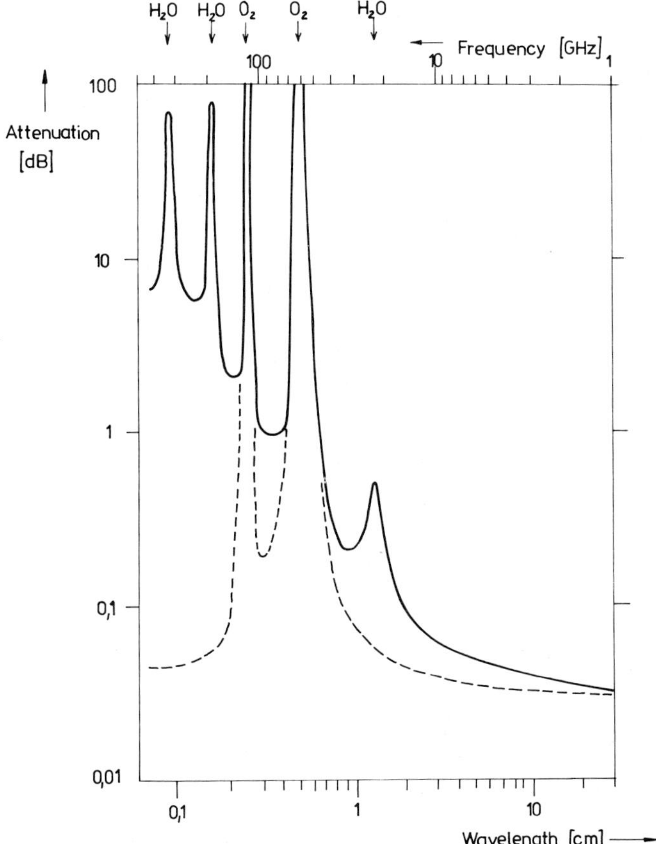

Fig. II.2. Attenuation of radio waves in the clear neutral Earth's atmosphere at vertical incidence (according to CCIR report cited by Blum, 1974). The full line corresponds to a water vapor concentration of 7.5 g m^{-3} (1.5 cm equivalent depth) at the Earth's surface. The dotted line refers to a dry atmosphere.

about 2 cm wavelength. An impression of these influences can be obtained from Figure II.2, which shows the atmospheric attenuation characteristics in the vertical ray direction for two particular cases of water vapor concentration according to Blum (1974). The knowledge of the properties of the absorption minima is especially important for the work of radio astronomy in the mm-wavelength region.

2.1.2. Radiation Quantities

The electromagnetic radiation received by any observer is principally determined by its intensity including the state of polarization. Additionally, information about the spectral dependence, time variations, and the angular distribution of the observed radiation can be obtained. Each measurement is restricted by its characteristic limits of sensitivity which are fixed for any given receiving equipment. In particular we distinguish: (a) the intensity resolution (temperature resolution); (b) the spectral resolution (receiver bandwidth); (c) the time resolution (receiver time constant; and (d) the angular resolution (half-power beamwidth).

The *intensity* (≡ specific intensity, cf. Chandrasekhar, 1960) of the radiation is defined as a scalar quantity determining the amount of energy ΔE_ν ('power') of a radiation field which is transported across a unit area $\Delta\sigma$ per unit solid angle $\Delta\Omega$ declining by an angle ϑ from the observed normal direction of $\Delta\sigma$ in the unit frequency interval $(\nu, \nu + \Delta\nu)$ and time unit Δt $(\Delta t \gg 1/\Delta\nu)$

$$I_\nu(r, \vartheta, \varphi; t) = \Delta E_\nu(r, \vartheta, \varphi; t)/(\Delta\nu\, \Delta\sigma \cos\vartheta\, \Delta\Omega\, \Delta t)\,[\mathrm{erg\,cm^{-2}\,sr^{-1}}]. \qquad \text{(II.1)}$$

This definition generally used in astronomy is not to be confused with another definition of intensity (or radiance) sometimes used in experimental physics by the time-averaged Poynting flux. In order to avoid misunderstandings in radio astronomy the intensity defined by Equation (II.1) is often called *brightness*.

An Integration of I_ν over the solid angle of a source region yields the *flux density*

$$S_\nu = \int_{\Omega_{\mathrm{source}}} I_\nu(\vartheta, \varphi)\, \mathrm{d}\Omega' \qquad \text{(II.2)}$$

which is usually adopted for radiometric measurements of spatially unresolved sources (point sources). For solar radio astronomy, using the mks units,

$$1 \text{ Solar Flux Unit (SFU)} = 1 \times 10^{-22} \quad \mathrm{W\,m^{-2}\,Hz^{-1}}, \qquad \text{(II.3)}$$

has been introduced, which is sometimes also called 1 'Jansky'.

A comparison between the fields of applications of the terms flux density and intensity is given once more in Table II.1.

It may be further remarked that the quantity

$$u_\nu = \int_{4\pi} (I_\nu/v_g)\, \mathrm{d}\Omega' \qquad (\mathrm{erg\,s\,cm^{-3}}) \qquad \text{(II.4)}$$

(v_g = wave group velocity) is called (spectral) energy determining the radiated energy per volume unit.

TABLE II.1
On the use of S_ν and I_ν

Object	*Whole distributed* source	*Single points* of a distributed source
Observed as	*Point source*	Brightness *distribution*
Relevant radiation quantity	S_ν, T_e (flux density, effective temperature)	$I_\nu(\vartheta, \varphi)$, $T_b(\vartheta, \varphi)$ (intensity, brightness temperature)

An integration over the time yields the following expressions

$$I'_\nu = \int I_\nu(t')\,\mathrm{d}t' \tag{II.5}$$

and

$$S'_\nu = \int S_\nu(t')\mathrm{d}t', \tag{II.6}$$

which are employed e.g. for the estimation of the (temporal) mean values of solar burst fluxes $\bar{S}_\nu$

$$\bar{S}_\nu = (1/t_d) \times S'_\nu \tag{II.7}$$

where t_d = burst duration time. Finally, an integration over the frequency leads to

$$I = \int_f I_\nu\,\mathrm{d}\nu \tag{II.8}$$

which is the *total brightness* ('integrated intensity') in a spectral region f.

Instead applying the quantities I_ν, S_ν, another representation of radiation measures in the form of equivalent temperatures is convenient using the concept of the black-body radiation (thermal equilibrium). At a fixed frequency the brightness I_ν measured in a special direction corresponds to a brightness temperature T_b, which is the temperature of a thermally radiating black body replacing the source of radiation according to Planck's law

$$I_\nu(T) = (2h\nu^3/c^2)/(\exp(h\nu/KT) - 1). \tag{II.9}$$

In the radio range, where $h\nu \ll KT$, i.e. $\exp(h\nu/(KT) \approx 1 + h\nu/(KT))$, Equation (II.9) can be replaced by the Rayleigh–Jeans approximation

$$\begin{aligned} I_\nu(T) &= 2KT/\lambda^2 = 2\nu^2\,KT/c^2 \\ &= 3.075 \times 10^{-28}\nu^2_{[\mathrm{MHz}]}T_{b[\mathrm{K}]}\,[\mathrm{W\,m^{-2}\,Hz^{-1}\,sr^{-1}}] \\ &= 2.761 \times 10^{-23}T_{b[\mathrm{K}]}\lambda^{-2}_{[\mathrm{m}]}\;[\mathrm{W\,m^{-2}\,Hz^{-1}\,sr^{-1}}]. \end{aligned} \tag{II.10}$$

It is evident, that $T_b = T$ is valid only for optically thick isothermal source regions.

Correspondingly one obtains for the flux density

$$S_\nu = \int_{\Omega_\odot} B_\nu(T_e)\,\mathrm{d}\Omega' \qquad \text{(II.11)}$$

$$= 2KT_e\Omega_\odot\lambda^{-2} = 2\nu^2 KT_e\Omega_\odot/c^2,$$

where T_e is named *effective temperature* or apparent disk temperature.

In the radio range the angular size of the Sun depends strongly on the frequency. Thus, for convenience, often the optical size of the Sun is used as a standard. With an average solar diameter of 1919″,3 one gets

$$\Omega_\odot = 6.80 \times 10^{-5}\,[\mathrm{sr}]$$

and consequently,

$$S_\nu = 1.88 \times 10^{-27}\, T_{e[\mathrm{K}]}\lambda^{-2}_{[\mathrm{m}]} \qquad [\mathrm{W\,m^{-2}\,Hz^{-1}}]. \qquad \text{(II.12)}$$

2.2. Radio Telescope Aerials

2.2.1. Fundamental Aerial Parameters

Equipment suitable for the detection and measurement of radio waves generally consists of two main parts: the antenna (aerial) and the receiver system.

Aerials are linear devices that extract power from incoming radiation from a certain direction in a given frequency range and deliver the collected wave energy to an amplifying receiver and the output detecting device.

In radio techniques, by the use of the reciprocity theorems, the definition of an aerial is also applicable for transmitter antennas. However, with the exception of astronomical radar, only the passive aspect of receiving of radiation is of interest for radio astronomy. Particular attention has to be paid to questions concerning calibration, size (collecting area and directional characteristics), mounting and driving techniques, control systems, polarization properties of aerials, etc.

Since we are dealing with transverse waves, all aerials are polarized. In principle this means that only half of a randomly polarized wave can be received. In order to convert the whole energy contained in all polarizations, two complementary aerials and two separate receiver systems must be employed.

After these general remarks some of the most important quantities characterizing radio astronomical antennas will be considered.

a. *Aerial Patterns*

The directional response of an aerial is called the *aerial pattern* which is represented by a function $A(\alpha, \beta)$ and expressed either in terms of the intensity (*intensity pattern*), the Poynting vector (*power pattern*), or the field strength (*field pattern*).

Strictly speaking, we are dealing here with radiation fields sufficiently distant from an aerial to be referred to as *far-field patterns* (*near-field patterns* are also a function of the distance). As already mentioned above, by reciprocity the patterns are the same for both receiving and transmitting conditions, which is of great theoretical and practical importance (*radiation pattern–reception pattern*).

Typically the reception pattern exhibits a number of lobes, where the largest or *main lobe* determines the main aerial direction. Since the patterns are three-dimensional, they can be reproduced as a closed surface in space by means of contour maps or by the representation of orthogonal cross sections, e.g. through the narrowest and widest part of the main lobe (*principal-plane patterns*). For antennas which are not linearly polarized, a more extensive description of the patterns is required.

The beam pattern (far-field distribution) represented in the form of a polar diagram e.g. of a filled aerial aperture is quite analogous to the well known Fraunhofer diffraction pattern of visible light for an aperture of corresponding shape. For one dimension this is given by the Fourier transform of the aperture-field distribution $F(x)$

$$A(\alpha) = \int_{-\infty}^{\infty} F(x/\lambda) \exp(2\pi i(x/\lambda) \sin\alpha) \, \mathrm{d}(x/\lambda). \tag{II.13}$$

Accordingly the polar diagram of an aerial of a rectangular shape would be the same as the diffraction pattern of a slit of the width x/λ. Hence it follows, that the criterion of the resolving power of optical telescopes originally introduced by Rayleigh can also be applied for radio telescopes. In this criterion it is stated that two point sources can be resolved, if the first maximum of the diffraction pattern caused by the one source lies on the first minimum of the pattern caused by the other when the polar diagram is swept in a suitable direction. If not a point source but an extended source of incoherent emission with an angular brightness distribution $I_\nu(\alpha)$ is observed, a pattern $S_\nu(\alpha)$ is observed

$$S_\nu(\alpha) = \int_{-\infty}^{\infty} p_\nu(\alpha - \alpha') \, I(\alpha') \, \mathrm{d}\alpha', \tag{II.14}$$

which results from the convolution of the brightness distribution with the beam pattern $p_\nu(\alpha)$.

From a given aerial pattern the directivity of an aerial can be expressed by means of different quantities. At first the *directivity* D is defined as

$$D = p_\nu(\alpha, \beta)_{\max}/\bar{p}_\nu \tag{II.15}$$

where $p_\nu(\alpha, \beta)_{\max}$ is the power radiated (or received) per unit solid angle in the maximum direction of the aerial pattern and

$$\bar{p}_\nu = \int_{4\pi} p_\nu \mathrm{d}\Omega/(4\pi) \tag{II.16}$$

is the power radiated (received) per solid angle averaged over all directions.

Another often used specification is made by determining the angular width of the main lobe measured at a certain level, e.g. the half-power points (HPBW – cf. Figure II.3). The directivity can be also expressed in terms of an effective solid angle (*beam*

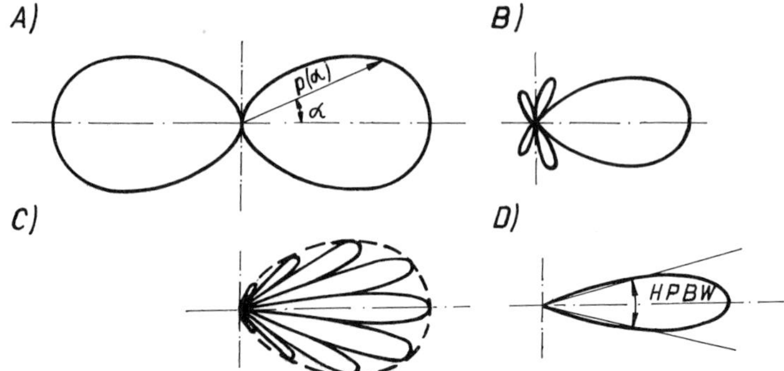

Fig. II.3. Schematic examples of different aerial patterns: A – dipole, B – Yagi-antenna, C – two-element interferometer, D – parabolic mirror (HPBW = half-power beam width).

solid angle, pattern solid angle) given by

$$\begin{aligned}\Omega_A &= 4\pi/D \\ &= \int_{4\pi} p_\nu(\alpha, \beta)/p_\nu(\alpha, \beta)_{\max}\, \mathrm{d}\Omega . \end{aligned} \tag{II.17}$$

b. *Effective Area and Gain*

The effective area of an aerial is defined as the quotient of the power p_ν available at the aerial terminals and the flux S_ν of a plane wave front of appropriate polarization ($S_\nu = (1/2)\, S_{\nu,\text{total}}$) crossing a unit area:

$$A_e = p_\nu/S_\nu . \tag{II.18}$$

The concept of the effective area (or effective aperture) is fundamentally related to the directivity and gain of an aerial and has much more significance than the simply projected area.

The available power p_ν is defined as the output power from the aerial terminals into an impedance which is adjusted to maximum power transfer:

$$p_\nu = \int_{4\pi} A_e(\alpha, \beta)\, I_\nu\, \mathrm{d}\Omega \quad [\text{erg s}^{-1}\ \text{Hz}^{-1}] . \tag{II.19}$$

In this way A_e is the area in the wave front through which passes a flux equal to p_ν. In general A_e depends on the wave direction (α, β) and the wave frequency. For isotropic radiation at a distance r from the aerial one obtains

$$p_\nu = S_\nu 4\pi r^2$$

which, however, is only an idealized case. For real aerials a factor $g \neq 1$ must be considered which, depending on the direction and frequency, is defined by

$$S_\nu = g p_\nu/(4\pi r^2) \tag{II.20}$$

or, taking $r = 1$,

$$g(\alpha, \beta) = 4\pi S_\nu / p_\nu . \tag{II.21}$$

In this way g is defined as 'gain relative to an idealized isotropic radiator'.

Taking

$$g(\alpha, \beta) = p_\nu(\alpha, \beta)/\bar{p}_\nu \tag{II.22}$$

there is also obtained

$$D p_\nu(\alpha, \beta)_{max} / p_\nu(\alpha, \beta) = g(\alpha, \beta) \tag{II.23}$$

yielding

$$D = g(\alpha, \beta)_{max} . \tag{II.24}$$

It is further evident that

$$\int_{4\pi} g(\alpha, \beta)\, d\Omega = 4\pi . \tag{II.25}$$

Thus, aerial patterns can be plotted also in terms of $g(\alpha, \beta)$. The level $g(\alpha, \beta) \equiv 1$ would correspond to the ideally isotropic (nondirective) aerial as a reference level.

There are some further useful relations between the fundamental aerial parameters and the wavelength:

$$A_e = \lambda^2 / \Omega_A \tag{II.26}$$

$$D = (4\pi/\lambda^2)\, A_e \tag{II.27}$$

$$= 4\pi / \Omega_A .$$

The ratio of the effective aperture A_e to the physical aperture A_p of an aerial is sometimes called the *aperture efficiency*

$$\varepsilon_A = A_e / A_p . \tag{II.28}$$

Besides that, a *beam efficiency* is defined as

$$\varepsilon_B = \frac{\Omega_M}{\Omega_A} , \tag{II.29}$$

where $\Omega_M \leq \Omega_A$ is the main-beam solid angle referring to the main lobe of an aerial pattern (Kraus, 1966):

$$\Omega_M = \int_{4\pi} \frac{p_\nu(\alpha, \beta)}{p_\nu(\alpha, \beta)_{max}}\, d\Omega . \tag{II.30}$$

There follows

$$\frac{\varepsilon_A}{\varepsilon_B} = \frac{A_e \Omega_A}{A_p \Omega_M} = \frac{\lambda^2}{A_p \Omega_M} . \tag{II.31}$$

c. *Effective Aerial Temperature*

For a loss-free aerial the *effective aerial temperature* T_a is defined as

$$T_a = \int_{4\pi} T_b A_e \, d\Omega \bigg/ \int_{4\pi} A_e \, d\Omega . \tag{II.32}$$

In the case of radiation received from a black body of the temperature T_s extended over a solid angle Ω this reduces to

$$T_a = \frac{D}{4\pi} \Omega T_s = \frac{\Omega}{\Omega_A} T_s . \tag{II.33}$$

If the aerial is not loss-free, a loss efficiency factor ε_L can be introduced according to

$$(T_a)_{\text{measured}} = (T_a)_{\text{loss-free}} \times \varepsilon_L + (T_a)_{\text{real}} \times (1 - \varepsilon_L) , \tag{II.34}$$

where $(T_a)_{\text{measured}}$ is the measured effective aerial temperature; $(T_a)_{\text{loss-free}}$ is the temperature which would be observed by a corresponding ideal loss-free aerial; and $(T_a)_{\text{real}}$ is the actual temperature of the aerial material.

2.2.2. Basic Types of Aerials

a. *Primary Antennas*

Two classical standard types of primary antennas can be distinguished:

- In the microwave region preferably rectangular or circular horn antennas are used in connection with waveguide techniques.
- At the longer wavelengths dipoles are employed in conjunction with coaxial cable techniques.

Horn antennas are funnel broadenings of the open end of a waveguide. Their gain can be calculated with a high degree of accuracy.

Dipoles are formed by conducting sticks of $2 \times \lambda/4$ length with the terminators in the center. In order to improve their low directivity, combinations with other elements are made in various ways, e.g. in form of Yagi aerials by placing a reflector behind and one or more director sticks in front of the dipole.

Several single dipoles or Yagis can be combined to broadside arrays or large antenna fields which are of some use in radio astronomy.

In contrast to dipoles, which primarily receive one linear component of polarization, helical antennas are suitable for receiving circular polarized radiation. According to the required sense of polarization, right- and left-handed helical antennas are to be used.

Horn antennas and dipoles are optimal only in a very restricted frequency range. In order to receive a broader frequency band, logarithmic periodic antennas and other types were developed.

b. *Reflectors*

Primary antennas ('feeders') are placed in the focus of a larger reflecting area, e.g. of a paraboloid, to obtain better directivities and greater collecting areas. The pencil beam of the parabolic mirror has obvious advantages in comparison to other aerial types.

In the Cassegrain system the focus is occupied by a small secondary reflector and the feeder is placed near the center of the parabolic dish.

Both the feed antenna and the reflecting area must be matched together in order to avoid either 'spillovers' or losses in the optimal utilization of the reflector area. For tracking the Sun or other cosmic objects parabolic aerials are preferably equatorially (paralactically) mounted. Larger instruments, however, have difficulties with the inclined polar axis and are therefore horizontally mounted, requiring a permanent transformation of the horizontal into equatorial coordinates.

Other aerials are designed as transit instruments steerable only in one (meridional) direction. The largest reflecting areas are unmovable, such as the great Arecibo telescope.

The above mentioned basic types of aerial are, in principle, also applicable to spaceborne radio astronomical equipments. But differences exist concerning the mounting and driving of the aerials. While ground-based aerials are commonly directly related to horizontal or equatorial coordinates, for space observations the movement and orientation of the artificial satellite or space probe must be taken into account, requiring additional efforts. Due to the largeness of space and the absence of interfering gravitational and meteorological influences, extraterrestrial experiments offer nice opportunities for radio astronomical and radio physical methods which are yet far from being exhausted.

c. *Special Constructions*

Apart from standard-type radio telescopes there exist a number of special large instruments, which will not be considered in very much detail here. Famous examples are:

The Pulkovo-type radio telescope RATAN 600 situated in the Caucasus (consisting of numerous steerable rectangular plane reflector plates placed on a circle around several feeder systems at different azimuths), cf. special issue of *Izv. Glav. Astron. Obs.*, Pulkovo, Nr. 188 (1972);

Kraus-type radio telescopes, situated at Ohio and Nançay (tiltable flat reflector in combination with a standing parabolic or spherical curved collecting area, cf. Kraus, 1966);

Diverse broad-band telescopes, such as

- the Clark Lake Array TPT (Erickson and Fisher, 1974);
- the Charkov UTR–1 and UTR–2 radio telescopes (Braude *et al.*, 1969);
- the Tasmania Llanherne low-frequency radio telescope (Ellis, 1972); and
- the large collecting array planned at Nançay (Aubier *et al.*, 1975).

A compilation of the angular resolving powers of different radio telescopes in dependence on size and frequency is given in Figure II.4.

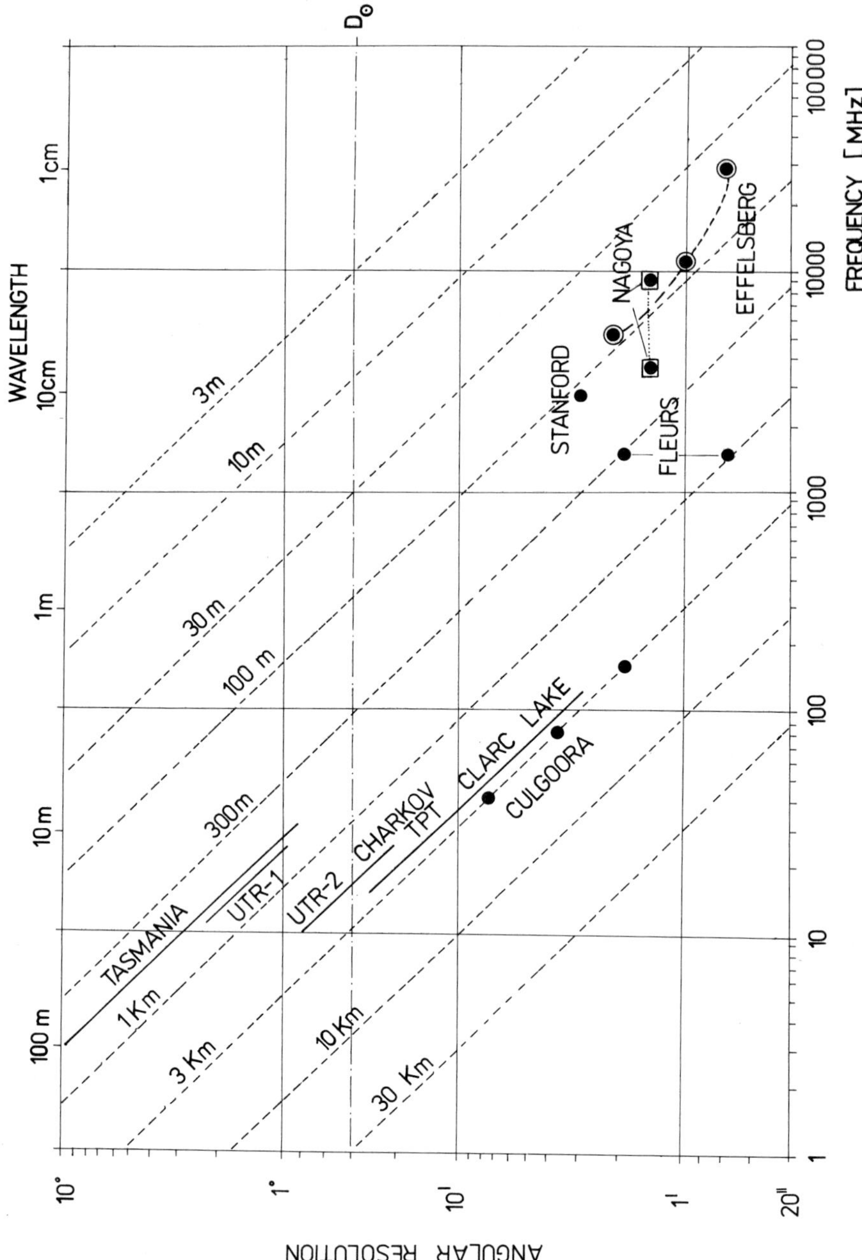

Fig. II.4. Angular resolution of different radio telescopes in dependence of aerial size and wave frequency.

A catalog of the world-wide distribution of all known radio telescopes has been published by the Committee on Radio Frequencies at the National Academy of Sciences and National Academy of Engineering, Washington/USA (1973).

2.3. Radio Astronomy Receivers

2.3.1. Fundamental Receiver Parameters

In contrast to the aerials, radio receivers contain nonlinear devices in general. It is their purpose to convert the high-frequency electromagnetic signals delivered from the aerial into a form convenient for detecting and measuring the incoming radio emission. For most astrophysical objects this emission consists of a large number of incoherently superimposed waves, which physically are of the same nature as the noise of the receiver itself or the background noise received from any thermal object. Owing to this fact and also to the small powers usually available from extra-terrestrial signals, sophisticated techniques are needed for the amplification and conversion of these signals. This applies especially to extrasolar radio astronomy, but also for solar research, regardless of the fact that the strongest radio signals reaching the Earth from astrophysical objects come from the Sun.

Each receiver can be characterized by a set of elementary properties such as:

(a) the sensitivity,
(b) the receiver noise temperature or noise factor,
(c) the amplification or gain factor,
(d) receiver band width, and
(e) integrating time (time constant).

It should be remarked that the sensitivity of a receiver, i.e. its property to detect signals of a certain limiting amplitude, is not a matter of mere amplification but it is essentially determined by the electromagnetic noise generated by the receiver itself. This random receiver noise is mainly due to thermal fluctuations in the resistors and related effects. It is therefore a basic problem in radio astronomy to keep the receiver noise as low as possible.

The receiver noise is characterized by the receiver noise temperature T_R which, according to Nyquist's fundamental theorem

$$p_R = KT_R \Delta\nu \tag{II.35}$$

(K = Boltzmann's constant), is related to the power p_R delivered to a detector and to the receiver bandwidth $\Delta\nu$.

The receiver bandwidth $\Delta\nu$ is generally defined as the difference between both frequencies at which the output signal falls to half of the value measured at the center frequency which is the nominal receiving frequency.

The noise temperature T_R would be equal to an effective aerial temperature T_A of an aerial delivering just as much noise as generated by the receiver.

The receiver noise can be specified also by another parameter, the *noise factor* or *noise figure* N_R taking an aerial at ambient temperature T_0 as reference. The noise factor is defined as the ratio of the total noise power to the noise power if the

Plate A: View from the air showing the Culgoora radioheliograph. 96 aerials equally spaced around the circumference of a circle of 3 km diameter are used for reception of 80 and 160 MHz, whilst 48 aerials around a circule of slightly less diameter are used for reception of 43 MHz. (Courtesy of J. P. Wild and K. V. Sheridan).

Plate B: The central Culgoora observatory building which houses the radioheliograph and radio-spectrograph equipments. (Courtesy of J. P. Wild and K. V. Sheridan).

Plate C: View of a part of the Pulkovo-type large radio telescope RATAN-600 erected in the Caucasus mountains. (Courtesy of Y. N. Parijskij).

Plate D: View of the 100m-Effelsberg radio telescope. (Courtesy of O. Hachenberg).

receiver would be ideally noise-free

$$N_R = (KT_0\Delta\nu + p_R)/KT_0\Delta\nu). \tag{II.36}$$

Hence it follows

$$p_R = (N_R - 1)\,KT_0\Delta\nu \tag{II.37}$$

and

$$T_R = (N_R - 1)\,T_0, \tag{II.38}$$

i.e. for an ideal receiver without noise it would follow that $p_R = 0$, $T_R = 0$, and $N_R = 1$.

The above quoted quantities are particularly useful for linear amplification. More general, if nonlinearities are to be taken into account, an *amplification* or *gain factor* G is defined by

$$G = \mathrm{d}P/\mathrm{d}p \tag{II.39}$$

where P is the output power delivered to a detector and p is the input power at the aerial terminals. Determining the sensitivity of a receiver which is limited by the r.m.s. fluctuations of the output receiver temperature $\Delta T_{\min}$, another important relation can be quoted

$$\Delta T_{\min} = aT_{\mathrm{sys}}/(\Delta\nu\cdot\tau)^{1/2}$$
$$\sim N_R/(\Delta\nu\cdot\tau)^{1/2} \tag{II.40}$$

where ΔT_{sys} is the system temperature consisting of the effective aerial temperature and the receiver noise temperature, $\Delta\nu$ is the receiver bandwidth, τ is the receiver integrating time or time constant, and a is a constant depending on the receiver type. From relation (II.40) it follows immediately, that even for an ideal noise-free receiver an increase of the sensitivity (smaller $\Delta T_{\min}$ values) must be purchased with an increase of the receiver bandwidth or of the integrating time.

2.3.2. Basic Types of Receivers

a. *General Principles*

The basic principles of the receivers used in radio astronomy are generally the same as of those used in other fields of radio science and radio engineering, although the development of radio astronomy initiated great progress in the whole field. A very common standard type is the superheterodyne receiver which is known to be widely used for many purposes in radio engineering.

Figure II.5 displays the schematic diagram of a superheterodyne receiver to be regarded as an example for one of the simplest receivers applied for radio astronomical tasks. Here the signal power delivered by the aerial is usually preamplified in a RF (radio frequency) amplifier (preamplifier). Then the signal is mixed with a power at different frequency ν_0 produced in a local oscillator resulting in a signal at an intermediate frequency (IF) which is proportional to the input RF signal power.

Next the IF signal is amplified in an IF amplifier making the largest contribution to the gain of the receiver. This stage is followed by a detector, converting the ampli-

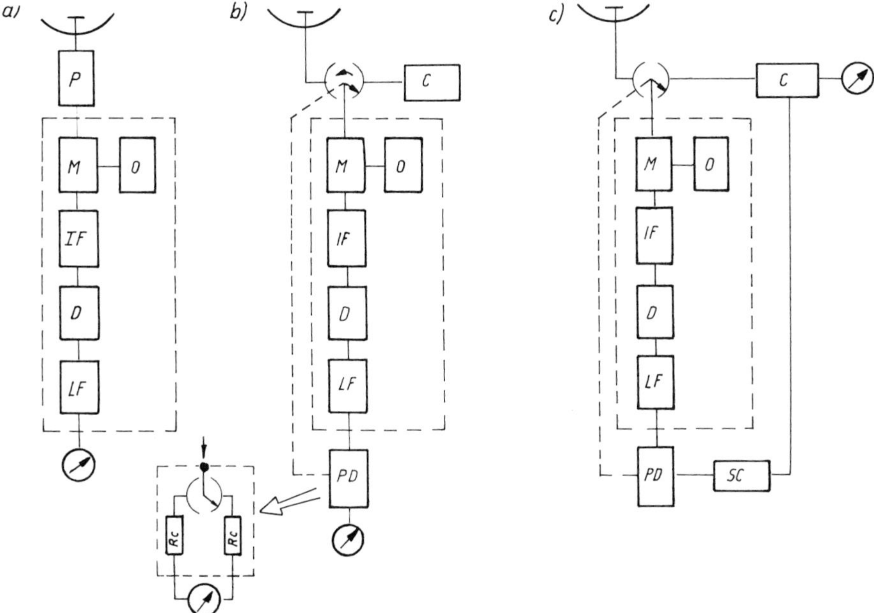

Fig. II.5. Basic elements of different receiver types: (a) – superheterodyne, (b) – Dicke's principle, (c) – Ryle's and Vonberg's principle, where P = preamplifier, M = mixer, O = local oscillator, IF = intermediate-frequency amplifier, D = (square-law) detector, LF = low-frequency amplifier, PD = phse-sensitive detector, C = calibration noise generator, SC = servo control.

fied signal by rectifying it into a low-pass signal. Most detectors have a square-law characteristic, so that the output voltage is directly proportional to the input power of the detector.

The stages of a receiver placed before the detector are called the predetection section or high-frequency part of the receiver. The part of the receiver following the detector is called the postdetection section or the low-frequency part of the receiver. It mainly consists of a low-frequency amplifier, the integrator, and a recording system. The choice of integration time results from a compromise between the required sensitivity and time resolution according to Equation (II.40). In most cases of applications in radio astronomy the simple superheterodyne receivers suffer from high receiver noise. Historically, a first step to overcome this difficulty consisted in applying the compensation method. Here the receiver noise was compensated by an equal d.c. voltage in the postdetection section. But this method has the great disadvantage that unwanted variations of the amplification are not compensated. Such gain variations can be principally eliminated, if the modulation method introduced by Dicke (1946) is applied. In this method the receiver input is continuously switched between the aerial output and a reference noise source, whereas the switch frequency is sufficiently high in relation to the occurring gain variations (switches employed in solar radio telescopes are ferrite or semiconductor switches, sometimes also mechanical or photo-mechanical modulating switches are used; as noise generators gas discharge tubes are used at shorter wavelengths ($\lambda < 10$ cm), while at longer wavelengths diode noise generators are employed). Simultaneously in the postdetection stage the rectified noise signal is switched (multiplied) into two identical integrators (RC-combinations with a time constant much larger than the reciprocal switch frequency, $\tau \gg 1/\nu_s$) formed by the phase-sensitive detector (synchronous demodulator). An output voltage is measured which corresponds to the difference ΔT of the noise temperatures of the aerial T_A and of the reference noise source T_C which for not too strong signals ($T_A \approx T_C$) is nearly independent of gain variations.

It is to be noted that due to the comparison with the reference noise source only about one half of the measuring time is exploited and the same fraction of the available signal power is used by the receiver. A better efficiency of signal power conversion can be achieved by switching the aerial between two identical Dicke-type receivers. Such a device has been described by Graham (1958).

Since in the Dicke-type receivers, gain instabilities are optimally eliminated if the receiver is balanced, it is convenient to introduce an automatic balance. This principle of null-balancing Dicke receivers was first employed by Machin *et al.* (1952) (cf. also Figure II.5c). By means of a servo-control the output of the noise generator is instantaneously maintained at the noise level entering from the aerial and can be directly recorded.

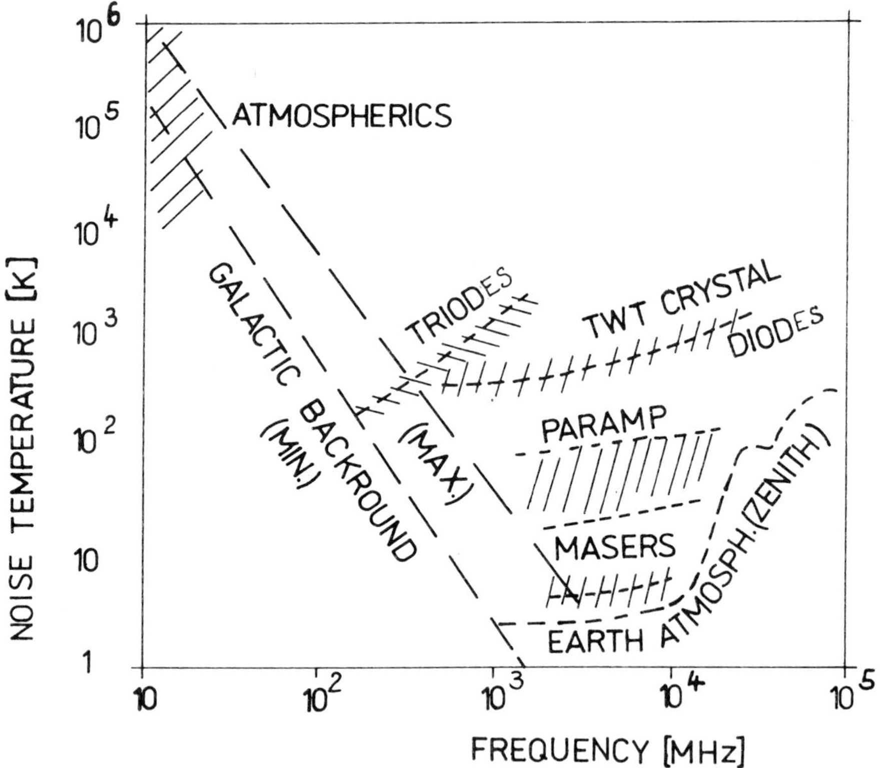

Fig. II.6. Noise temperatures of different types of amplifiers and sky background levels.

b. *Low-Noise Amplifiers*

If higher sensitivity is required, special low-noise amplifiers are used in radio astronomy. Such devices are acting as negative resistors. Special realizations are known employing masers, parametric amplifiers, tunnel diodes, and traveling-wave amplifiers. The former two are typical narrow-band amplifiers, the latter two are also suitable for broad-band amplification. Low-noise amplifiers are mainly used as RF preamplifiers. Their introduction was of vital importance especially for the development of nonsolar radio astronomy and has led to a new era of electronics in general. Concluding these remarks we give a short summary of noise temperatures attained by some typical devices as a function of frequency in comparison to the background temperatures of the sky and the Earth's atmosphere demonstrated by Figure II.6.

2.4. Polarization Measurements

2.4.1. Fundamentals of Polarized Radiation

In general extraterrestrial radio waves are partly polarized, i.e. they contain a polarized and a 'nonpolarized' ($\equiv$ randomly polarized) part, $I_{\nu,\mathrm{pol}}$ and $I_{\nu,\mathrm{ran}}$, respectively. The degree of polarization is defined as

$$\rho = I_{\nu,\mathrm{pol}}/I_{\nu} \tag{II.41}$$

where

$$I_{\nu} = I_{\nu,\mathrm{pol}} + I_{\nu,\mathrm{ran}} \,. \tag{II.42}$$

It follows that $0 \leq \rho \leq 1$; for totally polarized radiation we have $I_{\nu,\mathrm{ran}} = 0$ and $\rho = 1$, for totally 'unpolarized' radiation there is $I_{\nu,\mathrm{pol}} = 0$ and $\rho = 0$.

As a matter of fact the state of polarization refers to waves received in a given (more or less narrow) frequency band, time interval, and direction angle. In the general case of elliptically polarized electromagnetic waves the electric wave vector traces out ellipses in the wave-normal plane. Limiting cases of the elliptic polarization are the circular and linear polarization.

Determining right- and left-handed circular or elliptical polarization several definitions are used in the literature, which, in order to avoid confusion, are listed together in Table II.2. In the following we are using the definition generally applied in radio astronomy, according to which the polarization is called right-handed if the electric wave-field vector rotates clockwise in a fixed plane perpendicular to the wave normal viewed in direction of the wave propagation. The opposite case of anti-clockwise rotation refers to left-handed polarization demonstrated by Figure II.7. It is also shown in Figure II.7, that a right-handed polarized wave traces out a left-handed screw in space and vice versa.

It is well known that any state of wave polarization can be uniquely determined by the specification of four parameters. Such parameters are, for instance, the axial ratio p, an orientation angle χ, the intensity I_{ν}, and the degree of polarization ρ,

TABLE II.2

Definitions of right- and left-handed polarization

Wave propagation	Gyration	Sense of polarization Definition a	b	c	Magnetic polarity on Sun	Case of Fig. II.7	Screw space dependence
↑↑ H_z	electronic	right	left	right	north	A	left
↑↓ H_z	electronic	left	right	right	south	B	right
↑↑ H_z	ionic	left	right	left	north	C	right
↑↓ H_z	ionic	right	left	left	south	D	left

a – Definition commonly used in radio astronomy (observer looks into the direction of wave propagation),

b – Definition used in crystal optics (observer looks at the source),

c – Definition used in laboratory plasma physics (observer looks into the direction of the magnetic field H_z).

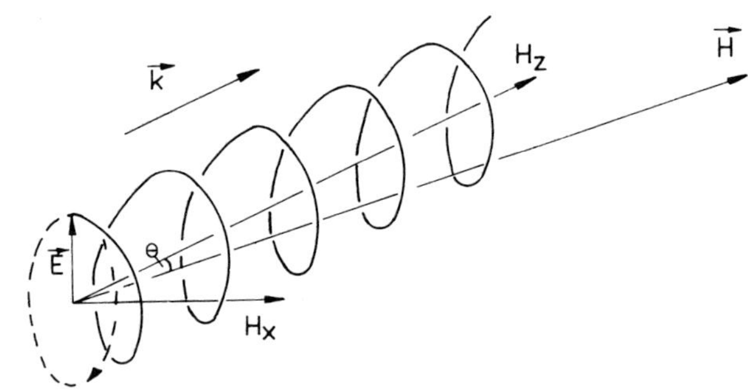

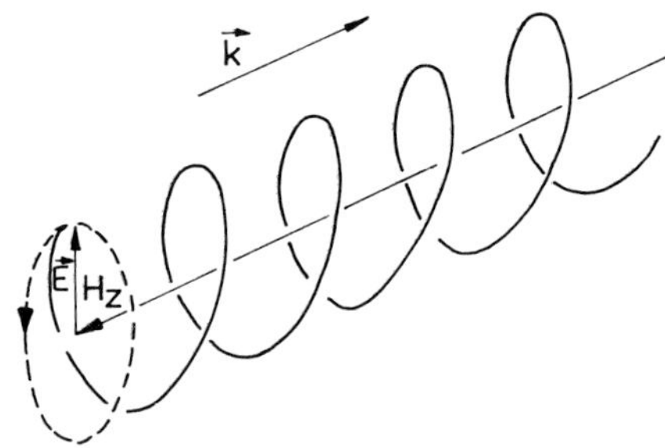

Fig. II.7. On the geometry of right- and left-handed polarization.

which can be transformed into the often used Stokes parameters I, Q, U, and V:

$$I_v = I$$
$$\rho = (Q^2 + U^2 + V^2)^{1/2}/I$$
$$\sin(2\sigma) = V/(Q^2 + U^2 + V^2)^{1/2}$$
$$\tan(2\chi) = U/Q, \tag{II.43}$$

where $p = \tan\sigma$ and $I_{v,\text{pol}} = (Q^2 + V^2 + U^2)^{1/2}$ (Chandrasekhar, 1960; Cohen, 1958). Hence it follows

$$I = I_{v,x} + I_{v,y} = I_v$$
$$Q = I_{v,\text{pol}} \cos(2\sigma)\cos(2\chi) = I_{v,x} - I_{v,y}$$
$$U = I_{v,\text{pol}} \cos(2\sigma)\sin(2\chi) = 2\overline{E_{0x}E_{0y}} \cos(\phi_x - \phi_y)$$
$$V = I_{v,\text{pol}} \sin(2\sigma) = 2\overline{E_{0x}E_{0y}} \sin(\phi_x - \phi_y), \tag{II.44}$$

where E_{0x} and E_{0y} are the time-averaged amplitudes of the wave E-vector:

$$E_{x,\text{pol}}(t) = E_{0x} \sin(\omega_0 t - \phi_x)$$
$$E_{y,\text{pol}}(t) = E_{0y} \sin(\omega_0 t - \phi_y) \tag{II.45}$$

and

$$I_{v,x} = I_{v,x,\text{pol}} + I_{v,\text{ran}}/2$$
$$I_{v,y} = I_{v,y,\text{pol}} + I_{v,\text{ran}}/2. \tag{II.46}$$

For linear polarization the degree of polarization is

$$\rho_l = \frac{I_{v,x} - I_{v,y}}{(I_{v,x} + I_{v,y})\cos(2\chi)}$$
$$= Q/(I\cos(2\chi)) = \rho\cos(2\sigma)$$
$$= U/(I\sin(2\chi)) = (Q^2 + U^2)^{1/2}/I. \tag{II.47}$$

Circular polarization is described by

$$I = I_{v,r} + I_{v,l}$$
$$Q = 2\overline{E_{0r}E_{0l}} \cos(\phi_r - \phi_l)$$
$$U = 2\overline{E_{0r}E_{0l}} \sin(\phi_r - \phi_l)$$
$$V = I_{v,l} - I_{v,r}, \tag{II.48}$$

where $I_{v,r}$, $I_{v,l}$, E_{0r}, E_{0l} denote the corresponding right- or left-handed circularly polarized components of I_v and E_0, respectively. The degree of circular polarization is given by

$$\rho_c = |V|/I = (I_{v,l} - I_{v,r})/(I_{v,l} + I_{v,r})$$
$$= \rho\sin(2\sigma). \tag{II.49}$$

In the general case we have

$$\rho = (\rho_l^2 + \rho_c^2)^{1/2}. \tag{II.50}$$

The intensities $I_{\nu,x}$, $I_{\nu,y}$, and $I_{\nu,r}$, $I_{\nu,l}$ are measurable by use of linearly and circularly polarized antennas, respectively. As is indicated e.g. by the set of Equation (II.44) the full determination of the polarization involves phase measurements and the execution of the cross correlation of the signals of oppositely polarized aerials.

2.4.2. Polarimeters

A principal possibility to perform polarization measurements delivering information about the four Stokes parameters or equivalent quantities is sketched in Figure II.8. Using two oppositely (linearly or circularly) polarized antennas, the outputs of two channels 1 and 3 are proportional to I_x and I_y (or I_r and I_l), respectively, if

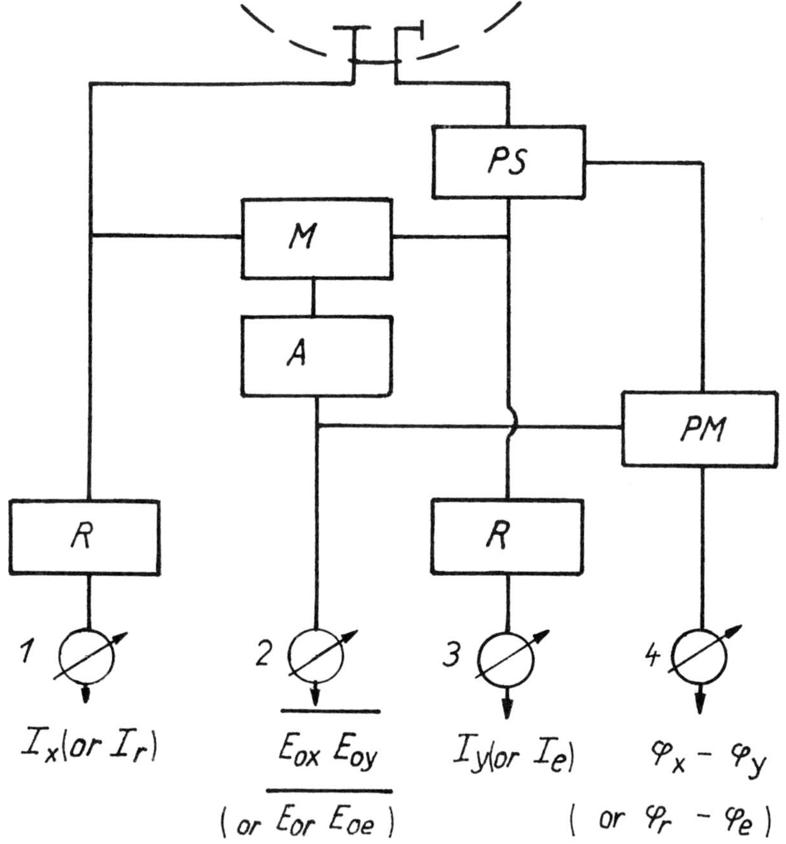

Fig. II.8 Block diagram of a polarimeter: PS – phase shifter, M – mixer, A – amplifier, PM – phase meter, R – receiver.

square-law detectors are applied. The signal from the second antenna is subjected to a continuous phase sweeping at a frequency $\omega_{\text{sweep}} \ll \omega$. Then by the action of a mixer, channel 2 displays the correlation of the polarized part of the emission at the sweep frequency of the phase shifter (in the case of random polarized emission this component becomes zero). Finally, the phase information is obtained in channel 4.

There are, of course, a number of different possible ways to get polarization measurements. In practice, for instance, especially when using linearly polarized aerials, the phase measurements are often omitted, in other cases only two Stokes parameters (e.g. I and V) are measured. Another practicable approach to polarization measurements is to measure only intensities by means of four independent antennas. This yields e.g. I_x, I_y, I_r, and I_l, replacing channels 2 and 4 of Figure II.8. Adding further aerials (e.g. crossed dipoles shifted by 45°), redundant information is obtained which can be used for checking the measurement determining instrumental errors. An example of this kind was given by Suzuki and Tsuchiya (1958).

Similar to optical devices the principle of using a combination of quarter-wave plates can be applied in radio astronomy. This technique was adopted for the study of microwaves. Some historically interesting solutions for putting a $\lambda/4$ plate before and behind the aperture plane of the aerial and other details were described by Covington (1951), Korolkov *et al.* (1960), and Akabane (1958).

Turning to a general description, the transmission of (quasimonochromatic) radiation through a receiver can be expressed in terms of a transmission matrix

$$(T) = T_{mn}, \tag{II.51}$$

so that

$$\begin{aligned} E'_x &= T_{11}E_x + T_{12}E_y \\ E'_y &= T_{21}E_x + T_{22}E_y \end{aligned} \tag{II.52}$$

if E'_x, E'_y and E_x, E_y represent the output and input waves, respectively (periodic factors omitted for simplicity).

Similarly the transmission of the Stokes parameters can be written

$$\begin{pmatrix} I' \\ Q' \\ U' \\ V' \end{pmatrix} = \begin{pmatrix} \alpha_{11} & \alpha_{12} & \alpha_{13} & \alpha_{14} \\ \alpha_{21} & \ddots & & \vdots \\ \alpha_{31} & & \ddots & \vdots \\ \alpha_{41} & \cdots & \cdots & \alpha_{44} \end{pmatrix} \begin{pmatrix} I \\ Q \\ U \\ V \end{pmatrix} \equiv (M) \begin{pmatrix} I \\ Q \\ U \\ V \end{pmatrix} \tag{II.53}$$

where (M) is called the instrumental matrix. The connection between (T) and (M) is established in the following way

$$(T) = \frac{1}{2}\begin{pmatrix} [(\alpha_{11} + a_{12}) + (\alpha_{22} + \alpha_{11})] & [(\alpha_{11} - \alpha_{12}) - (\alpha_{22} - \alpha_{21})] \\ [(\alpha_{11} + \alpha_{12}) - (\alpha_{22} + \alpha_{21})] & [(\alpha_{11} - \alpha_{12}) + (\alpha_{22} - \alpha_{21})] \end{pmatrix} \quad \text{(II.54)}$$

and

$$(M) =$$

$$= \frac{1}{2}\begin{pmatrix} (T_{11}^2 + T_{12}^2 + & (T_{11}^2 - T_{12}^2 + & (T_{11}^* T_{12} + T_{11} T_{12}^* + & i(T_{11}^* T_{12} - T_{11} T_{12}^* + \\ + T_{21}^2 + T_{22}^2) & + T_{21}^2 - T_{22}^2) & + T_{21}^* T_{22} + T_{21} T_{22}^*) & + T_{21}^* T_{22} - T_{21} T_{22}^*) \\ (T_{11}^2 + T_{12}^2 - & (T_{11}^2 - T_{12}^2 - & (T_{11}^* T_{12} + T_{11} T_{12}^* - & i(T_{11}^* T_{12} - T_{11} T_{12}^* - \\ - T_{21}^2 - T_{22}^2) & - T_{21}^2 + T_{22}^2) & - T_{21}^* T_{22} - T_{21} T_{22}^*) & - T_{21}^* T_{22} + T_{21} T_{22}^*) \\ (T_{11} T_{21}^* + T_{11}^* T_{21} + & (T_{11} T_{21}^* + T_{11}^* T_{21} - & (T_{11}^* T_{22} + T_{11} T_{22}^* + & i(T_{11}^* T_{22} - T_{11} T_{22}^* + \\ + T_{12} T_{22}^* + T_{12}^* T_{22}) & - T_{12} T_{22}^* - T_{12}^* T_{22}) & + T_{12} T_{21}^* + T_{12}^* T_{21}) & + T_{12} T_{21}^* - T_{12}^* T_{21}) \\ i(T_{11} T_{21}^* - T_{11}^* T_{21} + & i(T_{11} T_{21}^* - T_{11}^* T_{21} - & i(T_{11} T_{22}^* - T_{11}^* T_{22} + & (T_{11} T_{22}^* + T_{11}^* T_{22} - \\ + T_{12} T_{22}^* - T_{12}^* T_{22}) & - T_{12} T_{22}^* + T_{12}^* T_{22}) & + T_{12} T_{21}^* - T_{12}^* T_{21}) & - T_{12} T_{21}^* - T_{12}^* T_{21}) \end{pmatrix}$$

(II.55)

The knowledge of these components is of practical importance, e.g. for absolute calibrations of polarimeters (cf. Priese, 1972).

Refined techniques can be developed to combine polarization measurements with interferometers and spectrographs. A detailed discussion of these matters, however, is outside the scope of this short introduction.

2.5. Absolute Calibration Experiments

2.5.1. General Aspects

In essence most observations in solar radio astronomy refer to relative measurements, i.e. the measured quantities are primarily obtained in a more or less arbitrary scale of units. In such a way e.g. the fluxes of solar radio bursts can be expressed in units of the undisturbed pre-burst level based on units of the receiver noise etc. For an exact determination of solar brightness temperatures, the derivation of spectra, and for many other purposes, however, the knowledge of absolute flux or intensity values expressed in well defined physical units (e.g. in the mks system) is needed. As in other fields of experimental physics the absolute measurements are much more difficult and more expensive than relative measurements only. For this reason an absolute calibration of radio astronomical receivers is carried out in very few observatories delivering reference values which can be easily adopted by other stations. Such reference fluxes are known e.g. in the microwave region for the quiet Sun + S-component and the Moon. At longer wavelengths reference fluxes are available by cosmic radio sources, e.g. the galactic radio source Cassiopeia A.

2.5.2. Standard-Radiometer and Standard-Field Methods

For an absolute determination of an unknown radiation at a fixed frequency, the gain of the aerial and the scale of the receiver must be calibrated. In the case of polarization measurements, the instrumental matrix (M) must be known. In practice two main methods of absolute calibrations have been developed, the standard-radiometer method and the standard-field method.

Considering the first method, a standard radiometer consists of a standard antenna and a standard receiver. For a standard antenna, the horn antenna is most conveniently used. For this type of antenna the gain can be properly calculated (Schelkunoff and Friis, 1952) and a sufficient conformity between theory and experiment has been proved (cf. Fürstenberg, 1966). Another problem is the exact determination of the noise-temperature scale. The fundamental calibration norm is obtained from the thermal noise of a resistor of accurately known temperature. The principal scheme of such a calibration experiment demonstrating the comparison between calibration source, horn antenna, and calibrated antenna as well as between the standard radiometer and the radiometer which is to be calibrated is sketched in Figure II.9. In this way, with a rather sophisticated treatment regarding both, the accurate operation of the receiver and precisely working noise norms, accuracies of about 1 % have been achieved (Findlay, 1966; Priese, 1972). It should be mentioned, that for wavelengths below 10 cm the attenuation of the Earth's atmosphere becomes noticeable and must be taken into account.

The application of the standard-radiometer method requires some efforts con-

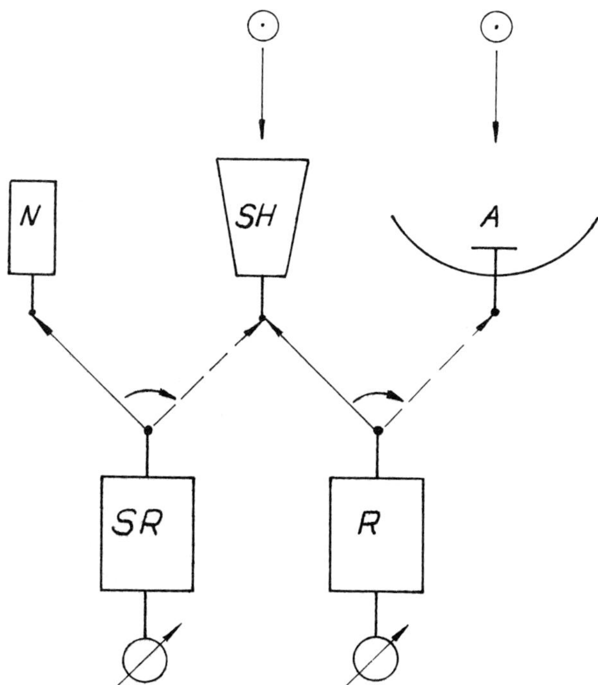

Fig. II.9. Scheme of absolute calibration by the standard-radiometer method: N – calibration norm (heated resistor), SH – standard-horn antenna, SR – standard receiver, A, R – antenna + receiver (radiometer) to be calibrated.

cerning the need of a standard receiver of high precision. To a certain extent these efforts are reduced by the 'standard-field method' (Mollwo, 1957). Here a calibrated transmitter connected with a horn antenna produces a known standard-radiation field which is received by the radiometer to be calibrated. The difficulty of the absolute calibration is then transformed to the problem of developing a sufficiently constant and reliable standard-radiation source which involves similar steps of comparison as in the standard-radiometer method. If the transmitter horn is mounted on a little tower, the aerial of the radiometer which is to be calibrated must be directed towards the calibration source measuring the incoming radiation. Advantages of this method are the possibility of quick repetitions and an overall calibration of the radiometer from the aperture plane up to the output meter. A certain inconvenience of the method lies in the necessary consideration of the near-field errors of the radiation field and in the elimination of the interference of ground reflections (Michel, 1965; Priese, 1969).

The history of solar absolute measurements especially in the microwave region and a summary of the most conclusive results were described by Tanaka *et al.* (1973). Further details and consequences of these results will be considered in Chapter III.

2.6. Spectrography

2.6.1. Swept-Frequency Spectrographs

Spectrographs are used for two main purposes in radio astronomy. The first goal is the study of narrow atomic or molecular line profiles which are currently of interest for nonsolar radio astronomy since line emissions from the Sun are below the limit of detectability in the classical radio range.

The second field of application of radio spectrographs is the observation of transient features and spectral fine structures in a broader frequency range which have a great relevance especially to the study of solar burst phenomena.

Radio spectrographs can be divided into swept-frequency and multichannel spectrographs. Furthermore one can distinguish between analog and digital spectrographs. The first solar radio spectrograph was initiated by Wild and McCready (1950) employing an analog swept-frequency technique. The equipment consisted of a nonselective aerial and a receiver which was motor-driven tuned over a frequency range of nearly one octave within about 0.1 s. Later on, the principle was applied to cover the frequency range of several octaves. The output of the spectrograph displaying a time-frequency pattern can be easily recorded by a moving 35 mm-film photographing the screen of a cathode-ray oscillograph tube. For this purpose the beam of the oscillograph is intensity-modulated by the signal strength and scanned across the width of the film according to the frequency sweep. In this way a plot of the brightness-modulated flux density versus time and frequency is obtained which is called a *dynamic spectrum*. During recent decades swept-frequency spectrographs have found a wide application for the continuous patrol or solar burst emissions in the m- and dm-region. Latest achievements provide a multicolor display allowing the recognition of polarized emissions.

2.6.2. Multichannel spectrographs

Swept-frequency spectrographs have the disadvantage of a comparatively poor intensity resolution and low sensitivity due to the fact that only one channel is received, tuned over the spectrum. Multichannel spectrographs are overcoming these restrictions. Their development leads to new generations of solar radio spectrographs, especially in combination with the digital data processing techniques. A prominent example for an analog multichannel solar spectrograph is the Utrecht radio spectrograph at Dwingeloo (van Nieuwkoop, 1971). This instrument contains sixty channels originally distributed at equal distances between 160 and 320 MHz. The simplified block diagram of the radio spectrograph is reproduced in Figure II.10. The output of the spectrograph is given in two ways: Firstly, in the form of a normal logarithmical flux display (dynamical range 15 dB) and secondly, as a representation of fast superimposed flux variations (dynamical range ± 1.7 dB) from a floating zero level derived with a time constant of 3 s for each channel. Due to this differentiating technique the spectrograph is well suited for the study of short-time fluctuations.

New facilities are opened by the recent advances of high-speed digital electronics technology. Apart from the inherent opportunity of a customary data handling by software the possibilities to include special blocks in hardware, e.g. adopting the

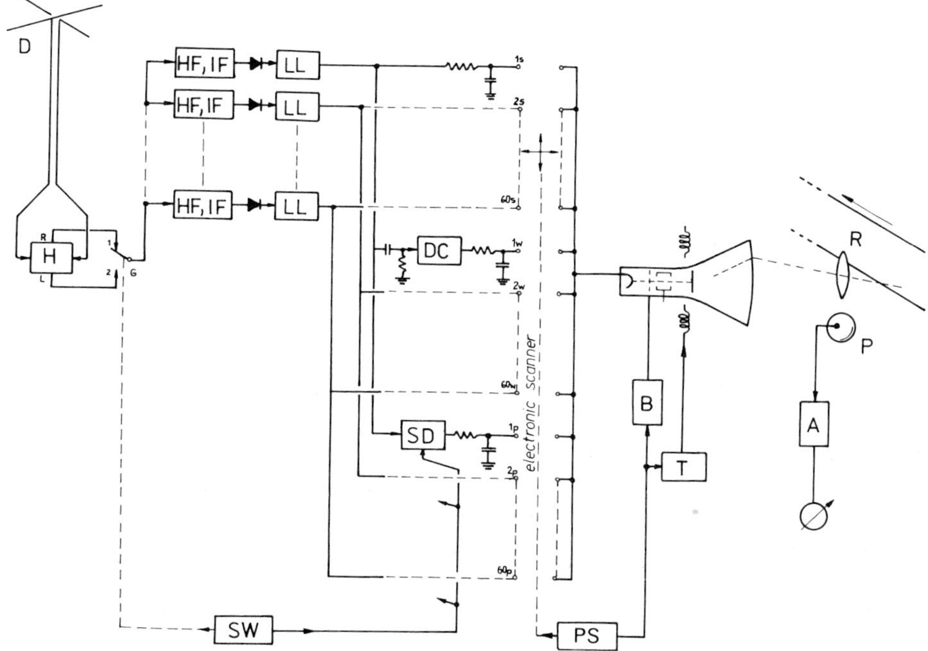

Fig. II.10. Simplified block diagram of the Utrecht 60-channel radiospectrograph (after van Nieuwkoop, 1971). D – log-periodic dipole, H – 90°-hybrid coupler, G – diode switch, HF, IF – HF and IF amplifying chain, LL – linear-logarithmic DC amplifier, SW – square-wave generator, DC – DC-amplifier, SD – synchron detector, B – blanking circuit, T – time base circuit, PS – programming scanner, P – photo multiplier, A – amplifier, R – record.

TABLE II.3

Characteristics of solar radiospectrographs (after Linscott *et al.*, 1975).

Type	Typical number of channels	Relative sensitivity	Characteristic dynamic range	Band pass [MHz]
Analog techniques				
Swept-frequency	1	~1/sweep speed	$\sim 10^3$	unlimited
Multichannel	~50	1	$\sim 10^3$	unlimited
Digital techniques				
FFT	10^3	~1	$\gtrsim 10^3$	10
Autocorrelator	10^3	~0.7	$\sim 10^2$	10

autocorrelation and fast Fourier transform technique, appear very attractive. The principle of a coherent radiospectrograph based on the fast Fourier transform (FFT) was described by Linscott *et al.* (1975). Using the digital techniques, the FFT is implemented into the hardware of the spectrometer. By maintaining coherence in signal processing predetection removal can be accomplished in real-time, whereas the spectrometer works either in a mode set for a specific dispersion or in a mode covering a broad range of dispersion. The above spectrometer system was particularly designed to operate in the microwave range in real-time for 10 MHz bandwidth by direct digital calculation of a 10 bit/1024 point spectrum in 0.1 ms in either the frequency or dispersion domain. In this way the time resolution can be improved by nearly 100 times in comparison to former spectrographs.

A compilation of some characteristics of different types of spectrographs is shown in Table II.3.

2.7. Interferometry and Heliography

Because of the limited angular resolution of single aerials special efforts are to be made to overcome this restriction. The question is invariably connected with the scale of wavelengths in the radio range requiring correspondingly large size antenna systems. It is quite evident that a simple increase of the aperture diameter of a single aerial (e.g. parabolic mirror) has technical and economical limitations and therefore other ways must be found to obtain the desired resolution power. Improved resolution has been achieved by interferometer techniques which play a prominent role in radio astronomy and have been grown in several variants.

2.7.1. Two-element (adding) interferometers

The simplest form of an interferometer is the (adding) two-element or twin-wave interferometer. Its principle follows from Young's well-known diffraction experiment in optics which was first applied to an optical telescope by Michelson in 1892 and adapted to radio astronomy by McCready *et al.* (1947) in the form of the cliff-interferometer; and by Ryle and Vonberg (1946, 1948) in its proper form. In essence the aerial system consists of two identical antennas spaced at a distance a, the inter-

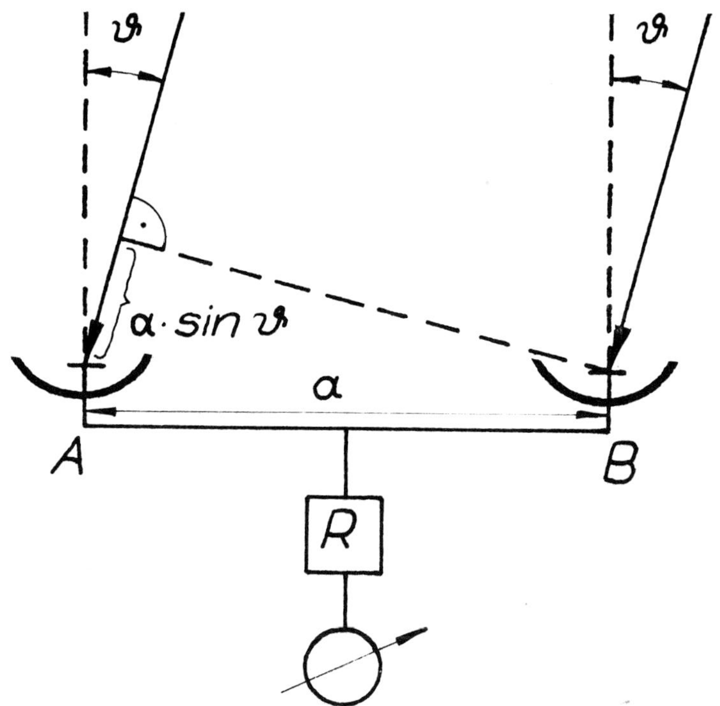

Fig. II.11. Principle of the two-element interferometer; a – interferometer base, R – receiver.

ferometer base, connected in parallel to the receiver input (Figure II.11). Observing a point source in direction ϑ the phase difference between the signals from the two antennas is $ka \sin \vartheta$ (where $k = 2\pi/\lambda$). The power at the receiver input is

$$P(\vartheta) = P_0(\vartheta)\left[1 + \cos\left(\frac{2\pi}{\lambda} a \sin \vartheta\right)\right], \tag{II.56}$$

if $P_0(\vartheta)$ is the power from a single antenna element. For larger antenna spacings a the number of lobes or fringes in the interferometer pattern increases, i.e. the fringe spacing decreases. The condition for the occurrence of the pattern minima and maxima is

$$\psi_{\min} = (2n + 1)\,\pi, \qquad \psi_{\max} = 2n\pi$$

where $\psi = (2\pi/\lambda)\, a \sin \vartheta$ and $n = 0, 1, 2, 3, \ldots$ is the fringe order.

In the general case of an extended radiation source the observed flux density is the convolution of the true angular distribution of the source's brightness $I_\nu(\vartheta)$ and the aerial pattern $P_\nu(\vartheta)$ for one direction:

$$S_\nu(\vartheta, a/\lambda) = \int_0^{\alpha} P_\nu(\vartheta)\, I_\nu(\vartheta - \vartheta')\left\{1 + \cos\left[\frac{2\pi}{\lambda} a \sin(\vartheta - \vartheta')\right]\right\} \mathrm{d}\vartheta' \tag{II.57}$$

(α = source diameter).

If the source extent is small in comparison with the single-element pattern of the interferometer, then it follows

$$S_\nu(\vartheta, a/\lambda) = P_\nu(\vartheta)\left\{\int_0^\alpha I_\nu(\vartheta - \vartheta')\,d\vartheta' + \int_0^\alpha I_\nu(\vartheta - \vartheta')\cos\left[\frac{2\pi}{\lambda} a \sin(\vartheta - \vartheta')\right] d\vartheta'\right\}. \tag{II.58}$$

If the observation is made in the main lobe, i.e. $P_\nu(\vartheta) = 1$, one obtains

$$S_\nu(\vartheta, a/\lambda) = S_{\nu,0}(1 + W), \tag{II.59}$$

where

$$W = (1/S_{\nu,0})\int_0^\alpha I_\nu(\vartheta - \vartheta')\cos\left[\frac{2\pi}{\lambda} a \sin(\vartheta - \vartheta')\right] d\vartheta'$$

and $S_{\nu,0}$ is the flux density of the source. In this way $S_\nu(\vartheta, a/\lambda)$ is expressed as the sum of a constant and a variable part.

For 'point sources', where $\vartheta - \vartheta' \ll \pi$, we have

$$W(\vartheta, a/\lambda) = (1/S_{\nu,0})\left\{\cos\left(\frac{2\pi}{\lambda} a\vartheta\right)\int_0^\alpha I_\nu(\vartheta)\cos\frac{2\pi}{\lambda} a\vartheta\,d\vartheta + \sin\frac{2\pi}{\lambda} a\vartheta \int_0^\alpha I_\nu(\vartheta)\sin\frac{2\pi}{\lambda} a\vartheta\,d\vartheta\right\}. \tag{II.60}$$

This can be expressed in the form of a cosine law

$$W(\vartheta, a/\lambda) = W_0(a/\lambda)\cos\left[2\pi\frac{a}{\lambda}(\vartheta - \Delta\vartheta)\right], \tag{II.61}$$

where $W_0(a/\lambda)$ represents the fringe amplitude and $\Delta\vartheta$ is the fringe displacement angle for a point source.

For illustration, some typical interferometer patterns are schematically shown for different cases in Figure II.12. From the foregoing equations it follows that

$$W_0(a/\lambda)\,e^{i2\pi(a/\lambda)\Delta\vartheta} = (1/S_{\nu,0})\int_{-\infty}^{\infty} I_\nu(\vartheta)\,e^{i2\pi(a/\lambda)\vartheta}\,d\vartheta \tag{II.62}$$

which corresponds to a Fourier transform of the source brightness. Then the brightness distribution itself is obtained, in principle, by the inverse Fourier transform

$$I_\nu(\vartheta) = S_{\nu,0}\int_{-\infty}^{\infty} W_0(a/\lambda)\,e^{-i2\pi(a/\lambda)(\vartheta - \Delta\vartheta)}\,d(a/\lambda). \tag{II.63}$$

This implies that for a practical application of this interferometric method, observations at various aerial spacings and/or at different wavelengths are required in order to derive the brightness distribution $I_\nu(\vartheta)$. Furthermore, no influences of other con-

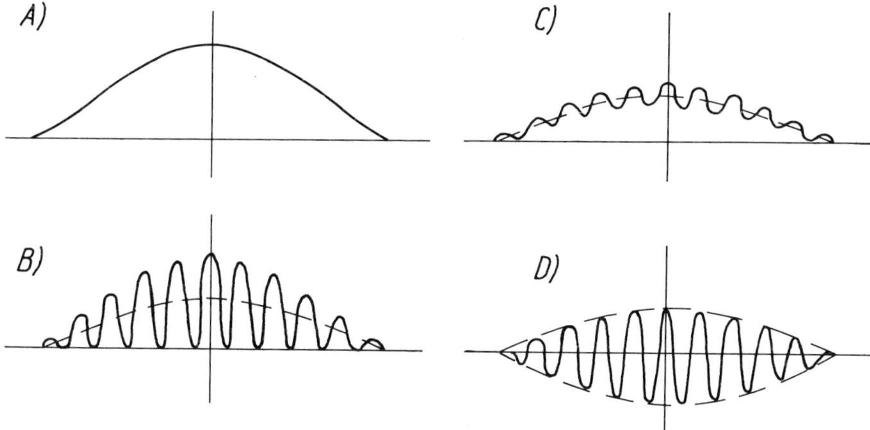

Fig. II.12. Idealized interferometer records: (A) – single element and (B) – two-element unswitched interferometer with point source; (C) – same as B with extended source; and (D) – phase-switched interferometer.

fusing sources should disturb the observed interferometer pattern and other technical interferences (effect of finite bandwidth, degree of stability of phase, amplitude, and frequency, the attenuation and matching of the aerial feeder system, etc.) must be taken into account.

For the above mentioned reasons, twin-wave interferometers are only employed in solar radio astronomy for more or less rough observations, e.g. as depending on the parameter a/λ to separate the whole Sun from the sky, to separate active regions from the background of the undisturbed Sun, or to eliminate burst centers.

Extensions of the interferometer method have been made in many directions. Some of them will be briefly discussed in the following subsections.

2.7.2. Phase-switched (multiplying) interferometers

The phase-switched interferometer, which was first introduced by Ryle (1952), operates according to the following principle: The phase of one of the aerials of a two-element interferometer is periodically (e.g. every 30 Hz) reversed and the difference of both patterns is recorded at the receiver output. The result is a record with fringes oscillating around the zero line (Figure II.12). The switching technique has considerable advantage through increased sensitivity and stability suppressing gain fluctuations of the receiver and eliminating variations of the background level. The switching process filters out only the high-frequency band of the Fourier components, which carries the essential information of the interferometer observations, whereas the low-frequency part, which is subject to unwanted interferences, is removed.

The received wave fields at the terminals of two identical aerials are for the case of phase coincidence (subscript 1 – upper sign) and phase reversal (subscript 2 – lower sign)

$$E_{1,2}(\vartheta) = E_0(\vartheta)\left[e^{i\pi(a/\lambda)\sin\vartheta} \pm e^{-i\pi(a/\lambda)\sin\vartheta}\right]. \tag{II.64}$$

Correspondingly the power pattern is given by

$$P_{1,2}(\vartheta) = |E_0(\vartheta)|^2 \left(e^{i\pi(a/\lambda)\sin\vartheta} \pm e^{-i\pi(a/\lambda)\sin\vartheta}\right) \times$$
$$\times \left(e^{i\pi(a/\lambda)\sin\vartheta} \pm e^{-i\pi(a/\lambda)\sin\vartheta}\right)^*, \qquad \text{(II.65)}$$

where the asterisk denotes conjugate complex quantities.

The recorded pattern is

$$P(\vartheta) = P_1(\vartheta) - P_2(\vartheta)$$
$$= |E_0(\vartheta)|^2 \left(e^{2i\pi(a/\lambda)\sin\vartheta} + e^{-2i\pi(a/\lambda)\sin\vartheta}\right). \qquad \text{(II.66)}$$

In the case of nonidentical aerials (a, b) one obtains

$$P(\vartheta) = E_a(\vartheta)\, E_b^*(\vartheta)\, e^{2i\pi(a/\lambda)\sin\vartheta} + E_a^*(\vartheta)\, E_b(\vartheta)\, e^{-2i\pi(a/\lambda)\sin\vartheta} \qquad \text{(II.67)}$$

In the case of symmetry $E_{a,b}(\vartheta) = E_{a,b}^*(\vartheta)$ there follows

$$P(\vartheta) = E_a(\vartheta)\, E_b(\vartheta) \cos 2\pi \frac{a}{\lambda} \sin\vartheta \qquad \text{(II.68)}$$

which gives the reason for the alternative name ‘multiplying interferometer’.

2.7.3. Swept-Lobe Interferometers

In the foregoing sections we have considered drift interferometers, that means the passage of the recepted radiation through the aerial lobe pattern is caused by the movement of the source or Earth rotation. In contrast to this the swept-lobe interferometer technique produces an artificial beam sweeping. It was introduced into solar radio astronomy by Little and Payne-Scott (1951) and has essential advantages for the study of short-lived phenomena like solar bursts, since beam-sweeping speeds with scanning rates of small fractions of a second (of time) can be achieved, which would otherwise not be possible by a mere beam passage due to the Earth's rotation. The beam sweeping is obtained by a continuous change of the relative phase of the signals applying a phase changing unit (phase shifter) in the local oscillator circuit. The application of this technique together with the phase switching system allows the possibility of applying preamplifiers behind each aerial. The usual difficulty of an inequality and instability of the preamplifiers does not make a great effect on the measurement because the long-periodic component of the output signal is not influenced.

2.7.4. Special Arrangements

a. *Receipt of Polarization*

If the two aerials of a twin-wave interferometer are oriented in planes perpendicular to each other, the state of the polarization of the received waves can be determined from the information of the phase and the amplitudes of the received pattern. This technique was first used for the observation of solar bursts by Payne-Scott and Little (1951) in combination with the swept-lobe interferometer.

b. *Swept-Frequency Interferometer*

The method of swept-frequency interferometry was introduced by Wild and Sheridan (1958) into solar radio astronomy. Here the receiver frequency is periodically swept over a certain frequency range producing a sinusoidal output pattern changing with time and frequency. A second interferometer with a different baseline can be used in order to eliminate ambiguities in the determination of the main lobe. By means of this method the position of radio sources (e.g. bursts in the corona) can be evaluated as a function of frequency and time.

c. *Wide-Band Interferometry*

In general the theory of interferometers is applied to conditions of the receipt of 'monochromatic waves'. The effect of finite bandwidth produces a progressive reduction of the fringe amplitude of the higher orders. In extreme cases the fringes are smeared over. This property has been applied e.g. by Vitkevich (1953) to construct a wide-band interferometer having only one central lobe of sufficient angular resolution.

d. *Long-Baseline Interferometry and Very-Long-Baseline Interferometry*

Because the angular resolution power of an interferometer depends on the linear extent of its baseline, an increase of the distance between the aerials is of high practical interest. Whereas in conventional interferometers with smaller aerial distances (up to a few kilometers) links by cables or waveguides are still practicable, for longer baselengths radio links have to be used. In the latter case the maintenance of the phase stability becomes a fundamental technical problem. A solution was offered by the post-detection correlation interferometer introduced by Hanbury-Brown and Twiss (1954). By this method the low-frequency signals resulting from the square-law detection are combined in a correlator.

Devices with separate square-law detectors for each interferometer channel are called *intensity interferometers*.

In contrast to the conventional and long-baseline interferometry, where the elements of an interferometer are linked by high-frequency techniques, i.e. cables (waveguides) or radio relays, respectively, the very-long-baseline interferometry makes use of a tape recording system which can be transported over great distances from one station to the other (cf. Figure II.13). This method is mainly applied to nonsolar radio astronomy providing resolutions of one-thousandth second of arc and better for intercontinental baselines. The signals from each aerial are separately recorded and later compared in a digital computer or special correlator. To do this, two requirements must be met. At first the recordings at the different tapes must be synchronized with a precision of about 1 μs. Secondly, the intermediate frequency signals, which lastly are to be correlated, must be produced by oscillators remaining coherent over the observing time. Thus the phase changes of the oscillators must be small enough so that the change in frequency is less than the reciprocal of the recording time. This can be achieved by means of atomic frequency standards. Favored by the properties of the Earth's atmosphere the very-long-baseline radio interferometry attains resolutions far better than those achieved by ground-based optical astronomy. More recently radio links by artificial satellites have been proposed.

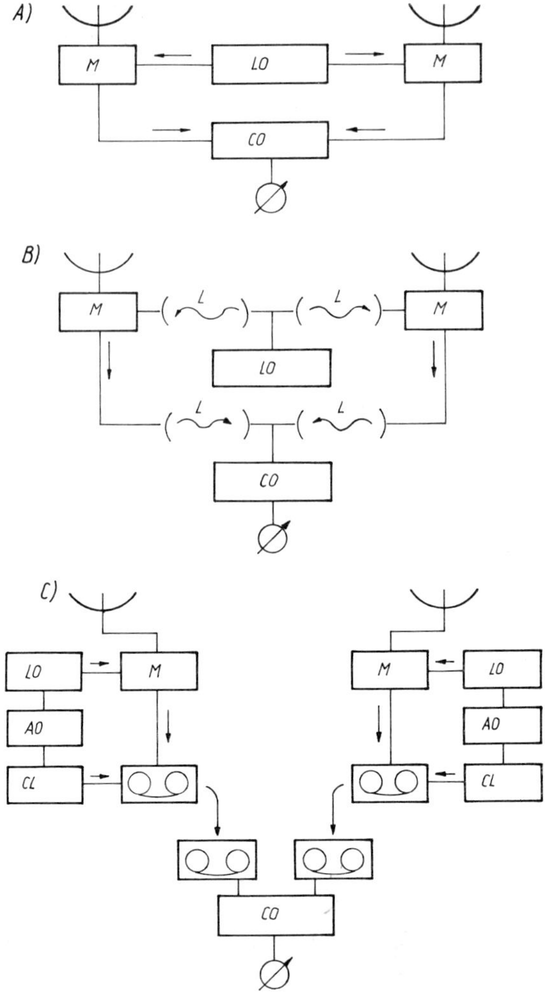

Fig. II.13. Interferometer systems: (A) – conventional interferometer, (B) – long-baseline interferometer, (C) – very-long-baseline interferometer, (M – mixer, LO – local oscillator, CO – correlator, L – radio link, AO – atomic oscillator, CL – clock).

2.7.5. The Grating (Multielement) Interferometer

In order to observe (nonstationary) brightness distributions, which is a common task in solar radio astronomy, two-element interferometers with a fixed baseline are not sufficient in general. Two-element interferometers could solve the task of the localization of a source, i.e. determine its position (and angular size). For a determination of the brightness distribution of a local radio source, however, the knowledge of a greater number of Fourier-components is required according to Equation (II.63). This can be achieved either by an interferometer of variable baselines, which requires a lot of observing time and analysis, or, better, by multielement interferometers.

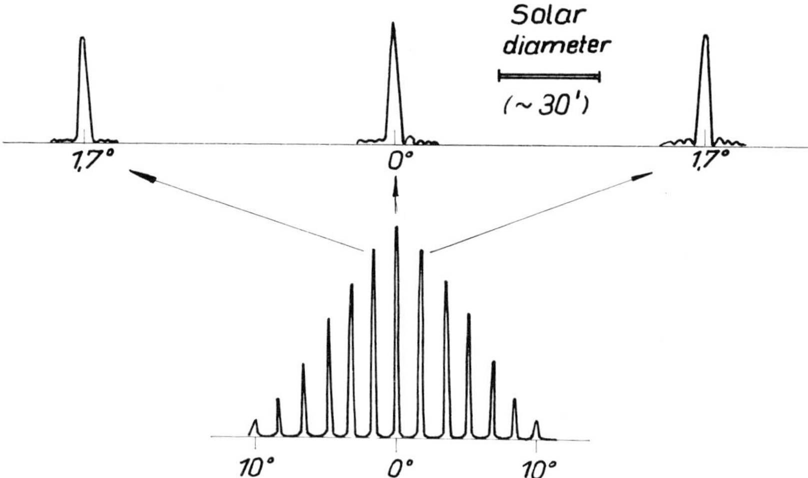

Fig. II.14. Example of an idealized 32-element grating interferometer record.

Taking such a multielement or grating interferometer (analogously to optical gratings) consisting e.g. of N identical aerials spaced by equal distances a between each other, the reception pattern is

$$A(\vartheta, \varphi) = A_0(\vartheta, \varphi) \frac{\sin^2\left(N\pi \frac{a}{\lambda} \sin\vartheta\right)}{N^2 \sin^2\left(\pi \frac{a}{\lambda} \sin\vartheta\right)} \tag{II.69}$$

where $A_0(\vartheta, \varphi)$ is the aerial pattern of an individual element. The pattern exhibits a number of clearly separated fan beams located at $\vartheta_k = 1/\sin(k\lambda/a)$ $(k = 1, 2, 3, \ldots)$ as sketched in Figure II.14. The first instrument of this type was built by Christiansen and Warburton (1953) using a system of 32 parabolic aerials. Because the distance between adjacent beams depends on the spacing between adjacent aerials it is convenient to hold this spacing as large as the beam distance is greater than the diameter of the Sun. Then it is possible to obtain scans of the Sun avoiding interferences between adjacent fan beam lobes.

2.7.6. Cross-type Interferometers

The linearly arranged multielement interferometer allows an angular resolution only in one dimension. In order to obtain a two-dimensional brightness distribution there are principally different possibilities, e.g. by using a series of one-dimensional fan-beam strip scans at different directions, scanning of pencil beams obtained by cross-type interferometers, heliographs, or large parabolic antennas, and synthesis of different kinds of unfilled apertures. In this way, cross-type interferometers are a particular means of achieving two-dimensional resolution. The cross-type interferometer was first introduced into radio astronomy by Mills and Little (1953). The instrument consisted of two aerials extended in perpendicular directions.

In solar radio astronomy the combination of the principle of Mills' cross with the grating interferometer techniques has found wide application. The first instrument of that type was Christiansen's cross consisting of two perpendicularly arranged grating interferometer profiles of 2 × 32 parabolic antennas operating at 21 cm wavelength (Christiansen and Mathewson, 1958). In contrast to Mills' cross this system has not one but several multiple pencil beams separated sufficiently widely from each other. A similar cross-type interferometer was constructed e.g. at Stanford/USA (Bracewell and Swarup, 1961) operating at 9 cm wavelength.

Sometimes grating or grating-crossed interferometers are connected with special single aerials to perform as compound interferometers. The advantage of the compound-interferometer technique is a reduced beam width as compared with a continuous aperture which is paid for by larger side lobes. A classical example for a compound interferometer was described by Covington (1960).

The first generation of pencil-beam interferometers obtained beamwidths of about 1–3 min of arc. Meanwhile the next generation has been designed aiming for an angular resolution of some tens of seconds of arc.

2.7.7. Heliographs

Solar multielement interferometry finds its perfection in the construction of radioheliographs. The most prominent example of this type is the Culgoora radioheliograph in Australia. The antenna system of this instrument consists of 96 steerable parabolic aerials (each of 13 m diameter) spaced uniformly along a circle of 3 km diameter. The number of the aerials results from the necessity of avoiding high-order diffraction patterns in the field of view of 2° diameter. The size of the individual aerials follows from the requirement of a sufficiently large collecting area which should be large enough to deliver a picture of the quiet Sun in the period of the order of 1 min. In order to monitor the fast development of solar bursts, a much more rapid sequence of pictures is needed. The Culgoora radioheliograph possesses a picture rate of one complete picture (two polarizations) per second.

The block diagram of the Culgoora heliograph is reproduced in Figure II.15 in the version as designed for operation at 80 MHz (Wild, 1967). The preamplified signals are coming in two polarizations to the central unit via separate broad-band transmission lines. Later on, a three-frequency operation was achieved, extending the observation to 160 and 43.25 MHz. For the latter frequency a set of 48 corner reflectors has been constructed because it proved impracticable to use the original aerials at a frequency as low as 40 MHz. The fields of view, beam widths, and picture formats are listed for the three operating frequencies in Table II.4.

Special care must be devoted to the process of beam shaping, because the response pattern of a circular array includes a number of unwanted rings around the pencil beam. The beam shaping is obtained by employing a special method for correcting the beam pattern known as 'J^2 synthesis' (Wild, 1965). The method makes use of the fact, that inserting appropriate phases into each transmission line from the aerials, any polar diagram with the form of the Bessel function $J_k^2(2\pi ar)$ can be generated

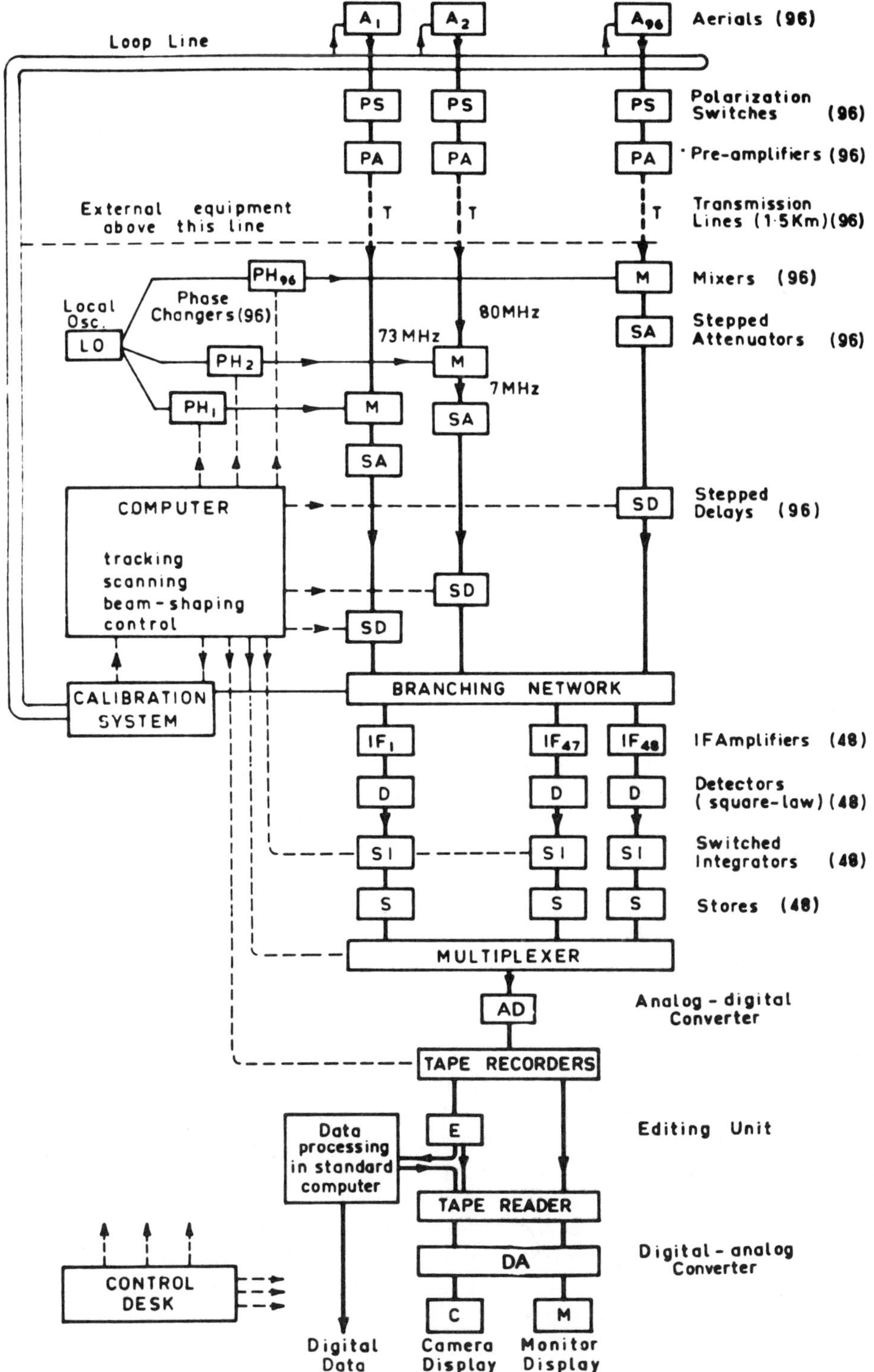

Fig. II.15. Block diagram of the Culgoora radioheliograph (after Wild, 1967).

TABLE II.4
Characteristics of the Culgoora radioheliograph

Op. frequency	43.25	80	160 MHz
Beamwidth	7.5′	3.8′	1.9′
Picture field	2° × 1.6°	2° × 1.6°	1° × 0.8°
Point number	30 × 24 = 720	60 × 48 = 2880	60 × 48 = 2880
Point spacing	4.2′	2.1′	1.0′

(where a = radius of the heliograph circle expressed in wavelengths, r = measure of the angular distance from the axis of the circle, and k = any integer). The realization of suitable phasing arrangements and all necessary switching operations are carried out by a computer which forms the 'heart' of the radioheliograph. The exact shape of the resulting pencil beam depends on the computer program. A beam was used which is 15% narrower than that of a corresponding mirror of 3 km diameter. The output picture of the Culgoora heliograph is formed by two intensity-modulated TV-tubes which can be photographically recorded but also digitally stored on a magnetic tape ready for subsequent data reduction computations.

Beside the grating-cross interferometers considered in Section 2.7.6 other projects for a two-dimensional mapping of the Sun have been designed in the microwave range. At Toyokawa, Japan, a 3-cm-radioheliograph has been established using a T-shaped antenna system composed of 32 + 16 parabolic dishes. The heliograph has an angular resolution of 1.6′ and 2.4′ before and after smoothing in direction near the zenith (Tanaka *et al.*, 1970).

A second radioheliograph has been completed also at Toyokawa operating at 8 cm wavelength and providing a maximum resolving power of 1.5′. The array consists of 32 + 17 T-shaped arranged parabolic dishes (3 m diameter) allowing a flexible operation in various modes in combination with a high-resolution compound interferometer (Ishiguro *et al.*, 1975).

2.8. Aperture-Synthesis Methods

2.8.1. General Principle

A radically new approach to the achievement of high angular resolution is the technique of aperture synthesis. This method was pioneered by Ryle and Hewish (1960), the first Nobel Prize winners of astronomy, and mainly used in nonsolar radio astronomy.

The principle of aperture synthesis is based on the fact that the radio signals from two aerials of a twin-wave interferometer yield an output which is related to certain Fourier components of the two-dimensional brightness distribution to be observed. Assuming a stationary radio source, it is possible to collect the needed information of the Fourier components by a series of measurements with one fixed aerial and the other movable in various positions thus subsequently filling the aperture of an imaginary giant telescope which is synthesized with the aid of a computer.

For applications to solar radio astronomy the attractiveness of the aperture-synthesis principle is somewhat restricted due to the requirement that the observed radiation has to remain constant throughout the full series of measurement. Evidently this makes an application to radio bursts more difficult than to observe the quiet Sun and certain slowly varying phenomena on the Sun.

CHAPTER III

PHENOMENOLOGY OF SOLAR RADIO EMISSION

A number of solar phenomena are observed at different radio and other wavelengths exhibiting characteristic time scales ranging from small fractions of a second to tens of years. These variations of the solar radiation are closely connected with solar activity and form a part of it. The variations can be expressed in terms of the ratio of 'disturbed' amplitudes to an 'undisturbed ground level'. Evidently such ratios are largest at the flanks of the solar electromagnetic spectrum (Planck curve), i.e. in the X-ray and radio regions.

According to the time-scale spectrum we can distinguish the following main components of the solar radio emission:

(1) The basic component and the quiet Sun;

(2) The slowly varying component and noise storms; and

(3) Various types of burst emissions.

Some details regarding this classification of solar radio waves are listed in Table III.1.

Besides textbooks there already exist a number of reviews on solar radio phenomena written from different points of view (e.g. Kundu, 1963; Maxwell, 1965; Wild *et al.*, 1963, 1972; etc.). Many details described in these reviews will not be repeated in this present survey. Instead, we emphasize the basic features which seem to be most important for a general orientation of the main phenomena and for an understanding of the essential physical processes which may take place in the source

TABLE III.1
Classification of solar radio waves

Phenomenon	Time scale	Place of origin	Wavelength range	Associated optical feature
Quiet Sun	steady	whole Sun	whole spectrum	solar disk
Basic component	~ 11 years	whole Sun	$S_{\nu,\max}$ at dm waves	solar disk
Slowly varying component (S-component)	months–days	active regions (A.R.) ('radio plages')	mm–cm–dm-waves	spots and plages
Noise storms	days–hours	restricted areas inside A.R.	dm–m–Dm-waves	large spot groups
Various types of bursts	minutes–seconds	restricted areas inside A.R.	whole radio range	flares (often)

regions of the radio waves. Also, we discuss the results of some recent developments not contained in the reviews. These developments are:

- Improvement of the angular resolution power, application of heliography and long-baseline interferometry;
- Extension of the solar radio spectrum towards longer wavelengths (Dm-, Hm-, and km-waves) by extraterrestrial measurements;
- Application of mm- and sub mm-waves;
- Adoption of high-resolution spectrography, investigation of fine-structure effects; and
- Application of polarization measurements.

3.1. The 'Quiet' Sun

3.1.1. Spectrum

Observations of the radio emission from the total Sun in sunspot minimum during the absence of bursts and noise storms yield the basic spectrum represented in Figure III.1. In principle, this spectrum can be interpreted as thermal radiation originating in different heights of the solar atmosphere depending on wavelength. At shorter wavelengths the radiation originates at deeper levels, correspondingly towards sub mm-waves the spectrum approximates the Planck curve at the photospheric temperature. At metric and longer wavelengths the coronal temperature of more than 10^6 K is indicated by the spectrum. Between these two extreme regions there is a transition region of intermediate radiation temperatures in the cm–dm range indicating the influence of the chromosphere and the transition zone to the low corona. Normally in the latter region the slowly varying component of solar radio emission is superimposed on the spectrum of the quiet Sun (cf. Section 3.2). Therefore the spectrum referring to the maximum slowly varying component as observed e.g. during the IGY is also indicated in Figure III.1. On any day between solar maximum and minimum the actually observed spectrum lies in the hatched region between these two extrema. In that way the 'quiet' radio Sun is defined by means of the lowest flux values observed during a particular minimum of solar activity. In practice, however, the real Sun is almost never totally 'quiet' and therefore the conception of the quiet Sun is strictly to be regarded as an idealized limiting case.

Theoretically the term 'quiet Sun' is often used to determine the radiation from the whole Sun eliminating the influence of single active regions. But we know, that the solar corona outside active regions exhibits systematic changes during the solar cycle which then lead to apparent variations of the quiet-Sun spectrum. These variations are commonly ascribed to the 'basic component'. Both notations should be distinguished. In the following we will consider the quiet Sun as an extreme (minimum) case of the basic component.

In a first approximation the quiet Sun can be described by the radiation from an idealized star possessing a static, nonisothermal atmosphere with cylindric–rota-

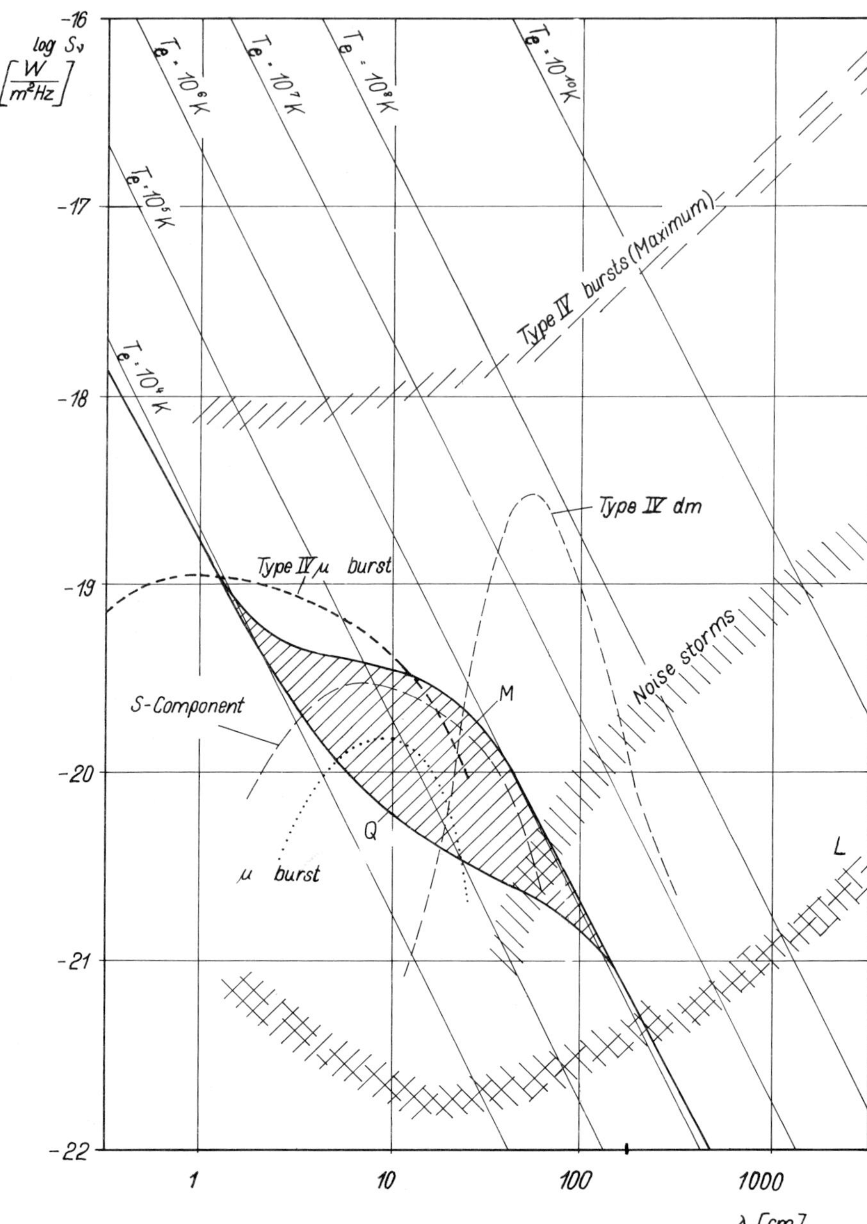

Fig. III.1. Solar radio spectra: Q – 'quiet Sun', M – maximum daily flux; L – limit of detectability of single-frequency patrol observations (after Krüger, 1968).

tional symmetry. The radiation comes from layers in local thermodynamical equilibrium (LTE). Influences of the magnetic fields can be neglected on the quiet whole-Sun spectrum. The theory of radiative transfer (cf. Section 4.2.4) is employed to calculate such models of the quiet Sun (cf. Section 3.1.3). The source height of the radiation at a given observing frequency or wavelength depends on the optical depth τ_ν along a ray path. The maximum contribution of radiation comes from regions of large gradients of τ_ν towards $\tau_\nu \gtrsim 1$. The refraction of the ray trajectories must be taken into account to calculate the ray path which becomes appreciable in the solar atmosphere for wavelengths greater than about 50 cm. For the quiet Sun the refractive index of the cold, collisionless, isotropic plasma can be used:

$$n^2 = 1 - (\nu_p/\nu)^2. \qquad \text{(III.1)}$$

Hence, at any given frequency ν the radio waves cannot propagate in regions where $\nu < \nu_p$ where $\nu_p \sim \sqrt{N_e}$ (N_e = electron density).

Since in the quiet solar atmosphere the electron density is increasing with height, a critical plasma level exists for each observing frequency below which practically no information can be gained at frequencies $\nu < \nu_p$. In this way the critical levels for increasing wavelengths are increasing with height in the solar atmosphere.

In calculating the ray paths in the quiet solar atmosphere for perpendicular incidence, due to the refraction at the plasma level, a turning point is derived which is denoted as a reflection point. For nonperpendicular incidence the height of the turning point is successively increased with increasing distance from the central meridian for geometrical reasons.

It should be further noted that the spectrum of Figure III.1 represents a smoothed curve. In the mm-range some authors suggested a nonmonotonic shape of the spectrum with a dip at 6 mm wavelength (for summary cf. Zheleznyakov, 1964/70). At present, however, it seems premature to draw any final conclusions. As the result of extensive absolute calibration experiments in the 1 to 10 GHz region, the flux values could be achieved with an accuracy of about $\pm 1\%$ (Tanaka *et al.*, 1973). On both sides of that region the accuracy of the flux measurements decreases. In the meter and decameter range the determination of the spectrum of the basic component or the quiet Sun is more problematical because the radiation is less intense than at shorter wavelengths and the influence of the Earth's sky background radiation comes into play. Additionally, principal questions arise concerning the definition of a quiet corona and the size of the radio Sun at long wavelengths.

Nevertheless, measurements carried out with large collecting areas may fix the spectral curve up to 10 m wavelength. The results show a certain decrease of the brightness temperature with wavelength, which cannot be well explained by current models of N_e and T_e for a homogeneously stratified corona. A certain reduction of this discrepancy could be achieved invoking the effect of scattering by an irregular corona (Aubier *et al.*, 1971).

3.1.2. Brightness Distribution

Spectra of the quiet Sun are not only obtained in terms of the whole-Sun flux density and effective temperatures, but also of intensity and brightness temperatures measured e.g. at the center of the solar disk. Beside that the intensity or brightness distribution over the whole radio Sun at different frequencies is of great interest yielding information on the distribution of temperature and electron density with height in the solar atmosphere. Such measurements can be obtained by solar eclipse observations and pencil-beam observations.

Summarizing the main results of the hitherto existing measurements of the brightness distribution of the Sun, we find (cf. Figure III.2):

(a) The diameter of the radio Sun approximates that of the optical disk at high frequencies (mm-range).

(b) In the mm-range there are contradictory results confirming or denying effects of limb brightening.

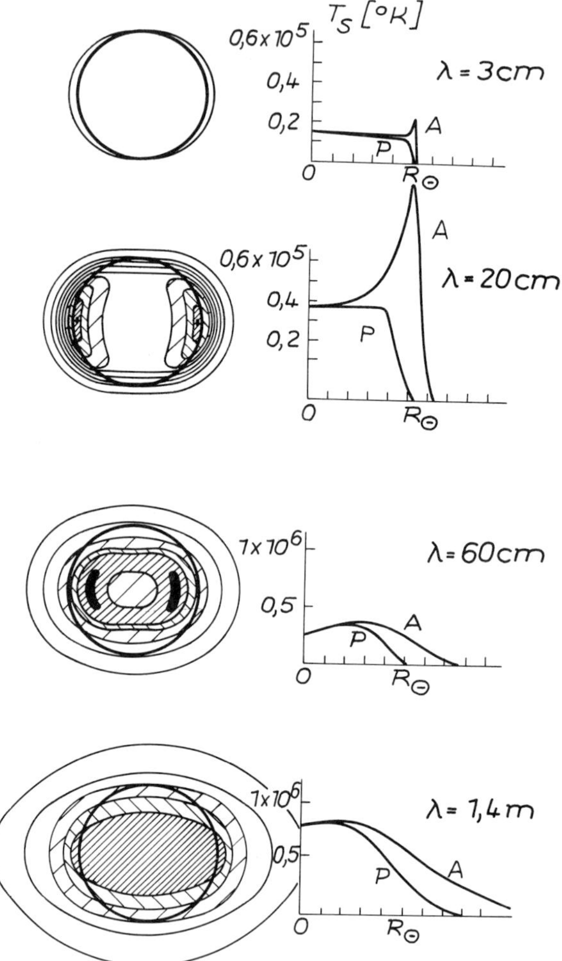

Fig. III.2. Characteristic brightness distributions of the quiet Sun at different wavelengths (after Hachenberg, 1959): *A* – equator, *P* – pole.

(c) At short millimetric waves, also limb darkening has been reported.

(d) In the cm- and lower dm-range a remarkable limb brightening is indicated at least at the equator.

(e) At longer wavelengths the size of the radio Sun increases steadily and deviations from the circular symmetry become remarkable.

(f) In the meter region a flattening of the brightness distribution is observed washing out a sharply defined solar limb.

3.1.3. Model Calculations

Postulating reasonable distributions of the temperature and electron density inferred e.g. from optical observations, model calculations of the quiet Sun assuming LTE, isotropy, and horizontal homogeneity already provide an understanding of the most important features. Applying the approximation of geometric optics, the solutions of the equation of radiative transfer along the ray trajectories taking height-depending temperature and density profiles can be easily achieved by numerical integration. A fundamental summary on these matters was given by Pawsey and Smerd as early as 1953. One of their excellent results which should not be omitted from any review on this topic is reproduced in Figure III.3 showing the height variation of the optical depth with frequency.

More refined calculations may take into account the inhomogeneous structure of the solar atmosphere at deep levels. The existence of a limb darkening effect in the short mm-region would require the use of models containing at least two components (Lantos and Kundu, 1972). Otherwise 'averaged' chromospheric models with temperatures increasing with height would lead to an increased brightness towards the limb because near the limb at same optical depths the radiation origi-

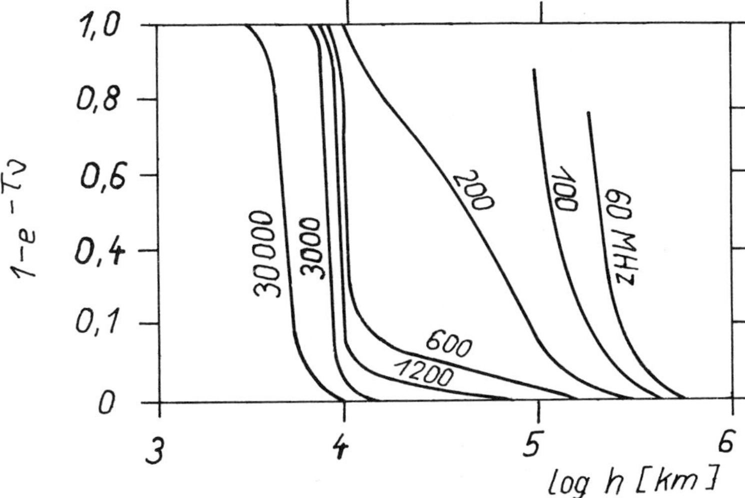

Fig. III.3. Quiet-Sun height variation of the optical depth for different frequencies (after Smerd, 1950).

nates at higher altitudes. This is due to the effect of increased geometrical lengths of the ray trajectories at greater central meridian distances.

In order to explain the limb darkening effect a two-temperature distribution corresponding to the spiculae and interspicular regions can be invoked. Then, among various models those must be rejected, for which the temperature at low altitudes is greater than the interspicular temperature (Fürst *et al.*, 1974).

From the observed radio spectrum compatible temperature or density profiles can be deduced applying the method of Laplace transform of the brightness temperature. This method has been used by several authors (e.g. Piddington, 1954; Moriyama, 1961; Chiuderi *et al.*, 1972). It provides a check of optical or UV models by radio data deriving temperature models from density models and vice versa. The results are of particular interest for the study of the transition layer between chromosphere and corona (cf. Athay and Newkirk (ed.), 1969; Athay, 1976).

The obtained distributions of T_e and N_e can in turn be applied to various problems like the derivation of the absolute abundances of elements in the corona (Chiuderi *et al.*, 1972). According to the relation

$$E \simeq C \times A \int F(T(h), \ldots)\, N_e^2(h)\, \mathrm{d}h \qquad \text{(III.2)}$$

the abundance A of an element cannot be derived from the optical flux of a resonance line E alone. The knowledge of the temperature T and electron density N_e as a function of height h is also required. The quantity C is a constant containing various parameters, e.g. the oscillator strength of the line, the Gaunt factor, and others (cf. Pottasch, 1964).

Turning to longer radio wavelengths the effect of scattering at macroscopic density fluctuations becomes important. The scattering of radio waves in an inhomogeneous corona has been already treated by many authors (e.g. Scheffler, 1958; Fokker, 1965; Steinberg *et al.*, 1971; Aubier *et al.*, 1971). For small scattering angles the definition applies

$$\langle \delta\Omega^2 \rangle = 2\sqrt{\pi} \int_{\text{ray}} \frac{\langle \delta n^2 \rangle}{n^2 h'}\, \mathrm{d}s \qquad \text{(III.3)}$$

(n = refractive index, h' = correlation radius of the density fluctuations, s = ray path) (Chandrasekhar, 1960; Hollweg and Harrington, 1968).

Since in the cold collisionless coronal plasma $\langle \delta n^2 \rangle \sim \nu_p^4/\nu^4$, the scattering angle varies as $(\nu n)^{-2}$ and hence the growing importance of the scattering at longer (e.g. decametric) wavelengths is easily explained. In the model calculations the scattering leads to a decrease of the brightness temperature (and also of the total emission) of the Sun. This fact is due to the effect, that on the average the turning point of the scattered rays is at greater heights than that of the unscattered waves, whereas the emissivity decreases with height in the corona.

3.1.4. The Basic Component of Solar Radio Emission

Since the solar corona exhibits systematic variations in the solar cycle, its radio emission also undergoes systematic cyclic variations. These variations are not only due to the direct influence of active regions (as it was sometimes believed) which is mainly expressed by the slowly varying component, but also to the physical state of the solar atmosphere outside of active regions. The latter is described by the term basic or B-component as indicated by Table III.1.

Several definitions and methods of a quantitative determination of the B-component have been used in the literature. One of these possibilities consists in a derivation of the lower envelope of a number of interferometric scans across the Sun over a period of one or more solar rotations. On the other hand, the B-component has been derived statistically as an extrapolated base level of a correlation analysis between the daily whole-Sun radio fluxes (exhibiting the influence of the slowly varying component) and a suitable standard parameter of solar activity, e.g. sunspot and plage areas or sunspot numbers. This method was first introduced by Christiansen and Hindman (1951) and in many variants used by numerous authors (for references cf. e.g. Krüger *et al.*, 1964).

A further statistical method was employed by Das Gupta and Basu (1965) extrapolating zero values of the variation amplitudes of daily solar radio fluxes of 27-day periods. Though different in their specific forms, all of the adopted methods aim to eliminate the direct influence of the active regions on the Sun and therefore give qualitatively similar results.

Adopting one of the above described methods of derivation of the B-component it is found that this component has a maximum contribution in the lower dm-region. This is in some contrast to the average spectrum of the slowly varying component of solar radio emission which has its spectral maximum in the 5–10 cm wavelength range (cf. Tanaka, 1964; Krüger and Olmr, 1973).

An explanation of the spectral distribution of the B-component can be given in terms of cyclic variations of the coronal electron density. Invoking thermal bremsstrahlung as the origin of the B-component, the increase of the optical depth would shift the source region of the main contribution to the radiation just in the lower dm-range from chromospheric into coronal heights, where the temperature is remarkably increased (cf. also Figure III.3).

3.1.5. Exploration of the Outer Corona

At greater distances from the photosphere than about one solar radius the quiet solar radio emission turns out to be an inappropriate tool to study the corona. Instead of this the coronal structure can be observed indirectly by the interplanetary scintillations and the effect of the angular broadening of nonsolar radio sources due to the diffraction and refraction of the radio waves by electron density fluctuations. This is an inexpensive and sensitive method for the investigation of the outer corona and the interplanetary space which are strongly affected by the solar wind (cf. Kakinuma and Watanabe, 1976).

The highly inhomogeneous structure as a permanent feature and the continuous merging into the interplanetary medium make a definition of an idealized quiet corona in the outer regions more difficult than at deeper levels; however, the wave-scattering properties of the corona appear to vary with the solar cycle especially at smaller distances to the solar surface.

The first studies of coronal scattering were made by Machin and Smith (1951) and Vitkevich (1951), who predicted that the corona would cause a deviation of the radio waves coming from cosmic objects close to the Sun–Earth line of sight. Subsequent interferometric occultation measurements, particularly of the Taurus-A radio source which passes the corona in June of each year at about 5 $R_{\odot}$ distance from the Sun, revealed effective solar occultation radii that are dependent on the observing frequency. Some details of that story have been reported by Kundu (1965). New developments in this field including theoretical implications and extensive references are summarized by Jokipii (1973).

By the occultation method information about the electron density, the shape of the corona, as well as its spatial and temporal variations, could be traced out up to distances of about hundred solar radii (Slee, 1961). Later on the discovery by Hewish *et al.* (1964) of the interplanetary scintillation from small-diameter quasi-stellar radio sources, as well as the observations of the dispersion rate of pulsar radio signals in comparison to optical light pulses (cf. Goldstein and Meisel, 1969) at positions near the Sun, have advanced the study of the structure and motions in the outer corona.

The interpretation, especially of scintillation power spectra, leads to the conclusion that the solar wind is turbulent near the Sun, supporting the idea that turbulent waves in the upper corona play an essential role as driving forces for the solar wind. In this way it appears likely, that the Alfvén waves also observed in the interplanetary medium at about 1 AU (Belcher, 1971) are of solar origin opening, as complementary to radio astronomy, a new field of 'Alfvén-wave astronomy'.

3.1.6. About solar line emissions

Up to now line emissions do not play an important role in solar radio astronomy. Following a suggestion by Dupree (1968) the high-quantum-number recombination lines of certain ions in the corona and chromosphere–corona transition region should be observable at mm-wavelengths. These high n-transition lines should arise, because dielectronic recombination in a medium of low density leads to departures from the LTE and, consequently, to an overpopulation of the high n-states. These lines are expected to appear in emission or absorption at shorter or longer millimetric wavelengths, respectively. Experimental detection of these line emissions is difficult because the lines are weak. This is due to several reasons, e.g. the occurrence of an intermediate region between emission and absorption, the rapid increase of the Stark broadening at longer wavelengths and local inhomogeneities of the radiating and absorbing medium (cf. Greve, 1974).

3.2. The Slowly Varying Component

3.2.1. General

Looking at 'slow-period' variations (in contrast to bursts) of the solar radio emission in the microwave region, two characteristic phenomena can be roughly distinguished. The first is the long-period variation of the B-component connected with the 11 yr cycle of solar activity, which was discussed in Section 3.1.4. The second phenomenon corresponds to the occurrence of individual active regions and represents their development and characteristic lifetimes. This component is called the slowly varying component or *S-component* of solar radio emission.

The S-component is best observed in the cm- and lower dm-range, but it is also detectable at mm- and m-wavelengths with instruments having sufficient spatial resolution. The enhanced radio emission originates in local regions above sunspots and chromospheric plages. For this reason sometimes also the terms 'spot component' and 'radio plages' have been used in the literature to denote the S-component. The emitting regions of the S-component are characterized by values of electron density, magnetic field, and temperature, which are enhanced with respect to the surrounding medium. The source regions are extending in height from the chromosphere up to the lower corona and were related to the occurrence of coronal condensations which are detectable in special emission lines of the visible light (e.g. 5303, 5694, 6374 Å).

It must be noted that a slowly varying component is also present in the solar soft X-ray emission exhibiting many similarities to and a good correlation with the S-component of the radio emission. But in contrast to the radio emission the amplitudes of the soft X-ray variations are very much higher and the X-ray flux level corresponding to the solar minimum is below the limit of sensitivity of the detectors used for solar patrol observations so that it is difficult to detect a basic component of the X-ray emission analogous to the radio emission. Thus, in general the observed soft X-ray flux corresponds to the slowly varying component (outside bursts) which in contrast to the terminology of the radio physicists sometimes has been called misleadingly the 'quiet component' in delimitation to burst emissions.

For solar physics the study of the S-component of the radio emission is of considerable interest especially from the following points of view:

(1) Investigation of the structure and physical processes in active regions, derivation of plasma parameters referring to layers in the solar atmosphere which are best accessible by radio methods with the goal to develop complex models of the active regions.

(2) Investigation of the interrelation between slow and fast acting processes (bursts), clarification of the role of the S-component as an indicator for energy-storage (and subsequent release) processes leading to flares with respect to a possible utilization for flare forecasts.

(3) Application of the measured flux values of the S-component as a general, physically well defined standard measure of solar activity (unlike sunspot numbers etc.), being available for different purposes of solar and space physics, as well as for geophysics, biophysics, and other scientific branches.

Since its detection in 1946 by Covington the systematic investigation of the slowly varying radio emission from active regions has been performed with a continuous improvement of the observational methods and physical models for its interpretation. Review papers dealing with different aspects of these matters were written in the past e.g. by Christiansen (1966), Swarup (1964), Pick (1965), and Hachenberg (1965).

3.2.2. Total-Flux Characteristics

The denotation 'S-' or 'slowly varying' component was originally derived from single-frequency, whole-Sun radio flux observations which exhibit a slowly changing level in the microwave region due to the day-to-day varying appearance of active regions on the visible hemisphere of the Sun. The flux variations are then due to the effect of the growth and decay of the active regions as well as to a center-to-limb variation resulting from solar rotation.

By means of low-resolution radio telescopes, various stations have established a relatively simple but effective service of daily solar patrol observations covering the spectrum by a set of single frequencies. These stations are distributed over different longitudes around the globe. Long-time series of such flux measurements have been performed; e.g. by the Astrophysics Branch of the National Research Council of Canada, Ottawa; the Radio Astronomy Observatory of the Research Institute of Atmospherics of the Nagoya University, Toyokawa, Japan; and the Heinrich-Hertz-Institute of the Academy of Sciences of the GDR, Berlin. In contrast to a mere watch for solar bursts the patrol of the S-component requires a greater stability and sensitivity of the receivers. High technical efforts are also demanded if absolute calibrations are tackled, forming a necessary but troublesome task without prospects of quick spectacular discoveries. Consequently this kind of work is carried out only by few solar observatories as mentioned in Section 2.5.

An example of a plot of daily mean fluxes at different frequencies is shown in Figure III.4 displaying the influence of the S-component for a half-year period. Such a time series provided (beside eclipse observations) the first extensive material about the slowly varying solar radio emission, stimulating a multitude of statistical studies.

The excellent correlation of the S-component with optical phenomena as sunspots and plages and also with soft X-ray fluxes has challenged the attention of the observers for a long time. Thus, numerous statistics are concerned with these features, separating them into a core, halo, and basic component. In the simplest case of a regression analysis one gets

$$S_\nu = a_\nu R + b_\nu, \tag{III.4}$$

where a_ν is the regression coefficient and R is a reference quantity of solar activity, e.g. the Zürich sunspot number. b_ν is the extrapolated base level of the flux density S_ν. An extension of this method involves a partial or multiple correlation analysis e.g. by putting

$$S_\nu = A_\nu R_1 + B_\nu R_2 + C_\nu, \tag{III.5}$$

where R_1 and R_2 may represent spot and plage areas, respectively. Naturally, the resulting regression coefficients and base levels depend on the test frequency and the selected time interval (phase of the solar cycle). Furthermore remember that different lifetimes of the compared phenomena (e.g. longer duration of the S-component emission in comparison to the sunspots) introduce systematic errors so that, for instance, the obtained base levels can be regarded as upper limits.

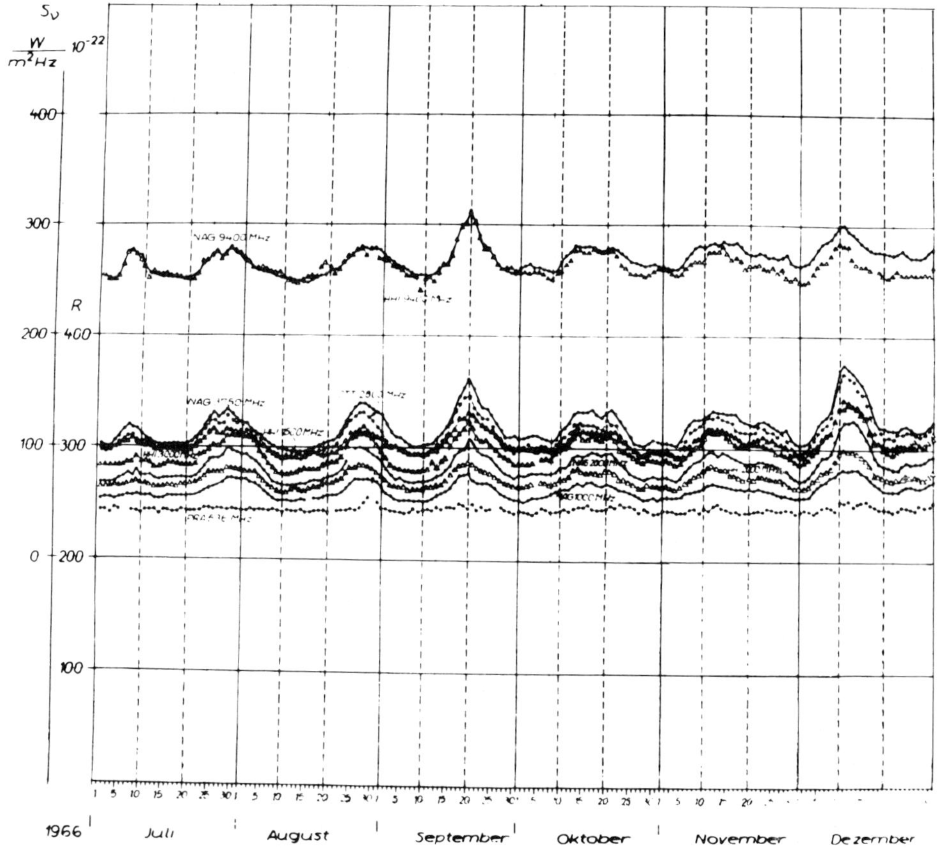

Fig. III.4. Example for daily means of solar microwave fluxes of a selected half-year period exhibiting the S-component variations.

Autocorrelation studies and spectral analyses of the time series of S-component fluxes display the influence of the 27-day period of the synodic solar rotation. The evaluation of the whole power spectrum which should reflect the distribution of lifetimes of the individual source regions is difficult without having a simultaneous patrol of the far side of the Sun.

The flux spectrum of the S-component has a maximum in the region between 5 and 10 cm wavelength. Its shape shows a remarkable stability in comparison to the more variable microwave burst spectra. The flux spectrum is smoothed out by the influence of the superposition of all active regions simultaneously present on the visible hemisphere at a given time. According to the measurements of Ottawa and Toyokawa during the 11 yr solar cycle a slow shift of the spectral maximum towards 10 cm wavelength near the solar maximum is observed indicating a certain influence of solar cycle temperature and density variations in active regions.

3.2.3. The source regions

a. *Observations*

Apart from mere flux observations more information is gained from measurements with higher spatial resolution. The source regions of the S-component can be studied by means of solar eclipse observations, interferometric and heliographic observations, single-dish observations, and application of the long-baseline interferometry and aperture-synthesis method.

From such measurements, various parameters as position, size, and height of the sources, brightness temperatures, center-to-limb variation, and directivity, etc. can be deduced. Results of the measurements are as follows:

The position and structure of the source regions of the S-component at mm- and low cm-wavelengths correspond well with those of the accompanying Hα plages. On the other side, Hα filaments are correlated with brightness depressions on the radio maps of the Sun (Buhl and Tlamicha, 1970). The radio absorption features sometimes appear larger (2 to 3′) than the corresponding dark Hα filaments. The depressions are of the order of 100 K but can reach up to 400 K below the background temperature level at millimetric waves. These temperature depressions can be interpreted by assumption of free–free absorption of the background mm-radia-

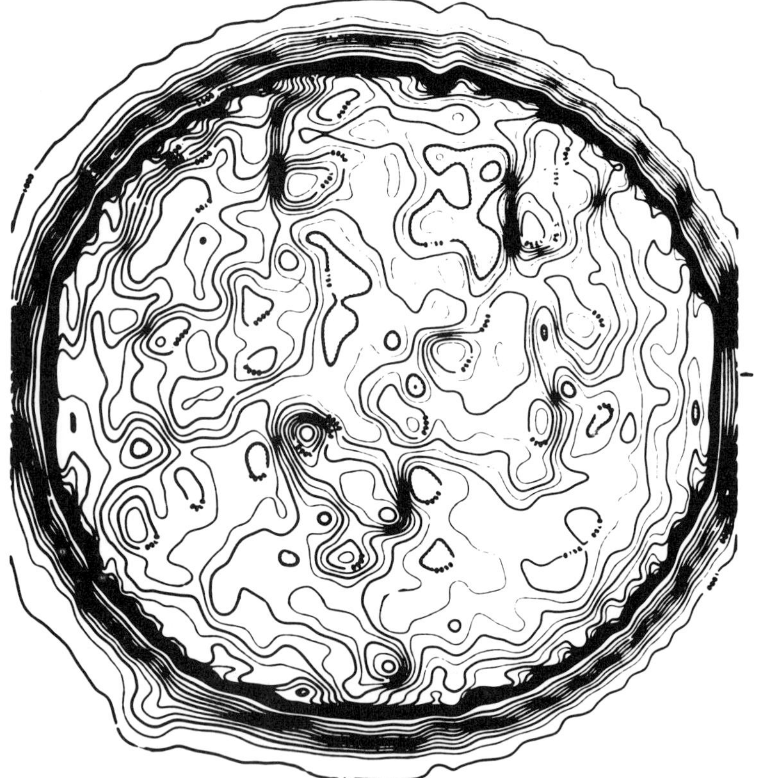

Fig. III.5. A contour map of the Sun at 3.5 mm wavelengths (March 14, 1970) (after Kundu, 1972).

tion by the filaments which are characterized by a higher electron density (typically 5×10^{10} cm^{-3}) and lower temperature (e.g. 6000 K at 3.5 mm wavelength) than the surrounding chromospheric matter (Kundu, 1970, 1972). Beyond the limb the prominences tend to correspond to emissive regions in the mm-region (cf. Figure III.5).

First attempts using very-long-baseline interferometry at 1.3 cm wavelength failed to detect fine structure in active regions with an angular resolution of 0.01 of a second of arc (Kundu, 1974). Fine structure with a scale of a few seconds of arc, however, is quite a common feature and can be revealed by long-baseline interferometer observations (Kundu *et al.*, 1974).

Typically the source regions of the S-component at centimetric wavelengths consist of bright points or cores (of about 20″ diameter) surrounded by a more or less diffuse halo. The cores correspond well to the positions of strong magnetic fields which are represented by the occurrence of sunspots. Bipolar structures are nicely seen in circular polarization, when by the different magnetic polarities the sense of the neighboring emission peaks is reversed. The circular polarization (R–L) maps of the Sun reveal that all prominent active regions which are displayed as uniform patches in the total power maps of lower resolution ($> 1'$), are really bipolar in nature if observed with sufficient resolution (Kundu and McCullough, 1972a, b). As a remarkable feature, the line of zero polarization passes near the peak of the total power emission (cf. Figure III.6). Comparisons with photospheric magnetograms exhibit an excellent correspondence even with regard to identical extensions with e.g. 9.5 mm-polarization features of the Sun. Positive (R) and negative (L) regions on these maps are corresponding to plus (magnetic vector towards the observer = magnetic north polarity) and minus fields on the magnetograms, respectively. Magnetic fields on magnetograms down to 5–10 G yield detectable effects in polarization of the solar millimetric emission.

The average excess brightness temperatures of the sources of the S-component at 1, 3, and 9 mm wavelength were estimated to be about 300, 500, and 1000 K, respectively. A poor correlation was found between the brightness temperatures in the mm-region and the magnetic fields of the related sunspots. However, the size and brightness temperature of the mm-regions are closely related to the size and magnetic field strength of the Ca-plage regions, respectively (Kundu, 1971b).

According to the above stated properties it seems likely that the main part of the emission of the S-component at mm-waves has its origin in a thermal free–free emission from local regions of enhanced density. The decrease of the average brightness temperature towards shorter wavelengths can be interpreted either as an effect of reduced optical thickness at shorter wavelengths and/or of a decrease of the electron temperature with decreasing source height in the chromosphere.

At mm-waves it is also indicated that new source regions appear to develop within time scales of the order of half an hour with a gradual changing of their structure over a few hours. Here, apparently, are interesting links to the development of the burst phenomena.

Prior to the utilization of high-resolution (pencil-beam and fan-beam) devices, the solar eclipses provided the unique means of studying details of the sources of the S-component. Observations of this kind are possible in relatively rare cases and

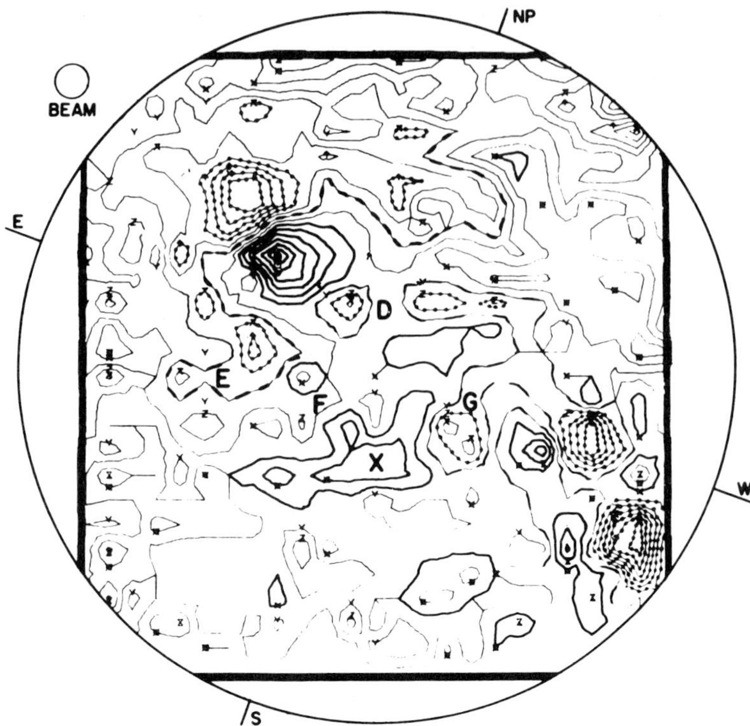

Fig. III.6. A circular polarization (R–L) map of the Sun at 9.5 mm wavelengths (February 27, 1971). Solid and dotted lines refer to − (L) and + (R) values, respectively. The dashed line corresponds to zero polarization (after Kundu and McCullough, 1972a).

in general require expensive expeditions. An extensive review of earlier solar eclipse radio observations was made by Castelli and Aarons (1965). A compilation of various observations of the particular case of the eclipse of May 20, 1966, by Soviet radio astronomers was edited by Gelfreikh and Livshits (1972).

b. *Interpretation*

In essence the radio emission of the S-component can be interpreted by thermal radiation, although a certain contribution of nonthermal energies cannot be fully excluded. The radiation can be generally explained by the two basic emission mechanisms of free–free Coulomb bremsstrahlung and gyro-synchrotron emission at low harmonics of the gyrofrequency. Hence three variants of the emission processes can be deduced, namely: (i) pure bremsstrahlung (in the absence of magnetic field influences); (ii) bremsstrahlung influenced by the presence of external magnetic fields (or also gyro-synchrotron radiation influenced by Coulomb collisions) as the intermediate case; and (iii) pure gyro-synchrotron (cyclotron or gyromagnetic) radiation of electrons in helical or circular orbits. As a matter of fact, the limiting cases of the emission mechanisms (pure bremsstrahlung and pure cyclotron emission) are easier to use in theoretical calculations than the intermediate case.

As to the history, the first mechanism treated with respect to the S-component was bremsstrahlung. Special models calculated in this direction were established by Waldmeier and Müller (1950) and Ikhsanova (1960). The effect of the gyroharmonic emission was introduced into the interpretation of the S-component by Zheleznyakov (1962) and Kakinuma and Swarup (1962). Models combining both mechanisms were presented by Zlotnik (1968) and Lantos (1968).

Presently it is believed that the strong bright-core component originates from the gyromagnetic emission, which refers to centimetric and longer wavelengths. The lower and denser (cooler) parts of the emitting volume and partly the halo region may preferably involve the bremsstrahlung mechanism (Felli *et al.*, 1975b).

The fundamentals and details of the emission processes will be outlined in Chapter IV. For the moment it is sufficient to note that the relative importance of both limiting emission processes is represented by the quantitative value of the optical thickness for each mechanism. In a magnetized plasma the contribution of the gyro-synchrotron emission exceeds that of the bremsstrahlung, if for a given electron density and propagation angle, depending on the harmonic number, a limiting temperature $T_{e,l}$ is exceeded. Hence for concrete models of a source region a height range can be calculated, where the one or the other emission mechanism is dominating.

3.2.4. Connection with X-rays and other phenomena

Apart from the S-component of the radio emission, influences of active regions can be observed in radiation of other spectral regions too. A striking feature is the connection with the slowly varying component of the solar soft X-ray emission. In the soft X-ray region the radiation consists of a continuum superimposed by discrete emission lines. The continuum is mainly due to thermal free–free and free–bound emissions. For the free–free emission the radiation flux is determined by

$$\begin{aligned} S_\nu &\sim N_e^2 \left(\frac{\chi_{\mathrm{H}}}{KT} \right)^{3/2} \mathrm{e}^{-h\nu/KT} \bar{g} V \\ &\sim N_e^2 V f(T, \nu), \end{aligned} \qquad \text{(III.6)}$$

where χ_{H} denotes the ionization energy of hydrogen and $\bar{g}$ is the average Gaunt factor (Elwert, 1954).

Analogously the free–bound (photo-recombination) emission flux is given by

$$\begin{aligned} S_\nu &\sim N_e^2 \left(\frac{\chi_{\mathrm{H}}}{KT} \right)^{3/2} \mathrm{e}^{-h\nu/KT} \sum_{z} \sum_{\substack{i \\ (h\nu \geq \chi_i(z))}} \mathrm{e}^{\chi_i(z)/KT} \frac{\chi_i(z)}{\chi_{\mathrm{H}}} f(N_e, N_i(z))\, V \\ &\sim N_e^2 V f(T, \nu, \ldots), \end{aligned} \qquad \text{(III.7)}$$

where $\chi_i(z)$ is the ionization energy of ions of the degree i and order z.

Thus both free–free and free–bound emissions can be regarded as indicators for the whole electron content and the average temperature of the emitting region. In contrast to the radio emission, the emitting volume appears typically transparent for X-rays, as a consequence of the small optical depth. Also different from the radio emission, no direct information about magnetic fields is contained in the X-ray emission mechanism.

Measurements of solar X-ray fluxes have been accumulated on a regular basis by means of satellites, such as the SOLRAD series, since the solar minimum period of 1964. In general the total flux from the Sun is monitored in fairly standardized wavelength (or energy) ranges, e.g. 44 to 60 Å, 8 to 20 Å, 8 to 12 Å, 0.5 to 3 Å, etc., according to special window materials used for the X-ray detecting equipment.

Time series of soft X-ray fluxes were used in several comparative studies with the solar microwave radio emission (White, 1963; Michard and Ribes, 1968; Krüger *et al.*, 1969).

In addition to the flux measurements, solar soft X-ray pictures reveal a wealth of details which are as yet difficult to achieve by radio. Valuable results were obtained by the Skylab mission. According to these observations a number of coronal features can be distinguished: active regions, active region interconnections, large loop structures associated with unipolar magnetic regions, coronal bright points, coronal holes, and structures surrounding filament cavities.

Evidently these findings bring completely new aspects to the phenomenon of solar activity. It would be a promising task for the future to look at corresponding radio phenomena. As yet radio pictures of the Sun with a spatial resolution comparable to the X-ray photographs are not available.

Special reviews informing in more detail about solar X-ray astronomy were published by de Jager (1965a, b), Mandelshtam (1965a, b), Goldberg (1967), Neupert (1969, 1971), Walker (1972), Culhane and Acton (1974), and Kane (1974).

3.2.5. Quasi-periodic oscillations

The investigation of fluctuations and quasi-periodic oscillations of solar emissions in different ranges of wavelengths is a relatively new and still developing branch of solar physics. The existence of quasi-periodic oscillations was first noted in optical line emissions with periods near 5 min (300 s) in the photosphere (Leighton *et al.*, 1962) and decreasing with height to about 150 s in the low chromosphere. Oscillatory motions are manifested by periodic Doppler shifts (indicating vertical velocity variations) and by periodic line intensity variations. There are different scales of velocity fields on the Sun which can be roughly divided into three groups: small-scale photospheric oscillatory motions, granulation, and a large-scale flow related to the supergranulation pattern (which according to Hα observations is extending into the chromosphere).

Furthermore other features on the Sun exhibit oscillations, such as sunspots (umbra and penumbra), plage regions, spiculae, quiescent and eruptive prominences, and coronal condensations, with different time scales between about 1 and 300 s.

Reviews on the solar oscillations were presented by Noyes (1967), Schatzman and Souffrin (1967), Gibson (1973), Bray and Loughhead (1974), Stein and Leibacher (1974), Lynch (1975), and others, providing many details regarding these questions relative to optical investigations.

Because the oscillatory patterns are not restricted to the photospheric level but also extend into the chromosphere and higher levels, it appears reasonable to ask whether the radio fluxes originating in these heights are influenced by modulations due to e.g. pressure (density) and/or magnetic field fluctuations. The occurrence of such quasi-periodic oscillations was reported by several authors (Yudin, 1968; Durasova *et al.*, 1971; Simon and Shimabukuro, 1971; Kobrin and Korshunov, 1972; Kaufmann, 1972; Bocchia and Poumeyrol, 1974) while other authors (Shuter and Cutcheon, 1973; Sentman and Shawhan, 1974) tried to demonstrate that the confidence of the above quoted results is marginal. In any case, the amplitudes of the oscillations in the microwave region are rather low (of the order of 10^{-3} of the undisturbed solar flux) and therefore hardly detectable with radiometers measuring solar fluxes on a routine basis. Special arrangements including the application of high-resolution spectrographic and interferometric equipment revealed several new features which still require more detailed studies. There are to be mentioned: (a) Quasi-periodic oscillations of the centrimetric flux from active regions (cf. Kobrin, 1976); (b) temporally changing fine structures in the spectrum of the S-component (Kaverin *et al.*, 1976); and (c) a small-scale quasi-periodic disk component (Lang, 1974).

The last point makes evident that quasi-periodic oscillations may not be only a phenomenon of the S-component and solar active regions. They can be also present in the 'quiet Sun', as well as in bursts and other phenomena on the Sun.

Thus the study of quasi-periodic oscillations seems suitable for contributing to various general interesting questions of solar physics, e.g.:

- Study of different phenomena in the solar atmosphere exhibiting fluctuations, such as the Sun as a whole, supergranulation, giant cells, spiculae, coronal condensations and magnetic cores, different kinds of prominences, etc.;
- Energy storage of active regions and the build up of flare energy;
- The heating of the corona and energy transport in the solar atmosphere; and
- Exploration of subphotospheric processes.

In this way, effects of fluctuating processes can be found in all layers of the solar atmosphere including the solar wind, with a wide scatter of the characteristic periods ranging from minutes to hours. It must be noted, however, that naturally not all of the fluctuations are always detectable but exhibit strong variations with time in intensity and other properties. This difficulty – added to the basic difficulties of the receiver stability and the elimination of fluctuations from the terrestrial atmosphere – may explain a number of contradictory results about the fluctuations, particularly of the S-component. The possible occurrence of enhanced fluctuations of the S-component with periods >20 min before the onset of proton flares (Kobrin *et al.*, 1976) bears interesting prospects for the forecast of strong solar flare events (proton flares).

3.2.6. Large-scale Patterns

Since the S-component of solar radio emission is closely related to the magnetic fields of active regions and since the active regions appear to be related to weaker long-lived large-scale magnetic field patterns on the Sun, it may be worthwhile to discuss some features of the S-component in connection with global structures. Such large-scale structures are indicated by the photospheric magnetic fields exhibiting the characteristic sector structures. These fields extend into higher levels up to the interplanetary space where often well defined sector boundaries can be observed by satellite measurements. These boundaries indicate the existence of well marked regions of opposite magnetic polarity of large extent. When comparing the photospheric and interplanetary magnetic fields near the Earth, it is to be borne in mind that due to the solar-wind transport the interplanetary magnetic field pattern

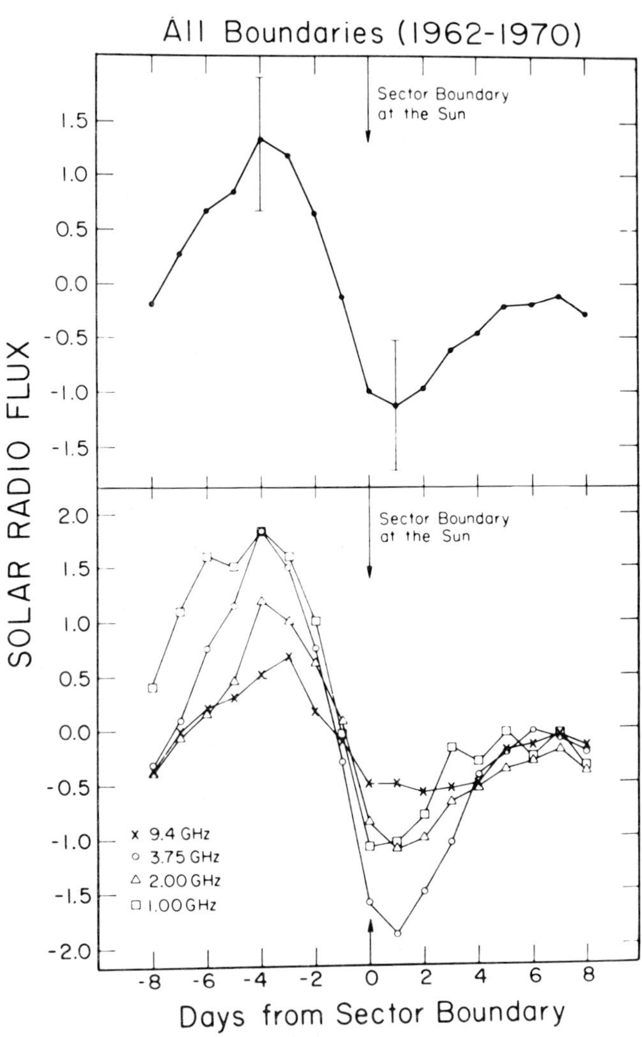

Fig. III.7. Superposed-epoch averages of daily solar radio fluxes from measurements of Toyokawa showing the effect of sector boundaries on the Sun. Top: Average plot for four microwave frequencies (after Scherrer and El-Raey, 1974).

reaches the distance of 1 AU about four days after its corresponding central meridian passage at the photosphere.

A superposed epoch analysis using the interplanetary magnetic field polarity inferred from tabulations of the interplanetary sector boundaries of Wilcox (1973) was made by Scherrer and El-Raey (1974) in order to examine the average flux level of the S-component near the sector boundaries. As indicated by Figure III.7 there is a peak of the emission several days before the sector boundary passage. In detail, the emission peak is higher within toward-field (−) sectors than if associated with away-field (+) sectors.

The above cited statements agree well with a result of Bumba (1972), who found more activity associated with large-scale regions of toward-magnetic-field polarity than with those of away-field polarity. These findings display nicely the existence of systematic relationships between the large-scale structures on the Sun and properties of individual active regions. The radio observations also suggest an asymmetric behavior of large-field patterns with opposite polarity. Similar connections are indicated for other phenomena of solar activity (e.g. flare activity).

3.3. Solar Continuum Bursts (A): Microwave Bursts

3.3.1. General remarks

Leaving the discussion of the quiet and slowly varying phenomena of the radio Sun, in the following sections we will deal with the more rapidly varying phenomena, viz. *solar radio bursts*. There are many kinds of solar burst radiation and generally solar burst activity is known to cover all regions of the radio spectrum occupying nearly all levels of the solar atmosphere. The literature about radio bursts is rather extended. During the recent years a number of reviews appeared (e.g. Wild *et al.*, 1963, Wild and Smerd, 1972; Kundu, 1965; Maxwell, 1965; Krüger, 1972; Boischot, 1974a).

It is a remarkable fact, that generally all types of solar radio bursts are more or less directly linked with the phenomenon of solar flares seen in visible light. In this way the optical flare appears as one part of a more general '*flare phenomenon*' which should be better called '*flare-burst phenomenon*' (cf. Section 1.2.3.d). The whole phenomenon comprises a multitude of very different features not only restricted to the optical and radio spectral ranges. The physical exploration of the flare phenomenon is one of the most fascinating tasks in modern astrophysics.

It must be noted that the various types of radio bursts can provide only special and limited aspects of the whole flare phenomenon, and the information coming from radio bursts is unlikely to replace that obtained by other methods. We start our discussion of the radio bursts with a special class of continuum bursts which shows the closest relation to the optical flares and herewith also to the question of the origin of the flare-burst complex in general.

The terms solar '*continuum*' or '*continuous*' radio burst emission were adopted for those kinds of radio burst radiation, which, in contrast to the narrow-band emissions of the drift-bursts (type II and III bursts) or single pips (type I bursts), exhibit more or less widely extended broad-band spectra at radio frequencies. It may be noted that the usage of the notation '*continuous*' *radiation* in r. dio astronomy has primarily nothing to do with the common definition of radiation continua applied in optics, where the contrast to the line emissions is stressed out, which plays a minor role in solar radio astronomy.

In essence, solar continuum radio bursts comprise two large groups of bursts, namely *microwave bursts* and *type IV bursts* including the *noise storm* continua.

The phenomenologically continuum-like appearance of type V bursts is misleading; physically it rather belongs to the class of drift bursts.

3.3.2. Morphology

Solar microwave bursts can be observed as short-period enhancements of the flux level when the whole-Sun flux density is monitored by single-frequency radiometer observations. In striking contrast to longer wavelengths, in the microwave region ($\nu \geq 1$ GHz) the burst spectra (as derived from single-frequency observations) and burst morphology (time profiles of flux-density records) are rather smooth as a prevailing feature.

In the past special attention has been paid to the shape of the time profiles of the burst records (the burst morphology). From an examination of a large number of microwave bursts classification schemata of certain morphological types can be established. The most general classification scheme distinguishes three basic types:

(a) Gradual bursts.
(b) Impulsive bursts.
(c) Complex and type IV bursts.
(cf. Figure III.8).

It can be conjectured that these basic morphological types reflect different complexes of physical processes and display three characteristic stages of the flare-burst development which will be discussed later.

The group of the *gradual bursts* comprises beside the characteristic bursts of the type 'gradual rise and fall of flux density' also the typical 'pre-' and 'post-burst increases' reflecting relatively smooth, quasi-stationary (quasi-thermal?) processes of energy dissipation in the source region.

As next *impulsive bursts* often develop as superpositions on a pre-existing gradual burst (precursor or gradual rise and fall) and indicate the onset of a rapid (evidently nonthermal) process of energy conversion. An extreme case of an impulsive burst is the so-called microwave 'spike burst' ('hair-pin' burst) which is defined by a comparatively high peak flux and short duration.

The study of the impulsive stage of a burst (which is displayed also in other spectral regions e.g. by the flash phase of the Hα flare, the hard X-ray phase, and the type III burst) is of particular interest from the point of view of flare physics. It is interesting to note, that if the impulsive increase of the flux density is observed with a greater time resolution (resolving seconds), often a complex form of the time profile is revealed.

The last large group, the *complex bursts*, may be complex also in their nature. So far as they are a mere multiplication of single impulsive ('simple') events, the justification of a separate group of 'complex' bursts would be physically doubtful.

But there is still another class of physical processes associated with the phenomenon of the microwave type IV bursts (type IVμ). Historically, the *type IV burst* has been defined as a 'long-period continuous radio burst emission accompanied by a major flare' without regard to the frequency range of the occurrence (cf. Section 3.6.1). Now, because the type IVμ bursts are generally complex in morphology, a precise distinction between complex and type IVμ bursts seems to be difficult to achieve.

The type IV burst evidently reflects a third stage of the flare development which is only attained for the largest events and is associated with the ejection of particles ('proton flares') during the explosive phase.

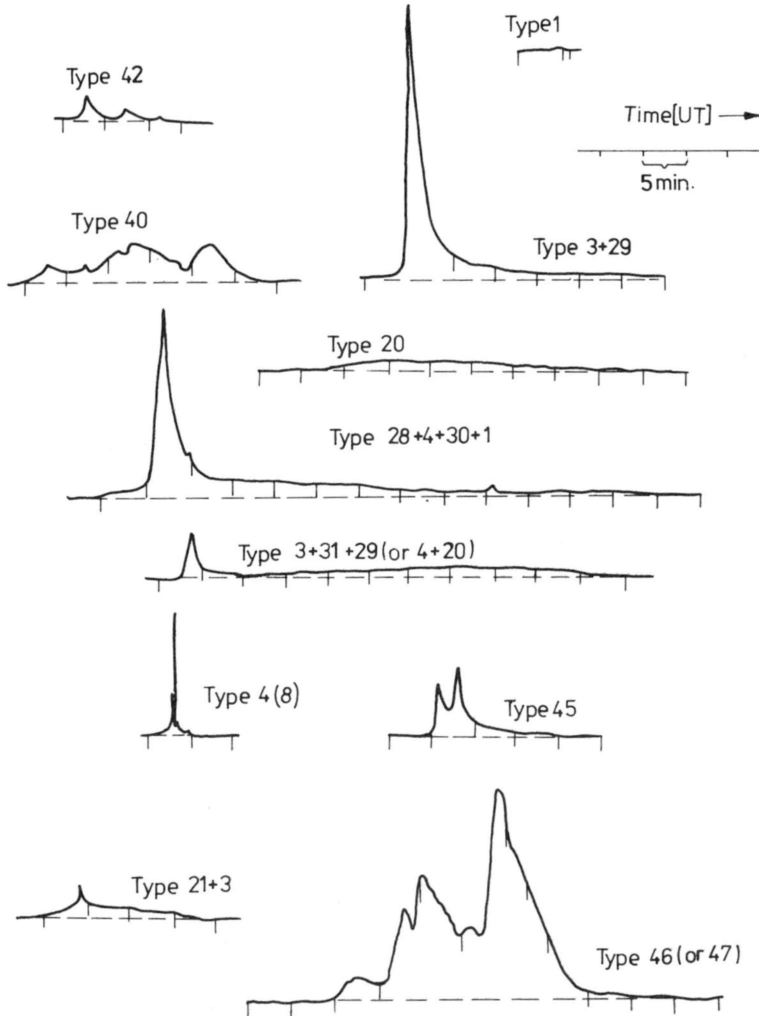

Fig. III.8. Examples for different morphology types (according the SGD-classification Table III.2) of solar microwave bursts (after Krüger, 1968).

TABLE III.2

Morphological classification of solar microwave bursts (according to the Instruction Manual for Monthly Reports prepared by H. Tanaka (1975)); phrases in parentheses are added by the author

SGD* Code	Letter symbol	URSIGRAM codes URANE	URANO	URANJ	Type	Remarks
1	S	1	1	3	simple 1	(small event)
2	S/F (S/A)	1	1	3	simple 1F (1A)	S + F (S + A), F means fluctuations, A means underlying
3	S	2	1	3	simple 2	(moderate event)
4	S/F (S/A)	2	1	3	simple 2F (2A)	S + F (S + A)
5	S	2	1	3	simple	(not specified)
8	S	–	1	3	spike	self-evident by duration (and intensity)
20	GRF	3	1	3	simple 3	gradual rise and fall
21	GRF (GRF/A)	3	1	3	simple 3A	
22	GRF (GRF/F)	3	1	3	simple 3F	
23	GRF (GRF/AF)	3	1	3	simple 3AF	
24	R	–	8	2	rise	
25	R (R/A)	–	8	2	rise A	
26	FAL	–	–	–	fall	
27	RF	3	–	–	rise and fall	(steeper than GRF)
28	PRE	9	–	–	precursor	
29	PBI	4	2	6	post-burst increase	
30	PBI (PBI/A)	4	2	6	post-burst increase A	
31	ABS	–	–	–	post-burst decrease	
32	ABS	5	–	–	absorption	(temporal fall of flux, 'negative burst')
40	F	7	4	4	fluctuations	
42	SER	8	4	(4)	series of bursts	a series of bursts occurs intermittently from base level with considerable true intervals between the bursts
45	C	6	3	5	complex	
46	C (C/F)	6	3	5	complex F	
47	GB	6	3	5	great burst	(type IVμ burst)

* SGD – Solar-Geophysical Data (Boulder).

Beside the above stated general lines, more detailed morphological classifications have been established in various manners more or less haphazardly adopting different systems of numerations. This is illustrated by the outline of Table III.2. More detailed information for practical use can be obtained e.g. in the Solar Radio Emission Instruction Manual for Monthly Report issued by the Toyokawa Observatory, Japan (ICSU–STP–IAU World Data Center C2); the Solar Geophysical Data issued by the National Geophysical and Solar-Terrestrial Data Center, Boulder, USA, the IAU-Quarterly Bulletin on Solar Activity published formerly by the Eidgenössische Sternwarte Zürich, Switzerland, and now by the Tokyo Astronomical Observatory; the Synoptic Codes for Solar and Geophysical Data issued by the International Ursigram and World Days Service, Boulder, USA, and other sources from particular observatories.

3.3.3. Spectral characteristics

In addition to the burst morphology which reflects quite well the dynamics of the burst processes, valuable information about the emission mechanism and plasma parameters of the emitting and propagating regions is contained in the burst spectrum. The spectral region covered by microwave bursts ranges from millimetric to decimetric wavelengths. In comparison to the S-component, the microwave burst spectra show a greater variability, i.e. the position of the spectral maximum as well as the frequency extent undergo more rapid variations from case to case.

Attempts to make a classification of the microwave burst spectra ready for special purposes of statistical and physical studies were made by different groups e.g. of the Toyokawa Observatory (Japan), the Sagamore Hill Observatory (USA), and the Heinrich-Hertz-Institute (GDR). In the following we review briefly the classification scheme of the Heinrich-Hertz-Institute:

The spectra can be distinguished primarily with respect to the position of the *spectral maximum* on the frequency axis (or sign of gradients in a restricted range) and with respect to the apparent *frequency extent* (band width) of the burst emission. Hence follows:

(a) Grouping according to the position of the *spectral maximum* (first letter(s)):

- A – Spectra starting in the cm-range increasing towards shorter wavelengths (higher frequencies) generally reaching a maximum in the mm-range;
- B – Spectra starting in the cm-range, increasing towards longer wavelengths, generally reaching a maximum at dm-wavelengths;
- C – Spectra exhibiting a maximum in the cm-range; and
- D – Extremely flat broad-band spectra.

Combinations between these types can be also observed, e.g. the combination of A and B indicates a minimum in the cm-range, etc.

(b) Grouping regarding to the *frequency extent* (second letter):

- s – Small-band continua, i.e. spectra of relatively small frequency extent, say, $\Delta\nu/\nu < 0.3$.
- b – Broad-band continua with, say, $\Delta\nu/\nu > 0.3$.

Furthermore sometimes a subgrouping into different *intensity classes* (according to the flux density at an observing frequency near the spectral maximum) seems to be useful. Following a proposal of Castelli (1970) three groups may be chosen within the following ranges:

(c) Grouping according to *peak fluxes* (third symbol):
1 – $I_{max} < 50$ s.u.
2 – $50 \leq I_{max} \leq 500$ s.u.
3 – $I_{max} > 500$ s.u.

Invoking the gyro-synchrotron emission mechanism (cf. Chapter IV) for the interpretation of the microwave bursts, different parameter ranges of the magnetic field, electron energies, source dimensions, etc. can be attributed to the above quoted groupings. Systematic variations of the statistical distribution of the microwave spectra seem to be indicated during the solar cycle (Krüger and Fürstenberg, 1969).

3.3.4. Polarization and Source Structures

Polarization (flux) measurements indicate, that in general the microwave bursts are partially circularly polarized. Typically the degree of polarization ranges from 0 to 50%. Occasionally, the existence of weak linear components has been reported.

The most prominent features of the microwave burst polarization are its statistical dependency on the heliographic longitude (illustrated by Figure III.9) and in many cases an observed reversal of the sense of polarization between cm- and dm-waves. Occasionally double or multiple inversions of the sense of polarization of microwave bursts were observed (Scalise *et al.*, 1971). Complex bursts and compound bursts mixed by several morphological types can also exhibit temporal changes of the sense of polarization.

In the cm-region the degree of polarization appears to increase statistically with decreasing wavelength. In the dm-range an increase of the degree of polarization with increasing wavelength is indicated, which perhaps is caused by the occurrence of new burst components at longer wavelengths not being typical for the microwave region (cf. Kakinuma *et al.*, 1969).

In the microwave region (i.e. at mm-, cm-, and lower dm-wavelengths) the sense of the polarization is assumed to refer to the 'extraordinary' electromagnetic wave mode of the magneto-ionic theory, corresponding to a gyration of radiating electrons in a magnetic field. Then the majority of the observations is compatible with the statement, that the dominating magnetic field influence is set up by the leading spot ('*leading spot hypothesis*').

High angular resolution measurements resolved the source regions of microwave bursts into two oppositely polarized patches according to the bipolar (or multipolar) magnetic structure of the source regions. Similar to observations of the S-component, the fine structure of microwave bursts has been found interferometrically to have diameters less than 10″ (7000 km on the Sun) and polarizations sometimes ranging up to 100% (Allissandrakis and Kundu, 1975). These highly polarized core regions are embedded in surrounding halo regions of weaker intensity and about 1 to 2′ diameter.

In principle, the polarization characteristics of the microwave bursts can be understood as an effect of the geometry of the magnetic field configuration on the Sun

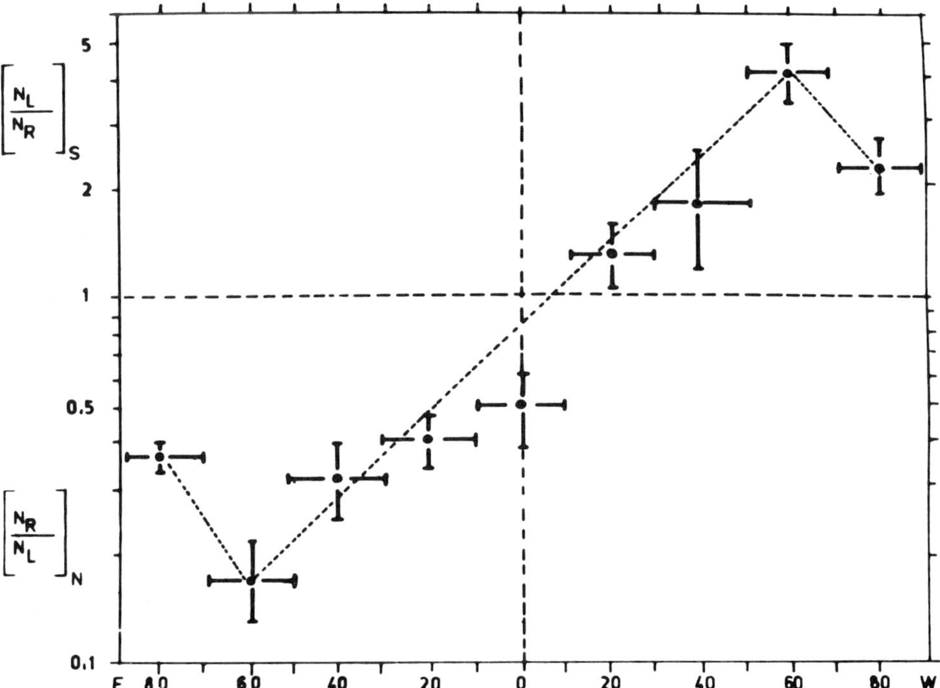

Fig. III.9. Distribution of the averaged ratio of left- to right-handed polarized bursts (numbers of events N_L/N_R) for the southern solar hemisphere and the inverse ratio N_R/N_L for the northern hemisphere in dependence on the heliographic longitude based on observations of the Heinrich-Hertz-Institute, at 9500 MHz and Toyokawa and Huancayo at 9400 MHz for the period 1969–1973 (after Krüger, 1976b).

with respect to the ray paths emerging to the observer determining the limiting polarization as was suggested earlier by Piddington and Minnet (1951) and Takakura (1961b). The polarization is right or left handed, if the angle between ray path and magnetic field is acute or obtuse, respectively. The same explanation may be valid even for the interpretation of the reversal of the sense of polarization with frequency (cf. Krüger, 1976b), though also alternative explanations invoking cutoffs at the reflection level for the extraordinary wave mode (Kakinuma, 1958), differential absorption by thermal electrons or gyro-resonance absorption (Takakura, 1960b, 1964; Ramaty, 1969a), or magneto-ionic wave coupling (Cohen, 1961a) offer a rich variety of alternative possibilities.

3.3.5. Association with Other Phenomena and Models

Depending on the range of the source heights and the electron energy distribution, microwave bursts are quite often associated with flares in Hα and other spectral lines (visible and UV) as well as with soft and hard X-ray bursts. In particular one finds sometimes an astonishing similarity of morphological details between the hard X-ray bursts and microwave bursts which challenged the scientific interests to reconcile these different components e.g. by the assumption of a common emitting volume and of a common energy distribution.

Even before direct X-ray measurements became available, a probable connection between the microwave and X-ray emissions was concluded by the study of sudden ionospheric disturbances (SIDs). The first attempts, however, to interpret simultaneously observed microwave and X-ray burst spectra by electrons of the same electron-energy population causing bremsstrahlung and gyro-synchrotron emission from one and the same emitting volume had led to an apparent contradiction: The number of electrons producing the X-rays should emit a radio emission three or four orders higher than actually observed (Peterson and Winckler, 1959). Attempts to remove this remarkable discrepancy invoked models, in which the main part of the X-ray emission was generated at heights below the plasma level of the microwave emission (Takakura and Kai, 1966) and the action of the gyro-synchrotron self-absorption process, which would significantly reduce the resulting radio emission (Holt and Ramaty, 1969). Later on the discrepancy was substantially removed by detailed model calculations using a nonuniform magnetic field distribution and a reduction of the high-energy part of the electron energy distribution, which is more sensitive to the amount of emitted radio waves than of the hard X-rays (Takakura and Scalise, 1970; Böhme *et al.*, 1976).

Applying a core–halo model of a common source region, a nonthermal core region contributes to the impulsive radio and (hard) X-ray components, whereas a quasi-thermal halo region may essentially apply to gradual microwave burst and soft X-ray burst components. During the lifetime of the burst an increase of the source dimensions can be possibly deduced.

3.4. Fast-Drift Bursts

3.4.1. General Properties

The 'drift' bursts owe their name to their most obvious characteristic property of a drift or time shift of certain morphological quantities (starting time, maximum, etc.) with the wavelength or frequency of observation. In a *dynamic spectrum* (i.e. frequency-versus-time diagram*), this feature is displayed by an inclined i.e. 'drifting' burst pattern. The distribution of a great number of observed drift rates shows two well separated peaks corresponding to the existence of two fundamental burst types, the slow- and the fast-drift bursts. It is widely accepted that the slow-drift bursts are due to the propagation of flare-related shock fronts (speed of the order 1000 km s^{-1}), and the fast-drift bursts are generated by fast electron streams expelled into the corona and the interplanetary space with speeds of the order of $0.5c$.

In particular the *fast-drift bursts* are very common features of the solar radio burst emission in the meter region. They were recognized as a special class of bursts soon after the first use of a solar radiospectrograph in Australia as communicated by

* In a dynamic spectrum on the ordinate scale the frequency is usually plotted in a downward direction. This relates to the fact that on the average, lower frequencies correspond to greater emission heights in the solar atmosphere.

Wild and McCready (1950). After considering storm bursts (type I) and slow-drift bursts (type II), the above authors classified the fast-drift bursts arbitrarily as *type III bursts*. This notation was generally adopted in the literature.

Excellent reviews presenting many details of the type III burst phenomenon are contained in the book of Kundu (1965) and in the papers of Wild *et al.* (1963) and Wild and Smerd (1972). Moreover, a special issue of Space Science Reviews of more than 300 pages was devoted in 1974 exclusively to the questions of type III radio bursts and solar electrons in space and is warmly recommended to the reader who is looking for further details.

From the point of view of solar physics the fast-drift bursts are of considerable value for several reasons: Firstly, fast-drift bursts are an attractive tool to investigate fast acceleration processes which must be considered to be the cause of the high exciter velocities. Secondly, fast-drift bursts can be used as natural plasma probes traversing the corona and yielding information about different plasma parameters. Thirdly, the fast-drift burst phenomenon provides a nice opportunity to study processes of different wave–particle (beam–plasma) and wave–wave interactions, which has stimulated a considerable progress in developing physical methods for a quantitative investigation.

In the following sections we will try to summarize the most prominent features of the fast-drift bursts with special regard to the more recently achieved developments.

3.4.2. Forms of Appearance (Gross Structure)

Fast-drift bursts comprise different forms of appearance. This fact is understandable if we remember that a great variety of physical conditions are covered in the wide spectral range extending from decimetric to hectometric wavelengths (and even higher) and different structures of the coronal magnetic fields may occur. In the following we discuss some of the main forms of fast-drift bursts which are most often observed.

a. *The 'Standard' Type III Burst*

An example of a dynamic spectrum of events which can be called a 'standard' case of a type III burst is shown in Figure III.10. The well expressed impulsive onset of the burst at each frequency defines a frequency drift which is approximately proportional to the observing frequency. Sometimes a separation of the emission pattern into two harmonics is displayed.

Adopting a model of the electron density distribution with height above the photosphere in the corona, the frequency drift can be translated into a height displacement of the exciting agent, if it is assumed, that the radio emission is generated in a region which corresponds to the local plasma frequency $\nu_p = (N_e^2/(\pi m_e))^{1/2}$ or to its second harmonic. There are many reasons to favor this assumption, which has been called the '*plasma hypothesis*' (Wild, 1950a, b). Its final proof, however, can be obtained only by simultaneous *in situ* measurements of both the electron density and the radio signal.

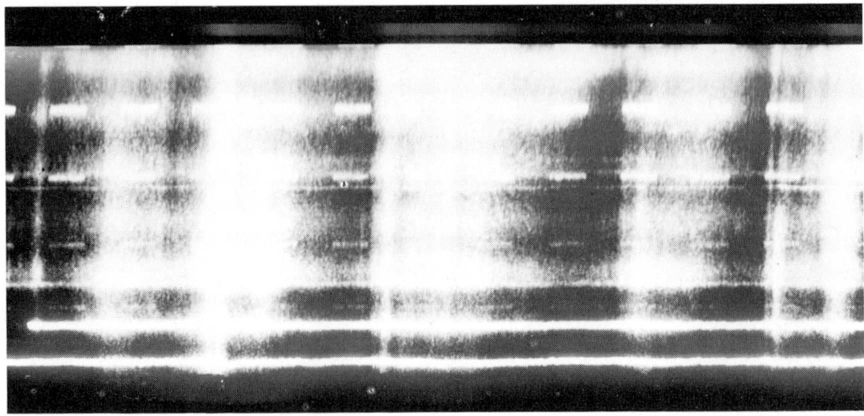

Fig. III.10. Examples of type III bursts in the 145–175 MHz range (frequency increases downwards at the ordinate scale, the markers have a frequency separation of 5 MHz and length of 1 s). Record of September 4, 1965. (Courtesy of Ø. Elgarøy).

b. *U-Type Bursts*

The U-type bursts (as first named by Maxwell and Swarup, 1958) are a special variant of fast-drift bursts. In contrast to the standard type III bursts, at a certain turnover frequency the drift rate is reversed thus exhibiting an emission pattern like the shape of an inverted 'U' in the dynamic spectrum (Figure III.11).

U-type bursts occur less frequently than standard type III bursts. Heliographic observations show that the source positions of the ascending and the descending branch of type U bursts are quite different (Wild, 1970c). This fact would be in accordance with the interpretation that U-type bursts are caused by exciters traveling up and down in the corona following closed magnetic field lines (loops). In this way, type U bursts could provide a means for the study of closed coronal magnetic field structures. Statistically, the turnover frequencies of U-type (and also of J-type bursts, see below) as well as possible stop frequencies of type III bursts are distributed over the whole frequency range of the occurrence of fast-drift bursts.

c. *J-Type Bursts*

J-type bursts are a modification of the U-type bursts, whereas the descending branch of the emission pattern in the dynamic spectrum is not well developed and therefore forms the shape of an inverted 'J' (Fokker, 1969). In this connection it should be noted that also in the case of well developed U-type bursts, usually the descending branch is somewhat fainter than the ascending one, so that J-type bursts can be regarded merely as a more extreme case of U-type bursts.

Different suggestions were made for an explanation of the weak trailing branch of the J-type bursts:

(a) The exciters of the J-type bursts may exert a pressure on the guiding magnetic field and thus distort it (Smith, 1970b).

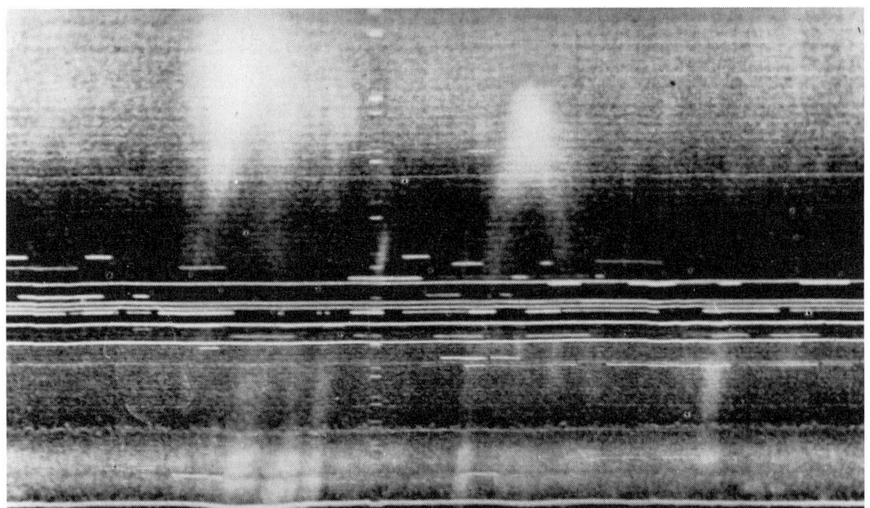

Fig. III.11. Examples of U-type bursts in the 93–186 MHz range. Record of November 5, 1973. (Courtesy of V. V. Fomichev).

(b) The magnetic field related to the descending branch of the J-type burst may have a diverging configuration successively causing a decrease of the density of the exciting electron stream (Daene and Fomichev, 1971).

(c) The exciting stream may be disordered at the top of a magnetic loop in the vicinity of a T-shaped neutral point configuration giving rise to a scattering of the exciter velocities (Caroubalos *et al.*, 1973).

d. *Type V Bursts*

Type V bursts exhibit a diffuse broad-band quasi-continuous emission of 0.5 to 3 minutes duration following some type III bursts at frequencies not higher than about 200 MHz. This peculiar phenomenon was first described by Wild *et al.* (1959b) and named type V burst providing an analogy of fast-drift burst 'couple' (type III/V bursts) to the formerly well known slow-drift/continuum burst association (type II/IV bursts).

Most properties of the type V bursts (such as source heights, polarization, peak fluxes, etc.) are quite similar to those of the associated type III bursts, so that they may be regarded rather as a special variant of a fast-drift burst emission occurring at greater source heights, than as an independent burst type. Differences from the type III burst emission arise mainly concerning the duration, the source position and source diameter, and a lack of definite evidence for the presence of harmonic structures, which, of course, is difficult to obtain because of the diffusiveness of the emission pattern (Weiss and Stewart, 1965; Daene and Krüger, 1966a; Wild, 1970c).

Considering the morphology of type V bursts, different forms of the time profile can be distinguished, e.g. the detached form showing a clear separation from the

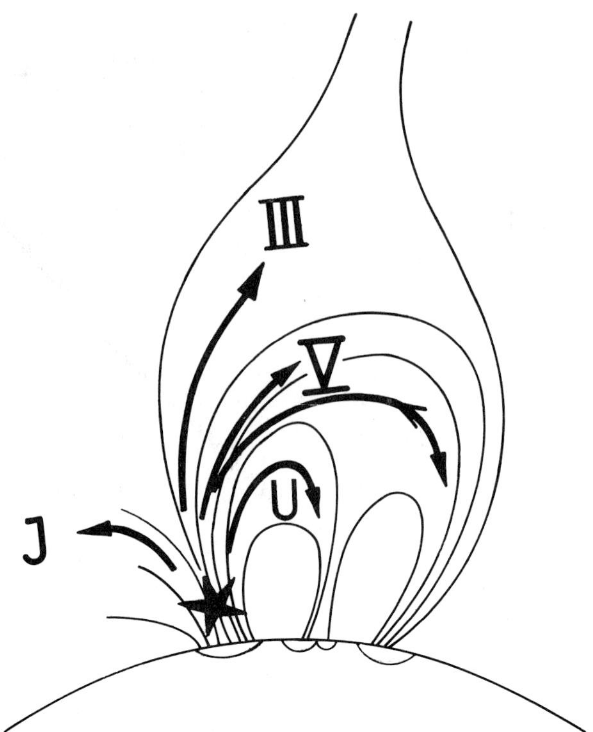

Fig. III.12. Proposed schematic picture of possible exciter paths of different kinds of fast-drift bursts and corresponding magnetic field structures.

preceding type III burst and the tail type consisting of a prolonged decay of the type III burst itself without interruption.

The cause of the type V burst must be sought in the fast electrons of the same origin as those producing the type III burst. Since heliographic observations show a displacement of both sources and greater sizes for the source of the type V burst emission, it can be assumed that the type V burst electrons are deflected into wide magnetic arcs and trapped there. A tentative picture comprising the exciter paths for the different main variants of large-scale fast-drift bursts is sketched in Figure III.12. An estimation of the particle energy rules out the initially suggested gyro-synchrotron emission mechansim. Therefore the emission mechanism of the type V bursts may be similar to that proposed for the type III bursts. A definite clarification, however, is still missing.

3.4.3. Fine-structure emissions

Beside the gross structure of fast-drift bursts, which is easy to recognize in dynamical spectra of moderate sensitivity and time/frequency resolution, a number of fine-structure drift bursts (or drift-burst-like phenomena) exist, for which a classification is less obvious. It appears that there is no longer a sharp boundary between the type III bursts and some features occurring in noise storms (type I bursts) and continuum bursts (type IV bursts). Apparently these features are of great interest for the understanding of the drift-burst phenomenon itself and provide nice opportunities to check special plasmaphysical theories. Unfortunately, due to the complexity of the phenomena to be observed, a satisfying interpretation concerning all involved processes is difficult to obtain and thus is still missing.

In the following we look into some details of the fine-structure emission features of the Sun in the meter region to be quoted here.

a. *Stria-Bursts (Split-Pair and Triple Bursts), Type IIIb Bursts*

As described by de la Noë and Boischot (1972) stria bursts are $\lesssim$1–2 s lasting, narrow-band emissions in the meter and decameter region at nearly constant center frequencies (negative drift rates between zero and about 0.1 MHz s^{-1}, bandwidth ~40–300 kHz). They occur singly or in groups sometimes forming chains in the dynamic spectrum. Very often they occur in pairs or triplets ('split-pair bursts', cf. Ellis and McCulloch, 1966) separated by intervals of 0.1–1 MHz.

There exist remarkable connections between the stria bursts and the type III bursts. The chains of stria bursts often exhibit a frequency drift in the same manner as type III or U-type bursts. In this way, the appearance of a 'broken' type III burst is formed (cf. Figure III.13). This phenomenon was first described by Ellis and McCulloch (1967) and called type IIIb burst by de la Noë and Boischot (1972). The latter was a result of a spectrographic study made with the large Arecibo ground reflector providing high sensitivity. It seems possible that stria structures can give a clue to a better understanding of the excitation of type III bursts and the exploration of coronal structures. It may be interesting to note that type IIIb bursts often occur as precursors of normal type III bursts. Further, chains of stria bursts can also show harmonic structures, though the recognition of these structures in fast drifting bursts is generally more difficult than in slow drifting bursts (Stewart, 1975). The circular polarization of the stria bursts can reach 100%, however, different elements of these bursts exhibit different degrees of polarization (de la Noë, 1975).

b. *Drift-Pair Bursts*

In contrast to the split-pair bursts, where the burst elements generally show a repetition at different frequencies (with zero or very slow drift rates), drift-pair bursts exhibit a repetition in time and fast frequency drifts (1–8 MHz s^{-1}). The pair consists of two almost identical elements, the second element has typically a delay of about 1–2 s. It is to be noted that single elements also occur without forming a pair. Both cases of forward drift (frequency decreases with time) and reverse drift (frequency increases with time) can be observed, but the reverse-drift pairs occur more frequently than the positive-drift pairs.

The drift-pair bursts were first described in detail by Roberts (1958). They form a characteristic phenomenon of the solar sporadic emission in the decameter range. Drift-pair bursts can be regarded as miniature fast-drift bursts, but they are also closely related to the type III bursts and noise storms forming a part of these phenomena. Special characteristics of drift-pair bursts were studied by Ellis (1969) and de la Noë and Møller-Pedersen (1971): The instantaneous bandwidth of one

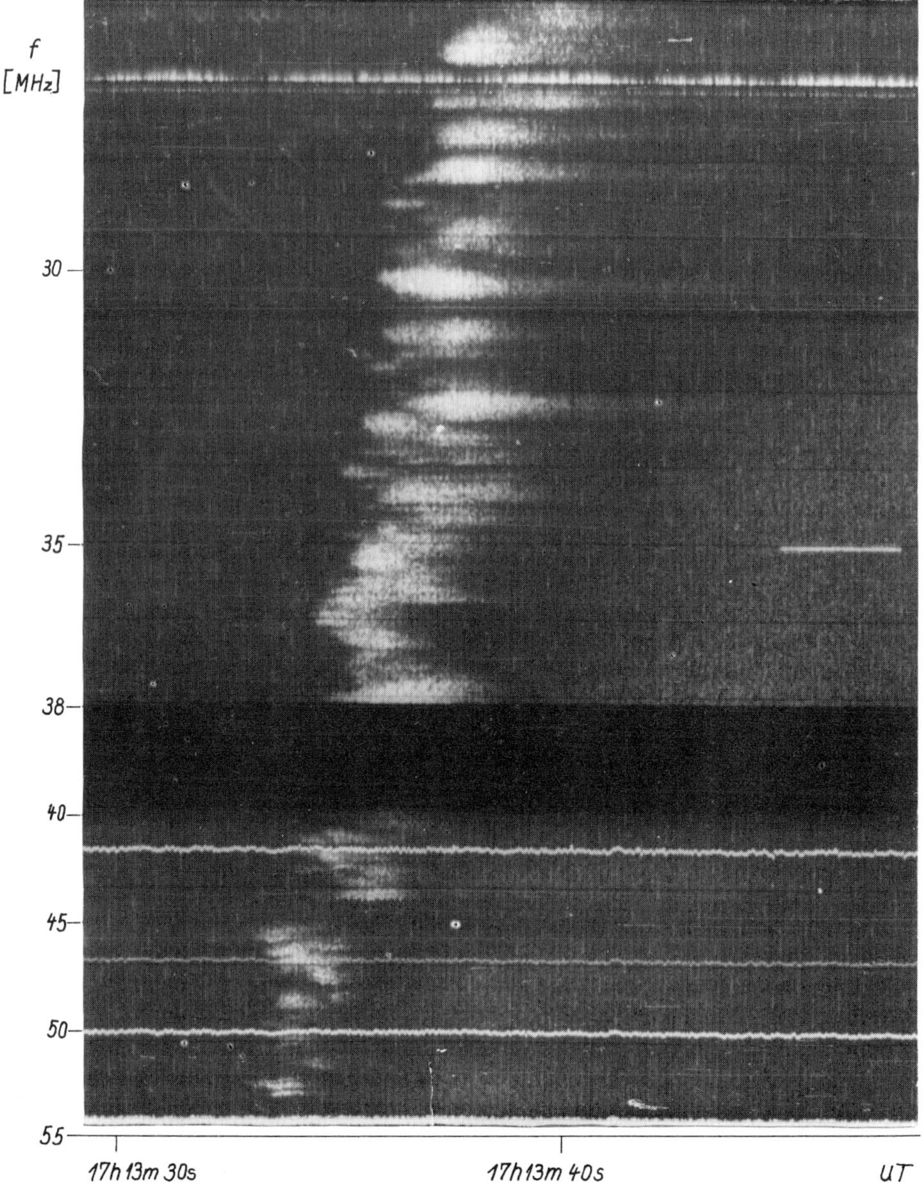

Fig. III.13. Example of a type IIIb burst (after de la Noë and Boischot, 1972).

element is about 0.3–4 MHz, its duration 0.5–4 s. The circular polarization is very weak, both elements were found to be polarized in the same sense. Since sometimes special fine structure of the first element can be found to be repeated at exactly the same frequencies in the second, it can be assumed that the second element is either an echo of the first one (reflected at lower levels in the corona), or that two modes of one and the same emission travel with different velocities. It can be supposed that drift pairs are formed by harmonic emission at the local plasma frequency (Roberts, 1958). Rare cases of single bursts at half the frequency of an associated drift pair can be interpreted as an emission at the fundamental frequency (de la Noë and Møller-Pedersen, 1971).

c. *Thread-Like Patterns and Shadow-Type III Bursts*

A pattern of fast drifting straight emission threads in the dynamic spectrum is often observed at different occasions (sometimes the threads are also called 'striations', but should not be confused with stria bursts) (cf. Solar Radio Group, Utrecht, 1974). The pulsating structure superimposed on some burst continua is of similar appearance. Sometimes single threads intensify and merge into type I burst-like or type III burst-like structures. All these phenomena are still insufficiently explored and may be different in nature.

It must be noted, that fast drifting thread-like patterns do not only occur in emission but also in absorption on the background of continuum or slow-drift bursts. An example of such a 'shadow'-type III burst was presented e.g. by Kai (1974). Quite often also negative thread-like patterns occur as juxtaposition of certain type IV burst continua. A possible explanation of these features may consist in a switching off of a plasma instability by high-speed particle streams in the solar corona (Zaitsev and Stepanov, 1974, 1975).

d. *Herring-Bone Structures*

Another example of fast drifting fine-structure emission patterns is given by the so-called herring-bone structure observed in certain slow-drift bursts (Roberts, 1959b). This pattern consists of a succession of both positive and negative fast-drift burst elements, emerging from both sides of the ridge ('backbone') of the slow-drift burst emission, respectively.

By proposing a shock wave as the primary cause of the slow-drift burst emission, the herring-bone structure can be regarded as evidence for the acceleration of particles (electrons) by such shock waves.

e. *Type III Storm Bursts*

Type III storms are a typical phenomenon of the disturbed solar emission in the decameter and hectometer wavelength range. Superimposed on a noise-storm continuum, instead of the type I bursts which are usually observed in the meter region, a great number of type III burst-like emissions appear. It should be noted that almost every decameter storm is associated with a type I storm in the meter region. The transition between both storm types is at about 60 MHz. In some contrast to the storms in the meter and decimeter range, storms in the decameter and hectometer range exhibit a greater variety of appearance: According to the preponderance of special burst elements, characteristic type IIIb storms, proper type III storms, type III DP (drift-pair) storms, and other forms can be found. Evidently their occurrence depends on evolutional and center-to-limb variations (Møller-Pedersen, 1974).

Observations made with the Australian 80 MHz heliograph indicate a characteristic displacement of the position of the sources of the type III storms in comparison to those of the normal type I storms (Stewart and Labrum, 1972). These authors suggest an explanation by disturbances originating in type I storm centers, which propagate nonradially outwards through a region of weak magnetic field and trigger an acceleration probably at the cusp of a coronal helmet structure leading to the type III storm. Thus the occurrence of type III storms can be regarded as a hint for the existence of local acceleration processes taking place even in the outer corona (cf. Section 3.7.4).

3.4.4. Ground-Based Observational Characteristics of Standard Type III Bursts

In the following we compile some basic properties referring mainly to standard type III bursts.

a. *Spectral Characteristics and Harmonic Structure*

The starting frequencies of the type III bursts range from about 600 MHz up to the decameter and hectometer region. It seems most likely that the exciters already exist at deeper levels before becoming visible. It can be supposed that many potential type III burst sources (electron beams) remain invisible because of the lack of transformation conditions into radio emission.

Adopting the plasma hypothesis (i.e. the emission occurs near the local plasma frequency) and taking a suitable electron density model, the frequency drift can be converted into exciter velocities. Thus, the derived velocities are distributed between about 0.25 and $\gtrsim 0.60c$ (corresponding to electron energies of 10 to 100 keV) with an average at about $0.33c$ (Stewart, 1965).

In the meter region the frequency-drift rate is found to be approximately proportional to the observing frequency. In a wider spectral range of 600 MHz $\gtrsim \nu \gtrsim$ 60 kHz an appropriate dependency of the form

$$d\nu/dt = -\alpha\nu^{\beta} \qquad \text{with} \quad \alpha = 0.01, \ \beta = 1.84 \tag{III.8}$$

was found by Alvarez and Haddock (1973a).

The exciters of the type III bursts are evidently moving without any significant deceleration up to distances of some tens of the solar radius (Alexander *et al.*, 1969). Later on at greater distances, they become decelerated (cf. also Section 3.4.5.d).

Against all doubts (cf. Daigne, 1975a) and observational difficulties arising from the fast drift rate and complex structure of type III burst groups there are examples which clearly demonstrate that in the meter region type III bursts may appear as a pair of fundamental and second harmonic emission.

The frequency ratio of the harmonic to fundamental emission is commonly slightly less than two (1.85–2.0) due to the cutoff of the fundamental radiation at the plasma level. In the case of missing harmonic structure, it is most likely that the observed emission corresponds to the harmonic radiation only.

No convincing indication has yet been found in favor of an occurrence of third (and higher) harmonics of the plasma frequency in type III bursts though the existence of such effects cannot be in principal excluded (cf. Section 3.4.6.c).

b. *Location*

Interferometric and heliographic studies do not show large differences between the positions of the fundamental and harmonic components in type III bursts. Minor displacements can be explained by the effects of scattering, refraction, and projection of the ray path in the solar corona (Stewart, 1972).

The angular diameters of the type III burst sources appear to be typically of the order of 4 min of arc in the meter region. The effect of scattering influences the observed positions (Riddle, 1974b).

The length and angular dimensions of the type III burst sources tend to increase with decreasing frequency.

The exciter paths are situated preferably in regions of relatively low magnetic fields as characterized by open field-line structures (Malitson and Erickson, 1966; Kai, 1970b).

c. *Morphology*

The time profile of the flux density of type III bursts recorded at single frequencies reveals a slightly asymmetric form. As typical for impulsive phenomena, a steep rise and a slower approximately exponential decay of the flux density records is to be observed. Moreover, high time resolution records especially in the decameter region often display a short precursor before the steep onset of the main burst (de la Noë and Boischot, 1972). The peak intensities tend to increase with wavelength through the whole frequency range of the drift bursts up to less than 1 MHz.

The duration t_d of standard type III bursts increases towards decreasing frequencies, on the average, according to

$$t_d = \alpha \nu^{-\beta} \qquad (\alpha \approx 60, \quad \beta \approx 0.7) \tag{III.9}$$

(Elgarøy and Lyngstad, 1972).

In the past electron temperatures and densities were deduced from the decay rate of the type III bursts invoking collision damping as the dominating process. However, this concept is no longer tenable since e.g. the recently observed fine structure in the time profile of the type III bursts make the validity of a collision damping extremely doubtful.

d. *Polarization*

Formerly type III bursts had been found to be moderately elliptically polarized. The reported degrees of polarization ranged from 0 to 70% with a certain preference for about 30 to 60%. These results depended on the observing frequency and the state of the receiving equipment (bandwidth) and other factors (cf. Chin *et al.*, 1971). Large uncertainties still existed concerning the content of linear polarization, whereas a number of authors reported different contributions of linear polarization in type III bursts (axial ratios $p \lesssim 2$). Recent measurements by Boischot and Lecacheux (1975), however, made with spectrographs displaying Faraday fringes when coupled with a linearly polarized antenna receiving linearly polarized radiation, did not reveal any amount of linear polarization from solar type III bursts in the range between 25 and 80 MHz and around 169 MHz within a time scale up to 0.02 sec and a bandwidth down to 20 kHz. A similar negative result of a search for linear polarization in type III bursts was published by Grognard and McLean (1973). In this way the content of linear polarization in solar radio bursts is highly questionable and all former investigations are to be considered with caution.

e. *Associated Phenomena*

Type III bursts are often associated with optical flares, microwave bursts, and (especially hard) X-ray bursts. As an important fact, the type III bursts preferably occur during the rising phase of these associated events. Impulsive events and strong flare-burst associations of different type as well as the occurrence of flare surges and sprays increase the probability of observing an accompanying fast-drift burst.

There is likewise a remarkable association with subflare activity and dark filaments without flare activity (Martres *et al.*, 1972; Tlamicha and Takakura, 1963). In other cases, however, a clearly detectable relationship between type III bursts and optical or microwave phenomena is missing. Nonrelativistic impulsive solar electron events ($E > 40$ keV) observed in the interplanetary space near the Earth's orbit by artificial satellites were found to correspond well with their radio response of the fast-drift bursts (Lin, 1970a).

3.4.5. Space observations

Since ground-based observations of the Sun are limited to frequencies above about 10 MHz, observations at lower frequencies require platforms at altitudes above the cut-off frequencies of the terrestrial ionosphere and plasmasphere. Such measurements became available by use of artificial satellites and space probes. Space radio astronomy emissions occur through all space between Sun and Earth and even beyond the Earth's orbit. In this way space radio astronomy provides a connection between solar physics, physics of the interplanetary space, and physics of the terrestrial atmosphere and magnetosphere. Much of the noise observed at the lowest frequencies is of terrestrial origin. Between the Sun and Earth the drift bursts are the most common radio phenomenon. With increasing distance from the Sun the coronal structure is determined more and more by the flow characteristics of the solar wind. Correspondingly open magnetic field configurations develop, favoring extended exciter paths of the type III bursts deep into the interplanetary space.

a. *Short History*

A compilation of the most important experiments concerning solar space radio astronomy during the first decade of its development is given in Table III.3. The start of radio astronomy experiments in space was, in the first half of the sixties, favored by the rapid development of the satellite techniques. The first successful observation of a solar radio burst from space was made in 1963 aboard the Alouette–1 satellite as reported by Hartz (1964). There followed burst observations aboard the space probes Zond–3 and Venera–2 in 1965 and Luna–11 and –12 in 1966/67 (Slysh, 1967a, b, c). The source directions were localized by means of the lunar occultation and the satellite spin modulation method. The latter method makes use

TABLE III.3

Compilation of some important solar space radio astronomy experiments

Platform	Reference	Remarks
Alouette–I	Hartz (1964)	First spaceborne detection of solar type III bursts, swept frequency 1.5–10 MHz.
Alouette–II	Hartz (1969) Hakura *et al.* (1969)	Swept frequency 0.1–15 MHz.
Zond–3 and Venera–2 Luna–11 and –12	Slysh (1967a, b, c)	Fixed frequencies: 0.02, 0.21, and 2 MHz; 1 MHz, 200, and 20 kHz; 30 kHz, 985 kHz, first utilization of spin modulation.
OGO–1	Dunckel and Helliwell (1969)	Still without burst observations,
OGO–3	Haddock and Graedel (1970) Graedel (1970)	2–4 MHz,
	Dunckel *et al.* (1972)	(0.02), 25–100 kHz swept frequency,
OGO–5	Alvarez *et al.* (1972) Alvarez and Haddock (1973a, b) Haddock and Alvarez (1973)	eight frequencies between 0.05 and 3.5 MHz, comparison with electron events.
ATS–II	Alexander *et al.* (1969)	Six frequencies between 0.45 and 3 MHz, comparison with streamer models.
RAE–1 (RAE–2)	Fainberg and Stone (1970, 1971)	First entire radio astronomy explorer, 0.2–5.4 MHz swept-frequency observations of type III storms.
IMP–6	Alvarez *et al.* (1974) Alvarez *et al.* (1975)	Eight frequencies between 0.05 and 3.5 MHz,
	Frank and Gurnett (1972) Kellogg *et al.* (1973)	64 frequenices between 23 and 230 kHz.
Mars–3	Caroubalos and Steinberg (1974) Caroubalos *et al.* (1974b)	STEREO–1 experiment, comparison with ground-based 169 MHz observations.
IK–Kopernik 500	Hanasz *et al.* (1976)	0.45–6 MHz, attempt of derivation of polarization by ionospheric refraction.
Helios A, B	Weber *et al.* (1977)	First measurement of type III burst directivity at low frequencies by interplanetary baseline observations.

of the rotation of an antenna with a nonisotropic diagram on a spin-stabilized spacecraft producing a definite modulation pattern of the received radiation.

Considerable progress in receiving a broad observational information was achieved by different experiments carried out e.g. aboard the ATS–II, OGO–3, and –5 satellites. The first satellite devoted exclusively to radio observations was the Radio Astronomy Explorer (RAE–1) which was launched in 1968 and provided, until 1972, much information on type III bursts and storms at frequencies up to about 200 kHz. Presently, the lowest frequency at which a type III burst has been reported was 10 kHz derived from a plasma-wave experiment aboard the IMP–6 satellite (Kellogg *et al.*, 1973). For further details the excellent review by Fainberg and Stone (1974) as well as the literature cited in Table III.3 is recommended.

b. *Hectometer Fast-Drift Bursts*

In comparison to the fast-drift bursts observed in the frequency range accessible to ground-based observations, the low-frequency type III bursts subtend larger distances in space being related to the structure of the solar wind and the interplanetary medium. The arrival of the exciters near 1 AU can be detected by spacecraft devices allowing one to observe both the electromagnetic radiation and the exciting particle streams. Such experiments support the assumption that electrons in the energy range between about 10 and 100 keV are indeed the exciting agents of the type III bursts (Lin *et al.*, 1973).

Another important feature of the fast-drift burst emission in the low-frequency range is the lack of the fundamental-wave emission. Postulating reasonable density distributions in the space between Sun and Earth, almost all low-frequency type III burst emissions should be regarded as belonging to the second harmonic of the local plasma frequency (Fainberg *et al.*, 1972; Smith, 1974b). According to Haddock and Alvarez (1973) the transition from the predominance of the fundamental to the second harmonic radiation tends to occur in the region between 0.1 and 4 MHz.

Similar to the high-frequency part of the fast-drift burst spectrum, variable cutoffs occur at both sides of the frequency range and only a very small part of all type III bursts is visible over the whole frequency interval extending from about some hundreds MHz to about 10 kHz.

Attempts to localize the emission level in space have been made by means of two methods using frequency-drift rates and directional measurements via spin modulation. Since the drift rates depend on the exciter speed, the density distribution, and the radio propagation velocity, which are not directly measured in general, the results are not free from the choice of suitable model assumptions. (Such models refer e.g. to different plasma parameters inside a streamer and the ambient solar wind.) Single models may differ by one order of magnitude as demonstrated in Figure III.14.

An impression of the typical drift rates observed over a wide range of frequencies is given in Figure III.15 which was compiled by Alvarez and Haddock (1973a). The variation of the observed drift rates with heliographic longitude can be obtained during type III storms lasting for periods of more than half a solar rotation. An example presented by Fainberg and Stone (1970) is reproduced in Figure III.16.

Apparently a deceleration of exciter speeds can be deduced for distances greater 10 $R_\odot$ which according to Fainberg *et al.* (1972) can be presented in the following form

$$\mathrm{d}\log v_e/\mathrm{d}\log R = -0.38, \qquad \text{(III.10)}$$

where v_e denotes the exciter speed expressed in units of the velocity of light and R is the distance from the Sun in solar radii.

c. *Directivity*

The first direct observations of the directivity of the radiation from type III bursts (as well as from other burst types) were achieved by the French-Soviet STEREO–1 experiment. Simultaneous ground-based and spaceborne observations of solar radio

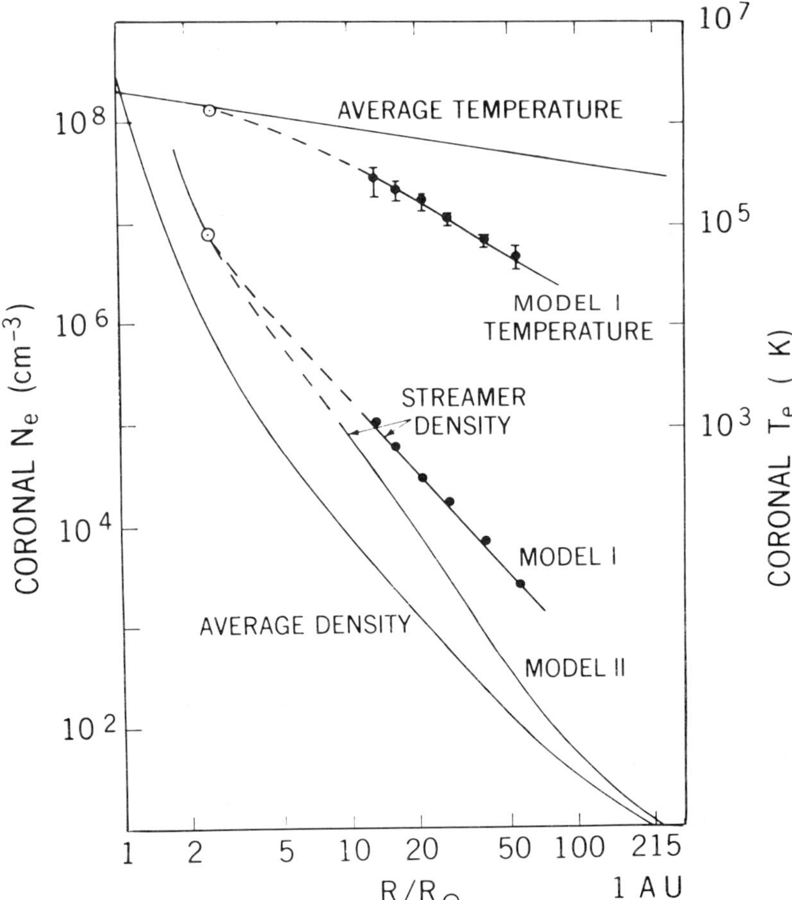

Fig. III.14. Models of the variation of electron density and temperature with distance from the Sun reported by Alexander *et al.* (1969) according to various references.

bursts at 169 MHz were compared to the dependence of the angle θ between the observing directions defined by the position of the moving space probe. The measurements were obtained at Nancay (France) and aboard the Soviet planetary probe Mars–3 mainly in 1971 (Caroubalos and Steinberg, 1974).

It was found that the shape of the time profiles of the fluxes of single type III bursts is widely independent of the observing direction according to a high density gradient in the related range of coronal altitudes. It may be concluded that the time profile of fast-drift bursts at the investigated frequency is a characteristic of the radiation source itself and is not essentially influenced by the propagation conditions in the solar corona.

The constant ratio of the flux in the space-probe direction to that of the Earth direction provided a characteristic for a given event called the directivity factor *R*

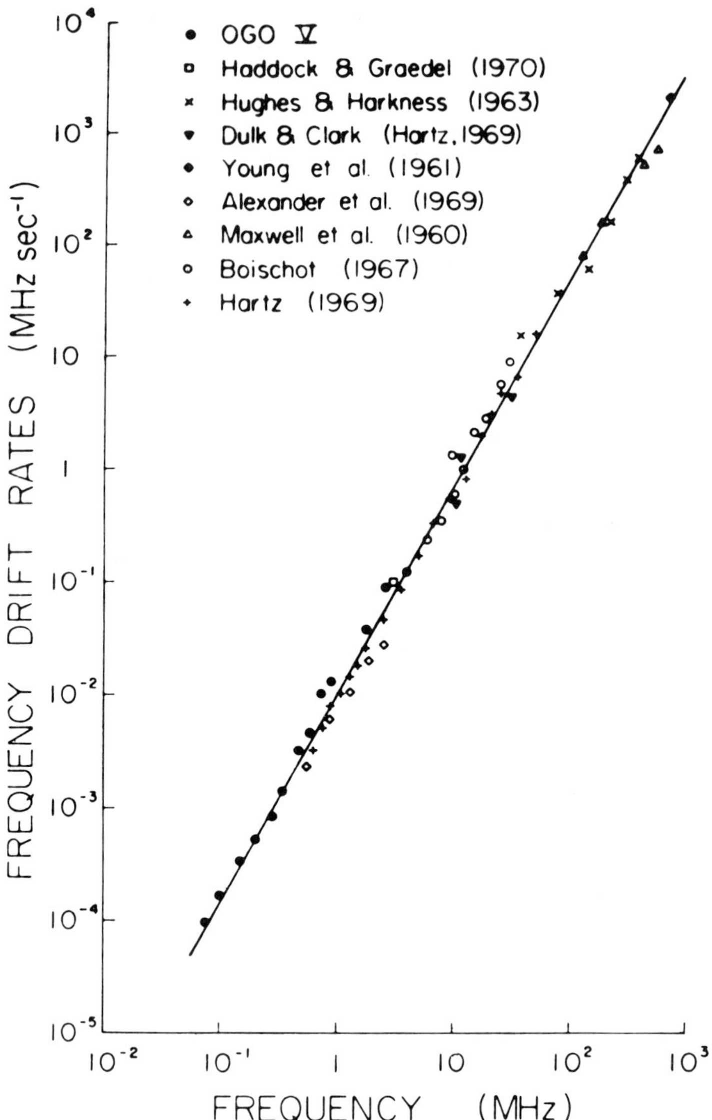

Fig. III.15. Typical frequency-drift rates of type III bursts over a wide range of frequencies (from Alvarez and Haddock, 1973a).

(Caroubalos *et al.*, 1974b). Characteristic double-humped bursts were interpreted by the authors as harmonic–fundamental pairs, whereas the directivity factor was systematically larger for the first (fundamental) component than for the second (harmonic) one. For events near the solar limb the directivity factor was found to increase with increasing central meridian distance of the source on the Sun. Apparently the source size also becomes larger with increasing R. According to the above cited

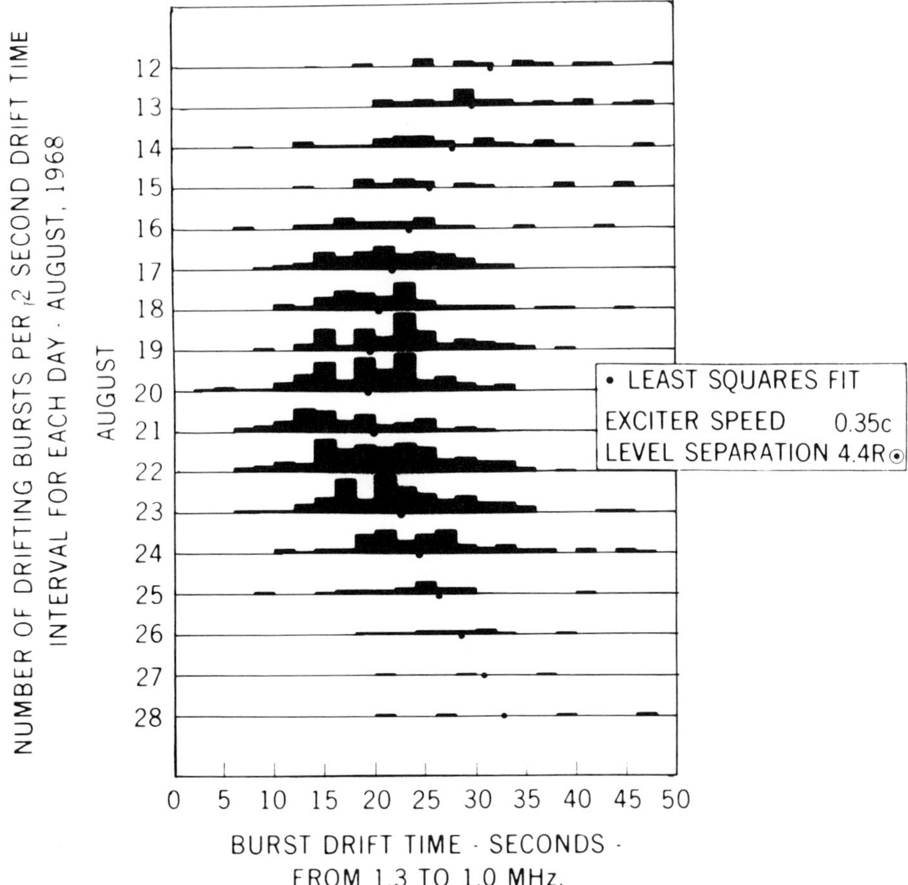

Fig. III.16. Histograms showing the dependence of the burst drift rate between 1.3 and 1.0 MHz on the heliographic longitude for a particular storm in August 1968 derived from RAE–1 measurements (after Fainberg and Stone, 1970).

authors, the results obtained are consistent with model calculations using the ray-tracing method in an inhomogeneous scattering corona for any value of θ independent of the longitude of the source. However, deviations from the spherical symmetry seem to be necessarily included in order to meet real conditions.

d. *Exciter Paths and Ray Trajectories*

The position of the type III burst radiation source (called 'emission level scale' by Fainberg and Stone, 1974) is determined by the path of the exciting particles (electrons) in space. Due to the relatively small gyroradii, the electrons are forced to follow the large-scale magnetic field direction. Therefore the type III bursts can be used as tracers of the magnetic field configuration and also other parameters in the corona and the solar wind. A nice example provided by an observation of an U-type

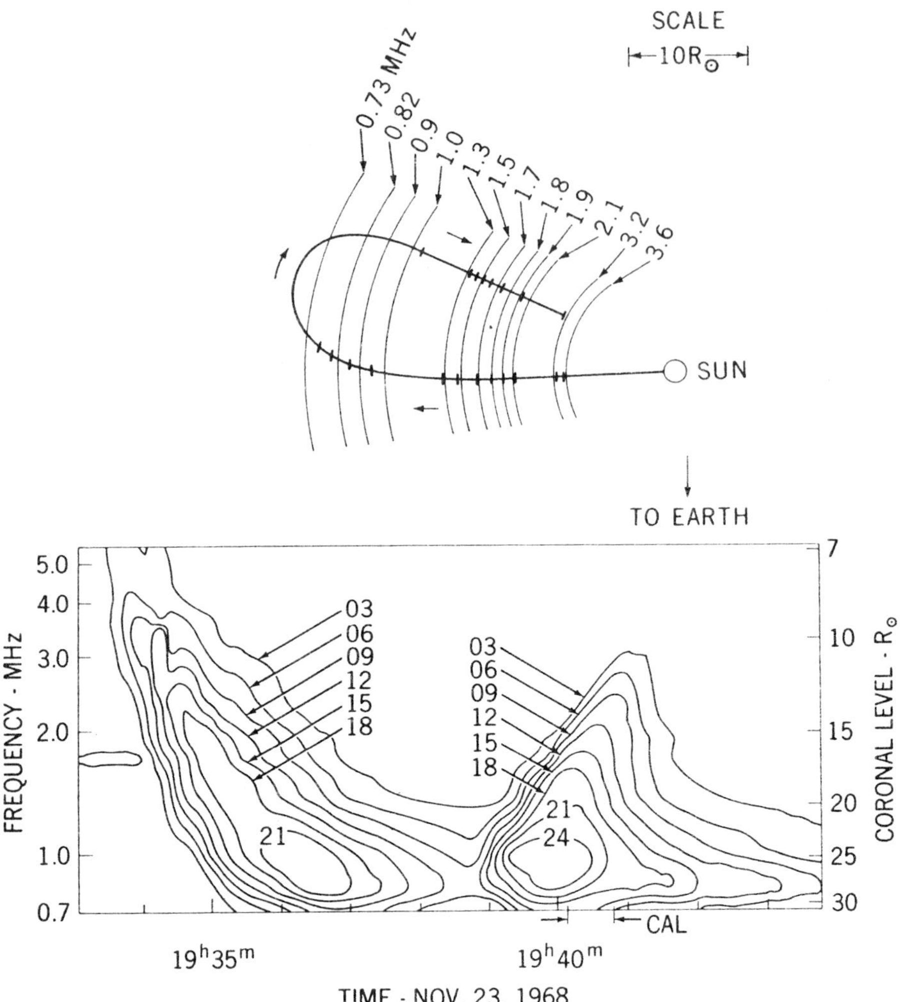

Fig. III.17. A computer developed dynamic spectrum of a hectometer U-type burst and the inferred model exciter trajectory determined from RAE–1 observations on November 23, 1968 (after Stone and Fainberg, 1971).

burst extending to a distance of about 35 solar radii by the RAE–1 satellite (Stone and Fainberg, 1971) is reproduced in Figure III.17.

Another possibility to derive information about the exciter path can be achieved by measurement of the low-frequency type III burst fluxes and the arrival directions from spin modulated data. Arrival directions in combination with a suitable emission level scale can be processed with a computer to follow the dynamic process of the passage of individual type III bursts through the interplanetary space up to the Earth's orbit. As an example a plot of data obtained from the IMP–6 satellite is shown in Figure III.18 displaying the traveling of the type III burst exciting elec-

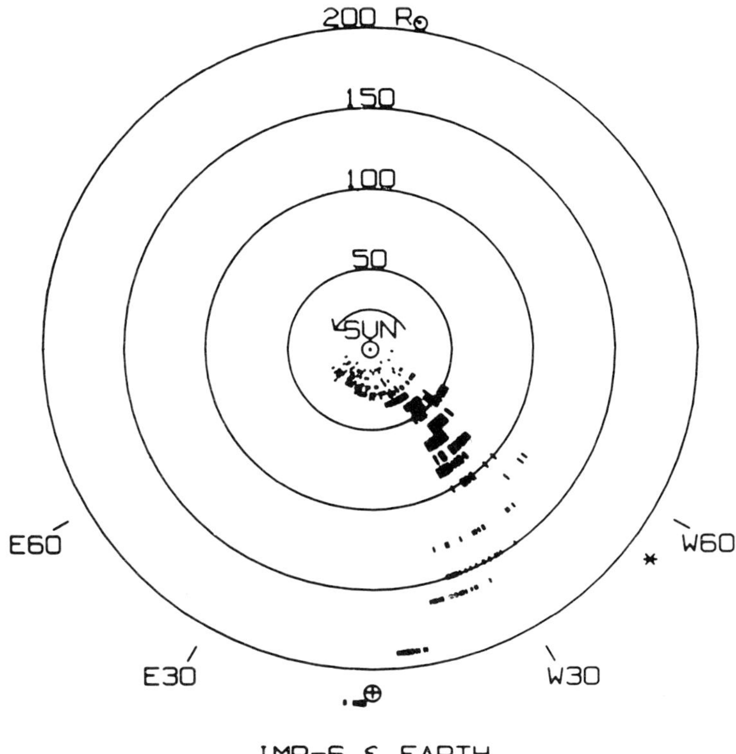

Fig. III.18. Example of a computer generated plot of data from IMP–6 tracing the exciter path of type III bursts down to the Earth. The plane of the figure is the ecliptic (Earth at bottom). For each minute of data, the intersection of the measured type III burst arrival directions with the spherical emission level is determined and a radial line segment with length proportional to log intensity is plotted (collected from data from 32 frequencies taken at 5 s intervals for a 2 min period) (after Fainberg and Stone, 1974).

trons as taken from a series shown at the IAU Symposium No. 57 on Coronal Disturbances in 1973.

The derived values of the exciter speed do not exhibit an appreciable deceleration over about the first 40 $R_\odot$ as already mentioned in Section 3.3.4 (Hartz, 1969). Nevertheless, the IMP–6 data clearly indicate a slowdown of the exciter velocities at greater distances from the Sun (Fainberg *et al.*, 1972). An increase of the pitch angles e.g. would produce the same effect. More detailed investigations are necessary here in connection with theoretical work.

A summary of some low-frequency radio emissions up to the vicinity of the Earth is contained in Figure III.19.

3.4.6. Interpretation

The current ideas on the generation of the type III bursts are based on the plasma (wave) hypothesis (cf. Section 3.4.2.a) which can be characterized by the following

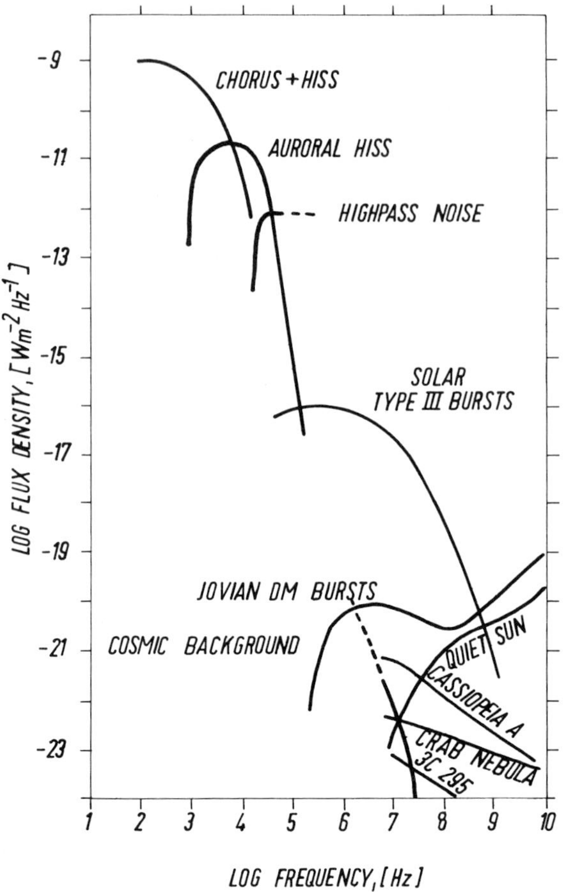

Fig. III.19. Characteristic flux densities of some low-frequency radio emissions detected by OGO 1 and 3 in comparison with other emissions (after Dunckel, 1974).

points:

(1) A fast moving exciting disturbance (electron beam) causes plasma waves at the local plasma frequency ω_p.

(2) A certain part of the resulting plasma waves is transformed into electromagnetic waves via scattering at ω_p and $2\omega_p$.

(3) So far as propagation is possible, the electromagnetic waves can reach the observer.

Each of these points contains a number of difficult problems, which are not yet fully solved as will be outlined below. Nevertheless, more and more convincing arguments have been collected in favor of the plasma hypothesis from the beginnings of type III burst studies (Wild, 1950b; Wild *et al.*, 1954a) up to the near past (Fainberg *et al.*, 1972).

a. *The Exciter Source*

As indicated by the above statements, a critical analysis of all possible exciter types (shock waves, ion and electron streams, temporal gradual changes of the exciting conditions, selective group-velocity retardation effects, etc.) led to the conclusion that fast streams of electrons are the most probable exciting agents of solar fast-drift bursts. In line with the plasma hypothesis electrons require less energy for acceleration than, e.g. protons and are more efficient in generating plasma waves. Moreover electrons were directly observed by space experiments (Lin *et al.*, 1973; cf. Section 3.4.5.b).

The problem of the exciter origin, i.e. of the acceleration process itself, is not yet solved satisfactorily. Obviously this question is intimately connected with the flare problem which will be treated in Chapter V. The short duration of the type III burst pulses and the occurrence of groups and storms of type III bursts should provide valuable hints for the investigation of the related acceleration processes. But unfortunately, since the starting heights of the type III bursts must not be regarded as the actual heights where the acceleration is initiated, the location of the exciter origin appears obscured. Independently from all considerations about the exciter origin, however, the mere fact of the existence of the exciting agents of solar fast-drift bursts must be accepted and their properties are to be deduced from their response to the radio emission.

For some time the question of the stabilization of the exciter beam puzzled a number of authors (cf. Smith, 1970a). It was thought that particle streams traveling over large distances, as the exciters of the type III bursts do, must be stabilized (Kaplan and Tsytovich, 1967b). But it could be shown that an electron-beam stabilization does not occur under the conditions relevant to the fast-drift bursts taking into account the effect of quasi-linear relaxation of the distribution function of the electron energies (Zheleznyakov and Zaitsev, 1970; Smith, 1970b).

An attempt to solve this obvious discrepancy was made by Zaitsev *et al.* (1972) who applied a dynamical theory of the type III bursts. They concluded that in a spatially bounded stream, in spite of the quasi-linear relaxation, plasma waves can be generated for a long time owing to faster particles escaping out of the front of the stream. This result was obtained by a solution of the relativistic quasi-linear equations for the temporal evolution of an electron beam by a similar theory under the conditions of a local explosion causing the bounded fast electron beam. An unchanged speed of the type III burst exciters over a wide range produces the illusion of a stream stabilization. An additional consideration of the effect of possible losses was made by Smith (1973).

b. *The Plasma-Wave Source*

The first approach to a detailed theory of the type III burst emission was made by Ginzburg and Zheleznyakov (1958a). Coherent (Čerenkov) plasma waves were invoked to be generated by a stream of fast particles and subsequently transformed into radio waves by Rayleigh and combination scattering at ω_p and $2\omega_p$, respectively. Meanwhile some details of this theory (e.g. concerning the scattering process) were replaced by newer concepts, but the basic idea remained tenable (Sturrock, 1964; Melrose, 1970; Smith and Sturrock, 1971).

In the following we restrict ourselves to some basic remarks on the theory. For more details the reader is referred to the review by Smith (1974a), the cited original papers, and partly to the next chapter.

Following Zaitsev *et al.* (1972) and Smith (1973), the dynamics of the electron stream producing the plasma waves of the fast-drift bursts are described by the approximation of quasi-linear equations:

$$\frac{\partial N_k}{\partial t} + \mathbf{v}_g \cdot \frac{\partial N_k}{\partial \mathbf{r}} = \frac{N_k}{m_e} \int w_k(\mathbf{v}, \mathbf{k})\, \hbar \mathbf{k} \frac{\partial f(\mathbf{v})}{\partial \mathbf{v}} \, \mathrm{d}\mathbf{v} + \int w_k(\mathbf{v}, \mathbf{k})\, f(\mathbf{v})\, \mathrm{d}\mathbf{v},$$

$$\frac{\partial f(\mathbf{v})}{\partial t} + \mathbf{v} \cdot \frac{\partial f(\mathbf{v})}{\partial \mathbf{r}} = \frac{\hbar^2}{m_e^2} \frac{\partial}{\partial v_i} \int w_k(\mathbf{v}, \mathbf{k})\, N_k(\mathbf{k})\, k_i k_j \frac{\partial f(\mathbf{v})}{\partial v_j} \frac{\mathrm{d}\mathbf{k}}{(2\pi)^3} +$$

$$+ \frac{1}{m_e} \frac{\partial}{\partial \mathbf{v}} \int w_k(\mathbf{v}, \mathbf{k})\, \hbar \mathbf{k} f(\mathbf{v}) \frac{\mathrm{d}\mathbf{k}}{(2\pi)^3}, \qquad \text{(III.11)}$$

where $N_k(\mathbf{k})$ is the wave distribution function, $f(\mathbf{v})$ the particle distribution function in the phase space, $w_k(\mathbf{v}, \mathbf{k})$ the emission probability of plasma waves with the wave vector $\mathbf{k}$ by an electron of the velocity $\mathbf{v}$, $\hbar$ denotes Planck's constant divided by 2π, and m_e the electron mass (cf. Section 4.5.3).

In practice for $N_k(\mathbf{k})$ suitable assumptions must be chosen since an exact determination meets with difficulties (Smith, 1974b).

The dispersion relation of the longitudinal electron plasma waves is

$$\omega_l^2 = \omega_{pe}^2 + 3 v_{Te}^2 k_l^2 \qquad \text{(III.12)}$$

(ω_{pe} = plasma frequency of electrons, v_{Te} = thermal velocity of electrons).

The emission probability for a plasma wave excited by an electron of the velocity $\mathbf{v}$ is given by

$$w_l(\mathbf{v}, \mathbf{k}) = \frac{(2\pi)^2 e^2 \omega_{pe}}{\hbar k^2} \, \delta(\omega_{pe} - \mathbf{k} \cdot \mathbf{v}) \qquad \text{(III.13)}$$

(Smith, 1974a).

The basic equations were solved for several special cases (Zaitsev *et al.*, 1972; Smith, 1974b). Nevertheless, the whole problem of the dynamics of an electron stream in the solar corona is not conclusively solved. In a more refined treatment of nonlinear effects, e.g. the induced scattering of plasma waves on ion polarization clouds, the coupling with ion-acoustic waves causing a pile-up of the plasma waves, charge neutralization effects, and others, must be considered.

c. *The Radio-Wave Source*

The generation of the fast-drift burst RF emission is due to wave-mode transformation, either by wave–particle interactions or by wave–wave interaction processes:

$$w' + P \rightarrow w + P',$$

$$w' + w'' \rightarrow w \qquad \text{(III.14)}$$

(w = 'wave', P = 'particle').

The four most probable processes of these kinds as relevant to type III bursts are schematically shown in Figure III.20. The processes (c) and (d) refer to the radio wave source where (c) represents the basic process for radiation near the fundamental plasma frequency described by a scattering of the plasma waves l' on the polarization cloud of an ion i (electron density fluctuations) producing a transverse electromagnetic radio wave t

$$l' + i \rightarrow t + i' \tag{III.15}$$

(cf. Smith, 1974b).

This process can be induced or spontaneous; it seems most likely, that the fundamental emission must be amplified (negative absorption) in order to account for the observed ratios of fundamental-to-harmonic emission in the meter region (Smith, 1974b; Heyvaerts and Verdier de Genduillac, 1974).

The transition probability of the process (III.15) is

$$u_{i,l \rightarrow t} = \frac{(2\pi)^3 e^4}{m_e^2 \omega_{pe}^2 \left(1 + \dfrac{T_e}{T_i}\right)^2} \frac{(\mathbf{k} \times \mathbf{k}')^2}{k^2 k'^2} \delta[\omega_t - \omega_l - (\mathbf{k} - \mathbf{k}') \cdot \mathbf{v}_i] \tag{III.16}$$

where T_e and T_i are the electron and ion temperatures, respectively (Tsytovich, 1967; Smith, 1970a) yielding the following emission and absorption coefficients

$$j(\mathbf{k}, \omega) = \frac{k^2 \omega_{pe}^2}{4(2\pi)^{5/2} v_g N_e \left(1 + \dfrac{T_e}{T_i}\right)^2} \int \frac{\hbar\omega_p(\mathbf{k}') N_p(\mathbf{k}')}{k' v_i} \frac{(\mathbf{k} \times \mathbf{k}')^2}{k^2 k'^2} \times$$
$$\times \exp\left\{ -\frac{1}{2}\left(\frac{\omega_t - \omega_l}{k' v_i}\right)^2 \right\} \frac{d\mathbf{k}'}{(2\pi)^3} \tag{III.17}$$

$$\alpha(\mathbf{k}, \omega) = \frac{(2\pi)^{1/2} \omega_{pe} \langle v_k \rangle}{4(3)^{1/2} N_e v_e c k T_i \left(1 + \dfrac{T_e}{T_i}\right)^2} \int \hbar(\omega_t(\mathbf{k}) - \omega_l(\mathbf{k}')) \frac{(\mathbf{k} \times \mathbf{k}')^2}{k^2 k'^2} \times$$
$$\times \frac{\omega_p(\mathbf{k}') N_k(\mathbf{k}')}{k' v_i} \exp\left\{ -\frac{1}{2}\left(\frac{\omega_l - \omega_t}{k' v_i}\right)^2 \right\} \frac{d\mathbf{k}'}{(2\pi)^3} \tag{III.18}$$

(Smith, 1974b) where $v_g = d\omega/dk$ is the group velocity, $k \ll k'$, and $\langle v_k \rangle$ is the average phase velocity of the plasma wave ($v_g = 3^{1/2} v_e c \langle v_k \rangle^{-1}$).

It is seen from Equation (III.18), that the absorption becomes negative for transverse waves, where $\omega < \omega_l$.

The basic process for the second harmonic radiation is the interaction (combination) of two plasma waves producing a transverse electromagnetic wave at the frequency $2\omega_{pe}$ (cf. also Section 4.6.2):

$$l + l' \rightarrow t(2\omega_{pe}) \tag{III.19}$$

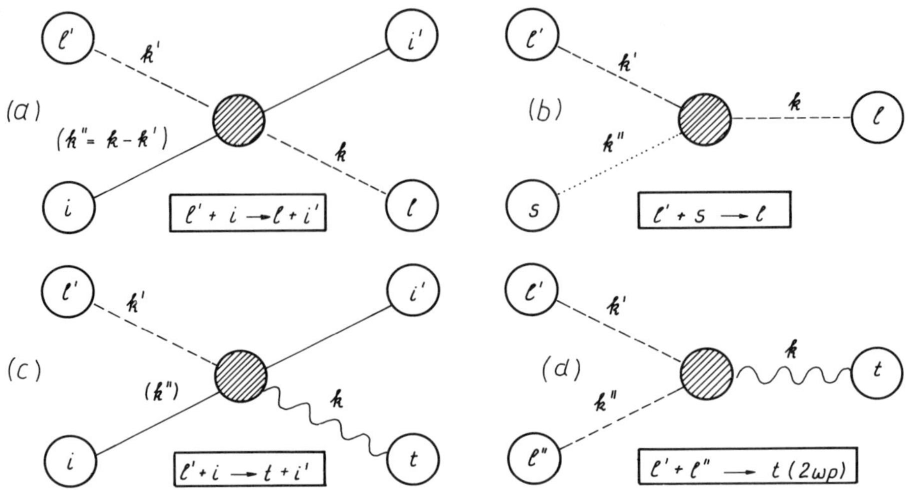

Fig. III.20. Schematic representation of nonlinear wave interaction processes a, c – scattering processes, b, d – combination processes.

with the transition probability

$$u_{l,l\to t} = \frac{(2\pi)^6\,\hbar e^2(k'^2 - k''^2)}{16\pi m_e^2 k^2 \omega_{pe}}\,\frac{(\mathbf{k}' \times \mathbf{k}'')^2}{k'^2 k''^2}\,\delta(\mathbf{k} - \mathbf{k}' - \mathbf{k}'')\,\delta(\omega_t - \omega_l - \omega_{l'}) \tag{III.20}$$

yielding

$$j(\mathbf{k}) = \frac{\pi\hbar e^2\omega_{pe}^2}{4(3)^{1/2}\,N_e m_e^2 c}\int N_k(\mathbf{k}')\,N_k(\mathbf{k} - \mathbf{k}')\,[\mathbf{k}'^2 - (\mathbf{k} - \mathbf{k}')^2]^2 \times \\ \times \frac{[\mathbf{k}' \times (\mathbf{k} - \mathbf{k}')]^2}{k'^2(\mathbf{k} - \mathbf{k}')^2}\,\frac{d\mathbf{k}'}{(2\pi)^6} \tag{III.21}$$

and

$$\alpha(\mathbf{k}) = \frac{\hbar e^2}{16\pi m_e^2\omega_{pe}}\int \frac{[\mathbf{k}' \times (\mathbf{k} - \mathbf{k}')]^2\,[k'^2 - (\mathbf{k} - \mathbf{k}')^2]^2}{k^2 k'^2(\mathbf{k} - \mathbf{k}')^2} \times \\ \times [N_k(\mathbf{k} - \mathbf{k}') + N_k(\mathbf{k}')]\,d\mathbf{k}' \tag{III.22}$$

(Smith, 1974b).

In contrast to the fundamental frequency, the absorption coefficient for the second harmonic can only be positive. In order for the above described process to occur, several conditions must be fulfilled: The combining plasma waves must meet together almost head-on and exhibit sufficiently high energy to guarantee energy and momentum conservation and to obtain nonthermal radiation. Beside that, secondary plasma waves occur mainly according to process (b) of Figure III.20.

Although the third harmonic is not generally observed in solar radio bursts, fast-drift bursts with a frequency ratio 2:3 and even, still more rarely, 1:2:3 are sometimes reported (cf. Takakura and Yousef, 1974), but it appears difficult to exclude purely random combinations. A possible mechanism to explain the occurrence of third harmonic radiation in solar radio bursts has been presented by Zheleznyakov and Zlotnik (1974).

d. *Radio-Wave Propagation*

On its way to the observer the radiation undergoes effects of scattering, refraction and polarization, absorption, etc., which modify the received radiation characteristics. Scattering and refraction effects have received considerable attention in the past. The method usually adopted consists of a computer simulation with a sufficiently large number of ray paths, which are subject to random scattering at electron-density inhomogeneities in the frame of special models (Fokker, 1965; Steinberg *et al.*, 1971; Caroubalos *et al.*, 1972; Riddle, 1972a, b).

Assuming a point source of radio emission with different directivities, the scattering may have an important influence causing either a focusing or a broadening of the received angular emission pattern dependent on the wavelength and the primary directivity of the source. In contrast to the conditions for a nonscattering corona, the scattering causes a scatter of the turning points of the ray trajectories too, so that the cutoff frequency becomes practically independent of the direction angle equal to the local plasma frequency of the source region (cf. also Section 4.2.4.a). Moreover, because scattering implies a multitude of different ray paths, the time profile of the received radiation is also influenced by an excess group delay (Wild *et al.*, 1959a).

3.5. Slow-Drift Bursts

3.5.1. General Observational Features

a. *Main Characteristics, Frequency Drift*

Slow-drift bursts are generally considered as the radio evidence of flare-induced collisionless magnetohydrodynamic shock waves in the solar corona. In contrast to the fast-drift bursts they are characterized by a 'slow' drift of the spectral features of the order 0.25 to 1 MHz s^{-1} from high to low frequencies (Figure III.21). The slow-drift bursts are outstanding phenomena connected with large flare events. Therefore, in contrast to fast-drift bursts, they occur comparatively seldom.

Historically, soon after the introduction of the solar radio-spectrograph observations the slow-drift bursts had been recognized as a special phenomenon and were classified as type II bursts by Wild and McCready (1950). It is interesting to note, that at the time of their detection (formerly contained in the label 'outbursts') type II bursts had not been distinguished from the type IV burst phenomenon.

An early comprehensive survey of the type II bursts was made by Roberts (1959b) and later supplemented by Maxwell and Thompson (1962). Newer accounts on the type II burst phenomenon were given e.g. by Wild and Smerd (1972), Dryer (1974), and McLean (1974).

The frequency drift of the type II bursts can be interpreted as a height variation

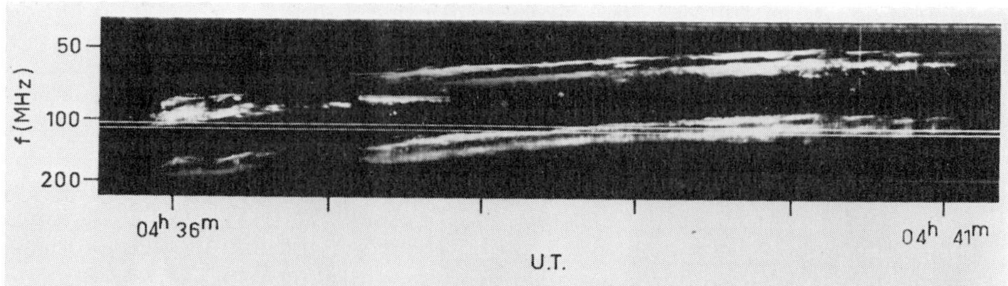

Fig. III.21. A type II burst (October 9, 1969) exhibiting harmonic and split-band structure (after Dulk, 1970a).

of a disturbance traveling outward through the corona, generally similar to the behavior of type III bursts. Inferring a reasonable electron density-distribution law in the corona and adopting again the plasma hypothesis, velocities of the traveling disturbances ranging from about 200 to 2000 km s^{-1} (and also beyond these figures) can be deduced.

These velocities are remarkably greater than the coronal sound velocity and Alfvén velocity, so that shock waves are to be anticipated.

There is a number of spectral features which may be of special interest for the interpretation of the type II bursts which will be discussed later: narrow bandwidths (> some MHz) of the drifting emission ridges, frequently simultaneous occurrence of fundamental and second harmonic features, and band splitting of both the fundamental and second harmonic components.

Due to the peculiar spectral features, the type II bursts provided the first observational hint for the existence of plasma oscillations in the solar corona, though the detailed theoretical proof was difficult. A direct argument for the validity of the plasma hypothesis came from position measurements with swept-frequency interferometers (Wild *et al.*, 1959b; Weiss, 1963a, 1965).

b. *Morphology and Other Properties*

Typically type II bursts start about 2–15 min after the beginning of a flare and last about 10 or 15 min in the meter region. After the onset the time profile is usually rather complex and exhibits very intense peaks of the flux ($T_b > 10^{12}$ K). The onset shows an abrupt cutoff in frequency. The starting frequency of the fundamental component is usually below 100 MHz, but can also reach higher frequencies, say, 250 MHz. Correspondingly the second harmonic features are observed at twice this frequency.

The polarization of the type II bursts is typically random; however, complex polarization structures can also be observed.

From ground-based patrol observations type II bursts can be traced up to decameter waves. Because of their rarity not very much is known about the frequency extent and the spectral behavior towards lower frequencies which is accessible to space measurements only. A good example of a type II burst observation at frequencies up to 30 kHz was obtained for the big solar flare event of August 7, 1972 (Figure III.22). In this case it was possible to track the path of the shock wave which was

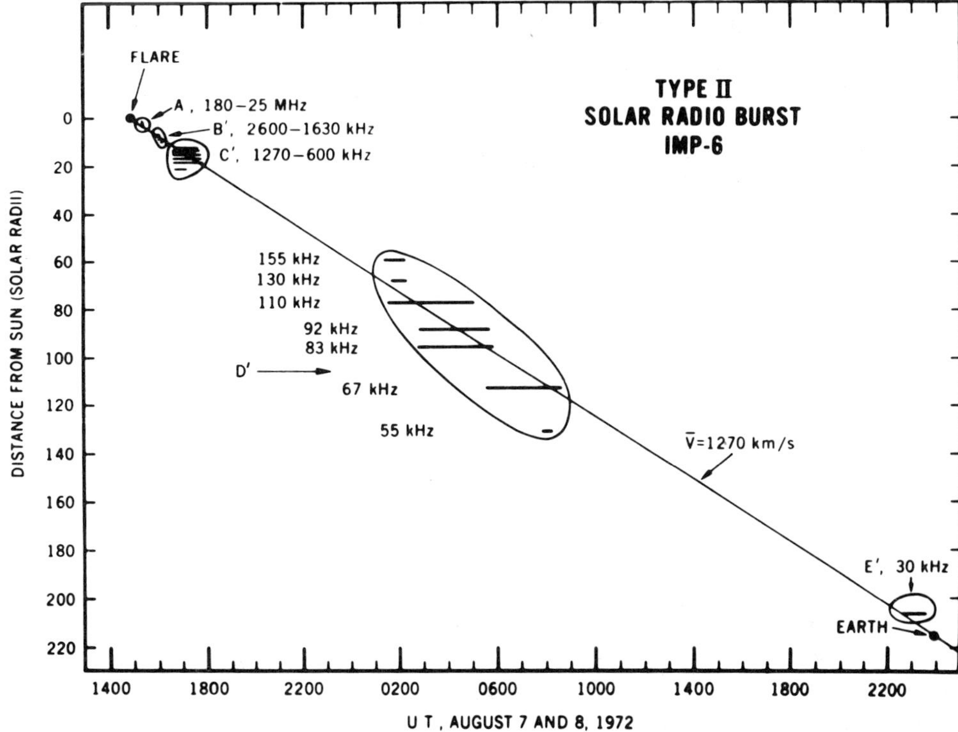

Fig. III.22. IMP–6 observations of a type II burst of August 7, 1972, in comparison with the average velocity of the shock front scaled under the assumption, that low-frequency type III bursts are observed at the second harmonic of the local plasma frequency (after Malitson *et al.*, 1973b).

marked by the type II burst by observations aboard the IMP–6 satellite from interplanetary space in to the Earth (where the arrival of the shock waves was indicated by a sudden commencement of a geomagnetic storm) by applying the RAE density scale, which was deduced inferring second-harmonic type III burst emission (Malitson *et al.*, 1973a, b).

c. *Heliographic Observations*

In recent years the main experimental information on type II bursts was collected from the Culgoora heliograph observations by displaying the spatial brightness distribution in a time sequence. Here the sources of the type II bursts are seen as extended emitting regions with a diameter of the order of $0.5R_{\odot}$, which is at first sight astonishingly large with regard to the comparatively small bandwidth of the emission. The sources often exhibit remarkable spatial and temporal variations of brightness (cf. Kai and McLean, 1968; Wild, 1969a; Kai, 1969c; Dulk, 1970a; Stewart *et al.*, 1970). The centroid of the source region exhibits small motions, even when observed at only one fixed frequency.

Sometimes the source consists of several distinct peaks situated on a large shell around the flare center (Figure III.23) (Smerd, 1970; Dulk and Smerd, 1971). The

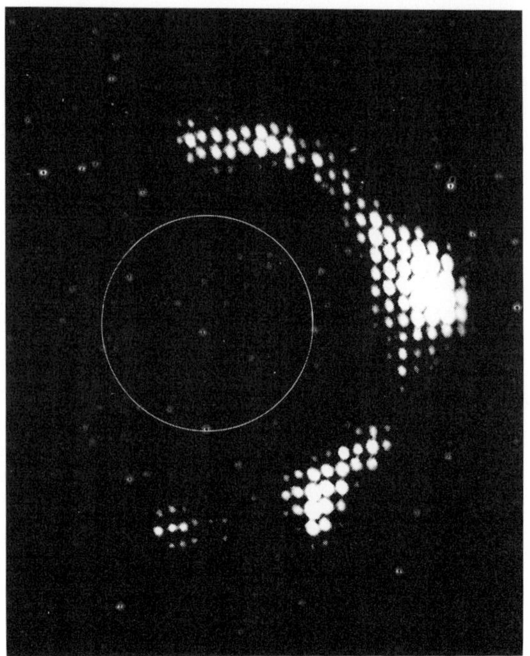

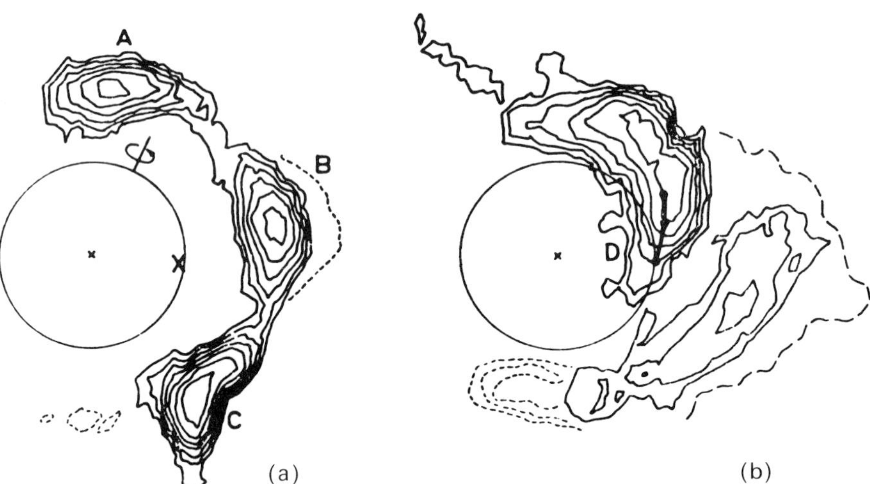

Fig. III.23. A very extended limb event observed by the Culgoora radioheliograph on March 30, 1969. Top: 80 MHz heliogram showing the great peripheral extent of the radio sources. Bottom: Computer-drawn brightness contours at two phases of the event, A, C, and D – type II burst sources, B – continuum source (after Smerd, 1970).

observations indicate a wide cone of the emission from the shock waves (i.e. up to about 200^0 if seen from the flare center). However, since the emission is not uniformly distributed over all angles, only special directions of the shock wave propagation are marked by the radio response. Apparently one single flare can produce several discrete sources of type II bursts, which – not necessarily simultaneously – can be distributed over an appreciable part of the solar hemisphere.

d. *Associated Phenomena*

Since the type II bursts are a typical phenomenon of large flare events, they are almost automatically associated, more or less directly, with the whole palette of single flare effects which commonly accompany big flares. Among them we would like to stress the relationship between some important features, e.g.: Hα flare waves,

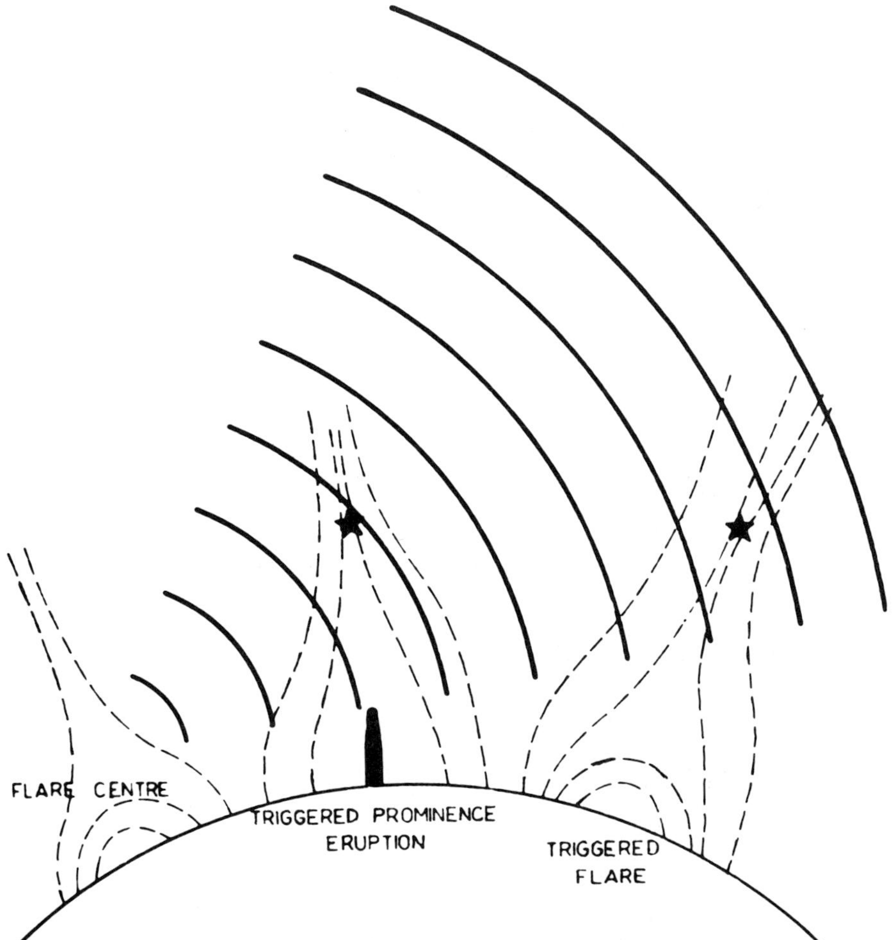

Fig. III.24. Schematical picture showing the possible trigger action of a flare-induced shock wave (after Wild, 1969a).

sprays and eruptive prominences, moving type IV bursts, and sudden commencements of geomagnetic storms.

The major optical evidence for the occurrence of shock-wave disturbances and for the ejection of matter in conjunction with flares and type II bursts comes from Hα-observations though it appears often difficult to discriminate by this method the true nature of the observed phenomenon. Some pioneering work was done in this field by Moreton (1964), Dodson and Hedeman (1964), and Smith and Ramsey (1966). A concise review on these matters was given by Bruzek (1974). Theoretical discussions are due to Meyer (1968), Uchida (1968), and Uchida *et al.* (1973).

There appears to exist a close connection between the occurrence of shock waves and the ejection of matter in the solar atmosphere. At radio frequencies this connection is indicated by the combination of type II bursts and moving type IV bursts. In the optical domain coronal observations reveal an association with flare sprays, but the statistical correlation with type II bursts is not exceedingly large (E. V. P. Smith, 1968; Smith and Harvey, 1971). Obviously not all optically visible disturbances come to regions favorable for a conversion into a type II burst emission.

At greater heights up to several solar radii coronographic observations in the white light from satellites display pictures indicating piston-driven shock waves moving outwards in connection with a slow-drift radio burst emission (Figure III.30) (cf. Stewart *et al.*, 1974).

Combined radio and optical observations also give a strong evidence for shock wave interactions between very distant centers around the Sun (Wild *et al.*, 1968; Wild, 1969a). Apparently the shock waves are capable of triggering eruptions of distant prominences in the corona and, moreover, can release high-energy particles as may be concluded from the appearance of the radio emission. Figure III.24 shows a sketch for such an interaction link, which was proposed by Wild (1969a).

Concerning the processes in the interplanetary space and planetary environments it may be briefly stated that in connection with type II bursts, flare-associated solar wind disturbances, as well as subsequent storms in the magnetospheres of the Earth and Jupiter, are marking reactions initiated by the disturbances traveling from the Sun.

3.5.2. Spectral Structure

a. *Harmonic Radiation*

A significant clue to the emission process of the solar slow-drift radio bursts is given by the existence of harmonics of the radiation in connection with narrow-band structures. This feature is recognizable in dynamic spectra and was first discussed by Wild *et al.* (1954a). A well expressed harmonic structure is discernible for more than half of all type II bursts in the meter region, whereas at longer wavelengths, e.g. in the hectometer region, harmonic structures are perhaps not so clearly visible. In the meter region the fluxes at both bands are usually of similar magnitude, though deviations exist in both directions. No clear evidence of components of higher harmonics could be achieved until now. Sharply defined details in the dynamic spectrum reveal a group delay between the harmonic and fundamental emission of about

1 s. The harmonic structure can be observed from nearly all central meridian distances and the harmonic ratio (usually a little bit smaller than two) does not seem to be seriously influenced by the position on the Sun. The question of the relative positions of the fundamental and harmonic emission demands still further work. A number of observations yields contradictory results, being partly inconsistent with simple ideas of a type II burst interpretation, putting the questions of whether (a) the harmonic radiation's position as measured at a fixed frequency is closer to the flare center than that of the fundamental and (b) the burst positions at fixed frequencies are really moving with time (cf. Dulk, 1970a, b; Riddle, 1970b).

b. *Frequency Splitting*

In several cases, as demonstrated by Figure III.21, the individual harmonic bands of type II bursts are duplicated once more by the effect of frequency or band splitting. The frequency separation is about 10% of the band midfrequency, i.e., of the order of 10 MHz in the fundamental and twice as great in the harmonic band.

The phenomenon of frequency splitting should give an important hint of the nature of type II bursts. At present, several explanations have been proposed but the interpretation is still open. The proposals fall into three main categories:

(a) Effects of the magnetic field:
Emission at ν_p, $\nu_p + \frac{1}{2}\nu_H$, $\Delta\nu = \frac{1}{2}\nu_H$ (Roberts, 1959a) or emission at ν_p and the upper hybrid frequency

$$\nu_{uh} = (\nu_p^2 + \nu_H^2)^{1/2}, \qquad \Delta\nu \approx \frac{\nu_H^2}{2\nu_p} \tag{III.23}$$

(Sturrock, 1961a; Tidman *et al.*, 1966).
In both cases the resulting magnetic field seems to be too large compared with the requirement

$$v_A \approx 7 \times 10^3 \frac{\nu_H}{\nu_p} \lesssim 10^3\,\text{km s}^{-1} \rightarrow \frac{\nu_H}{\nu_p} \lesssim 1/7$$

(Weiss, 1965).

(b) Doppler-effect interpretation (Zheleznyakov, 1965; Zaitsev, 1965; Fomichev and Chertok, 1967a): This approach is based on the Pikelner–Ginzburg model (cf. Section 3.5.3.b) yielding $\Delta\nu \approx (2v_y/c)\,\nu_p$ at the fundamental frequency.

(c) Geometrical interpretation (McLean, 1967, 1974). Basing on a reasonable model distribution of the electron density and the magnetic field a computer simulation shows a preference of the emission in different locally separated regions which can be directly attributed to the occurrence of split bands (cf. also Section 3.5.3.c).

c. *Herring-Bone Structure*

Certain type II bursts exhibit the so-called 'herring-bone' pattern in the dynamic spectrum consisting of a succession of fast drifting burst elements. First attention was paid to this phenomenon by Roberts (1959a,b). A short discussion of this matter was given already in Section 3.4.3.d. For illustration see Figure III.25.

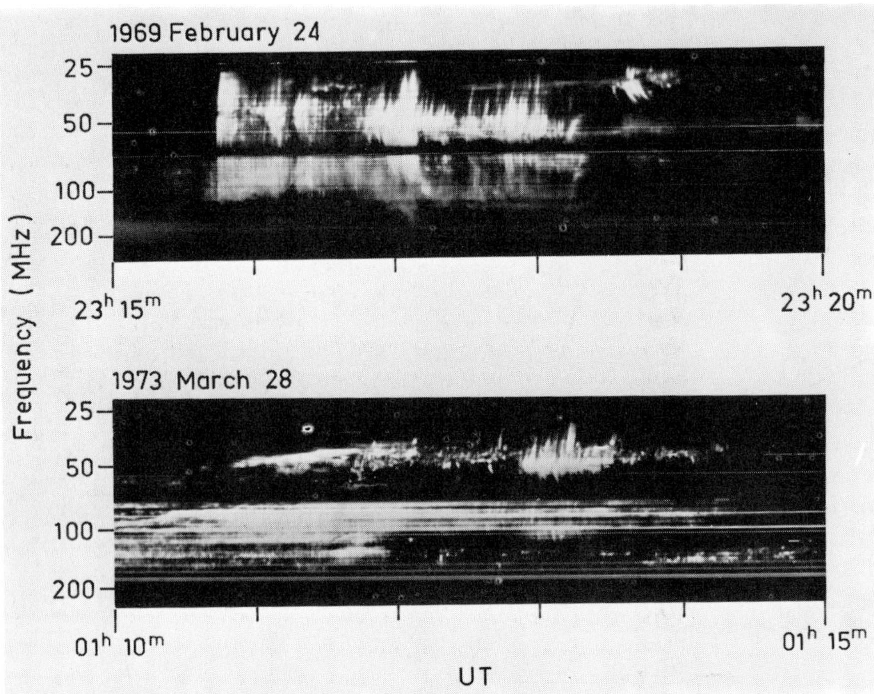

Fig. III.25. Examples of dynamical spectra of type II bursts showing forward- and reverse-drift herring-bone structure repeating itself in the fundamental and harmonic bands (after Smerd, 1976).

3.5.3. Interpretation

a. *History*

In 1946/47 remarkable time delays between the arrival of 'outbursts' at three different fixed frequencies were already noted (Payne-Scott *et al.*, 1947). Based on a suggestion by Martyn (1947) that the burst emission may originate near the critical level, where the refractive index vanishes, the time delay was interpreted as an upward motion of a certain physical agency passing the solar corona from high-frequency to low-frequency levels. In the same way the sharp low-frequency cutoffs of the type II bursts were interpreted by Wild (1950a) as an effect of the critical frequency ('critical-frequency hypothesis'). Later on Wild *et al.* (1954a) detected a narrow emission peak near the low-frequency cutoff implying the possibility of a resonance phenomenon near the critical frequency as it was formerly discussed by Martyn (1947) and Jaeger and Westfold (1949). In this way the *plasma* (*frequency*) *hypothesis* was developed from the earlier *critical-frequency hypothesis*.

b. *Type II Burst Models*

The exciter velocities v_{II} of the type II bursts are definitely less than the thermal velocities of the electrons v_{th} in the corona. Therefore particle streams with such low velocities cannot account for the generation of plasma waves as it is required for the type II bursts. One can estimate

$$v_s < v_A \lesssim v_{II} < v_{th} \qquad \text{(III.24)}$$

where $v_s = \sqrt{\gamma RT}$ is the sound velocity (≈ 120 km s^{-1}, if γ (polytropic exponent) = 5/3, $T = 10^6$ K, R = universal gas constant), $v_A = H/\sqrt{4\pi\rho} \approx H/\sqrt{4\pi N_e m_i}$ ($\approx 7 \times 10^3\ \nu_H/\nu_p$ [km s^{-1}]) is the Alfvén velocity, and $v_{th} = \sqrt{2KT/m_e}$ ($\approx 5.5 \times 10^3$ km s^{-1} if $T = 10^6$ K) is the thermal velocity. With regard to these figures an interpretation of the type II bursts in terms of shock waves seems most appropriate (cf. Uchida, 1960). Sound-shock waves in the quiet corona without any magnetic field influence would attain high Mach numbers

$$M_s = v/v_s \qquad \text{(III.25)}$$

and hence suffer a rapid dissipation of the wave energy by heating the ambient plasma. Run-away electrons with large free path lengths ($\sim 0.1\ R_\odot$) would be generated and penetrate into regions before the shock front (Welly, 1963).

For the above reasons it can be assumed that only MHD shock waves have a chance to propagate over long distances and generate plasma waves, which in turn can be transformed into the type II burst radio emission. The theory of these processes is still being developed; therefore, there is no final conclusion at this time.

Up to now there exist two major approaches to the problem of the interpretation of the type II burst emission. The first one is a model introduced by Pikelner and Ginzburg (1963) based on an investigation of the fine structure of MHD shock waves by Sagdeev (1962) and elaborated by Zheleznyakov (1965) and Zaitsev (1965, 1968). The model describes a collisionless shock wave propagating perpendicular to a magnetic field. An electron current $j = -eN_e v_y$ (where $y \perp x, z$; x = direction of the shock-wave propagation; z = direction of the external magnetic field) arises in the shock front as a consequence of the magnetic field gradient

$$\partial\mathbf{H}/\partial x = (4\pi/c)\,\mathbf{j} \qquad \text{(III.26)}$$

so that the drift velocity v_y is approximately proportional to the difference quotient $\Delta H/\Delta l$, when l is the thickness of the shock front, which is a function of the MHD Mach number

$$M_A = v/v_A. \qquad \text{(III.27)}$$

Plasma waves are excited, if the drift velocity v_y exceeds the thermal velocity v_{th}. For $v_y = v_{th}$ the condition for the 'critical Mach number'

$$M_C = 1 + (3/8)(8\pi N_e KT/H^2)^{1/3} \qquad \text{(III.28)}$$

can be obtained under the assumptions

$$H^2/8\pi \gg N_e KT \qquad \text{(III.29)}$$

and

$$\nu_H \ll \nu_p. \qquad \text{(III.30)}$$

For $M_A > M_C$ plasma waves are expected to be generated. However, it is quite obvious that the above mentioned conditions can be satisfied only for Mach numbers $M_A < 2$. In this range of Mach numbers the maximal drift velocity of the electrons in the shock front was calculated by Zaitsev (1965):

$$v \approx \frac{1.55 H \omega_H}{4\pi e N_e} \cdot \frac{M_A^2 - 1}{M_A^3} . \tag{III.31}$$

In the frame of this general model, estimations of the magnetic field from type II burst observations were proposed by Takakura (1964, 1966) (obtaining upper limits of the magnetic field strength by equating $v_{\text{II}} = v_A$) and by Fomichev and Chertok (1965) (using Equation III.28).

In general, there is some doubt about the validity of the assumption $\mathbf{v}_{\text{II}} \perp \mathbf{H}$, which seems to conflict with some observations indicating $\mathbf{v}_{\text{II}} \parallel \mathbf{H}$ (heliographic observations, correlations to the emitting regions of type III bursts which are thought to travel parallel **H**, and the extension of type II bursts well into the interplanetary space, where closed magnetic field structures are scarcely to be expected).

The other approach to a type II burst theory was done by Tidman (1965) and Tidman *et al.* (1966). These authors predict the occurrence of superthermal particles in MHD shock waves and refer to an analogous situation as the Earth's bow shock bouncing against the solar wind. The shock front is thought to be followed by a turbulent plasma region of disordered magnetic field confining superthermal electrons exciting incoherent Čerenkov plasma waves.

A great advantage of this approach seems to be that no restriction was made concerning the magnetic field direction relative to the shock front, which looks more realistic. A more detailed elaboration of the theory, however, is yet missing.

c. *Shock-Wave Propagation*

The propagation of flare-induced shock waves has very interesting aspects illuminating the structure of the solar corona above active regions.

By heliographic observations it was found that the path of the shock waves causing type II bursts can deviate remarkably from a direct straight line. For instance, Kai (1969c) interpreted one peculiar event as the observational evidence for a reflection of shock waves on a magnetic wall. In another case, Smerd (1970) described a behind-limb event and concluded a curved path for the type II burst exciters thus being either refracted or magnetically guided. Summarizing the factors, which possibly influence the shock-wave propagation, we have to take into account: (a) Refraction of shock waves; (b) channeling of the shock-wave energy by the structure of the magnetic field and density distribution; and (c) dissipation of the shock-wave energy at the lowest Alfvén velocity.

d. *Electromagnetic Radiation*

The transformation of the plasma waves excited by the shock waves into electromagnetic radiation (radio waves) can in principal be thought of as analogous to the case of the fast-drift bursts which was already discussed in Section 3.4.6.c.

Concluding our remarks concerning slow-drift bursts we want to stress the conditions necessary for the generation of a type II burst, which possibly could explain the rareness of these events:

(a) Existence of a sufficiently strong shock wave (i.e. conditions must be favorable for its propagation requiring not too high and not too low Mach numbers);

(b) Conditions must be favorable for a generation of plasma waves (the mechanism itself is not yet fully understood);

(c) Conditions favorable for a transformation into electromagnetic waves; and

(d) Conditions suitable for a propagation of the electromagnetic waves to the observer.

3.6. Continuum Bursts (b): The Type IV Burst Complex

3.6.1. Introductory Remarks

a. *General Features*

A *type IV burst* is defined as a 'long-period continuum event in any part of the radio spectrum following a flare' (Fokker, 1963a; Wild, 1962). Since this definition comprises the whole range of the radio frequencies, and the radio continua are rather complex in nature, a number of subtypes or special components of type IV bursts can be distinguished, which were classified by various authors in various manners (cf. Section 3.6.1.c). Hence, speaking about 'type IV bursts' one means either the totality of all possible type IV burst components (the '*type IV burst complex*') or simply one part of this, which is particularly developed in a really observed event.

Like the slow-drift bursts, type IV bursts are also comparatively rare and, in general, restricted to major flare events. Recent reviews discussing type IV bursts are due to Krüger (1972) and Boischot (1974a). Catalogs of type IV bursts, in particular in connection with associated proton flares, have been collected by different authors (e.g. Švestka and Olmr, 1966; Krüger *et al.*, 1971; Švestka and Simon (ed.), 1975).

b. *Brief History*

As already mentioned in Section 3.5.1, slow-drift and continuum bursts were not differentiated until 1957, though the latter phenomenon was already studied by a number of authors. The type IV burst as a new, distinct class of solar radio bursts was first recognized by Boischot (1957, 1958, 1959) with the aid of the Nançay multi-element interferometer operating at 169 MHz.

A first interpretation of the phenomenon was given by Boischot and Denisse (1957) in terms of synchrotron radiation from electrons spiraling in coronal magnetic fields.

Boischot's concept of type IV bursts (forming the only major spectral type of solar radio bursts which was not born in Australia) originally referred to only one component of the type IV bursts, namely to that which later on turned out to be named 'moving type IV bursts'. Soon after the introduction of the fourth burst spectral type a rapidly growing number of studies made around the globe (partly under the

TABLE III.4

Survey of type IV burst subclassifications

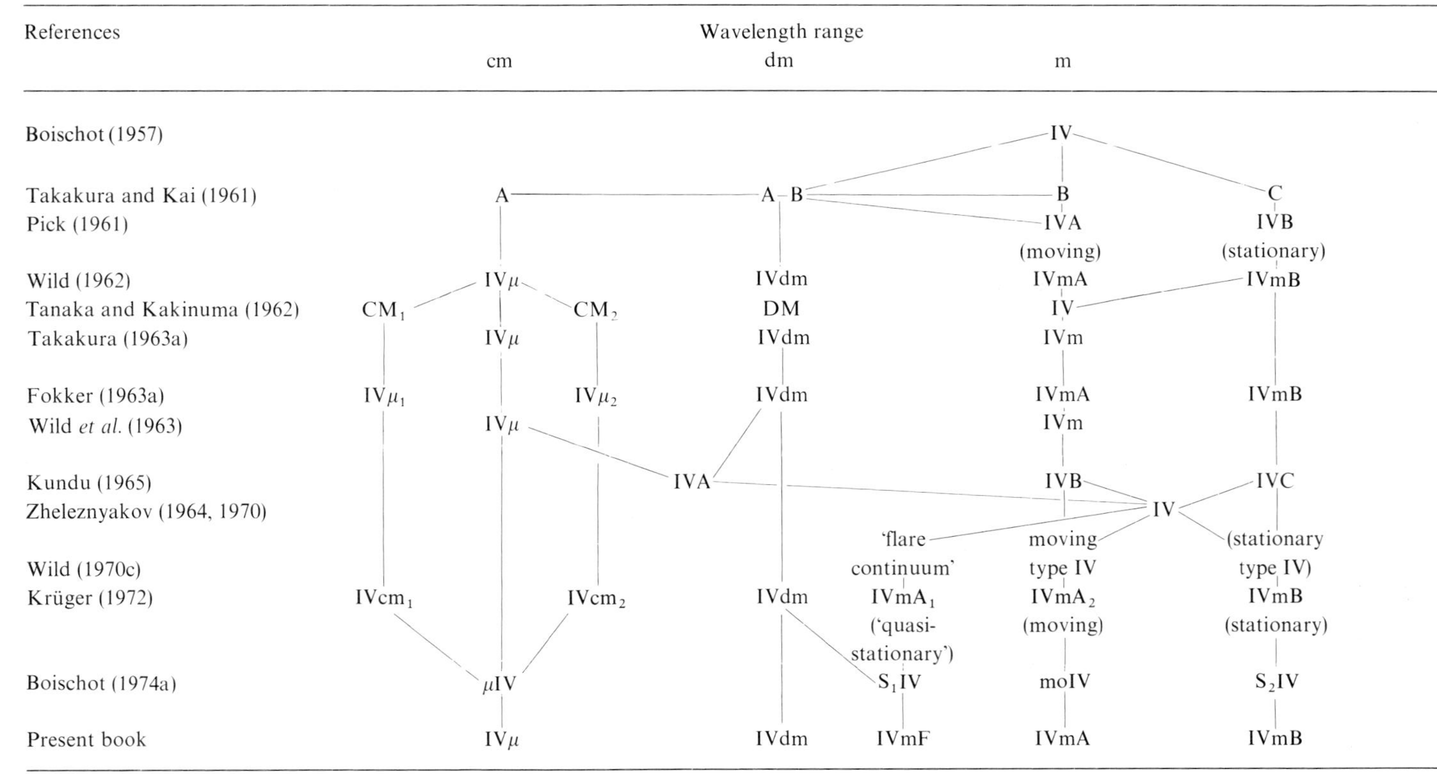

References	Wavelength range								
		cm			dm		m		
Boischot (1957)							IV		
Takakura and Kai (1961)		A			A B		B		C
Pick (1961)							IVA (moving)		IVB (stationary)
Wild (1962)		$IV\mu$			IVdm		IVmA		IVmB
Tanaka and Kakinuma (1962)	CM_1		CM_2		DM		IV		
Takakura (1963a)		$IV\mu$			IVdm		IVm		
Fokker (1963a)	$IV\mu_1$		$IV\mu_2$		IVdm		IVmA		IVmB
Wild *et al.* (1963)		$IV\mu$					IVm		
Kundu (1965)				IVA			IVB		IVC
Zheleznyakov (1964, 1970)								IV	
Wild (1970c)						'flare continuum'	moving type IV		(stationary type IV)
Krüger (1972)	$IVcm_1$		$IVcm_2$		IVdm	$IVmA_1$ ('quasi-stationary')	$IVmA_2$ (moving)		IVmB (stationary)
Boischot (1974a)		μIV				S_1IV	moIV		S_2IV
Present book		$IV\mu$			IVdm	IVmF	IVmA		IVmB

aspect of the eminent relevance to solar-terrestrial relationships) extended the initial conception which led to some confusion considering the variety of different forms of the appearance of this burst type.

In the mean time much work has been done enlarging the knowledge about the type IV bursts by investigation of the spectral, polarization, statistical, positional, and other characteristics, which will be considered in the next subsections. Nevertheless, much work is still to be done in the future on this burst class.

c. *Classifications*

The variety of large solar continuum radio bursts is reflected by a multitude of attempts to classify the main features resulting in a puzzling picture for those who do not actively work in this field. In order to make these things clear the major schemata of type IV burst subclassifications widely distributed in the literature are compiled in Table III.4. A look in this table shows that although there are no essential differences in single notations, some uncertainties exist especially with regard to the classification of the decimeter part of the type IV bursts and the subdivision of the type IV burst in the meter range.

It should be remarked that the classifications used hitherto are basically oriented on observational aspects. It is hoped that a deeper knowledge about the physics of the type IV phenomena will result in a well founded and generally accepted classification. For the purpose of the present book we suggest distinguishing between the following main components based on the essence of Table III.4:

(1) type IVμ bursts,
(2) type IVdm bursts,
(3) type IVmF bursts (flare continua, 'quasi-stationary type IV bursts'),

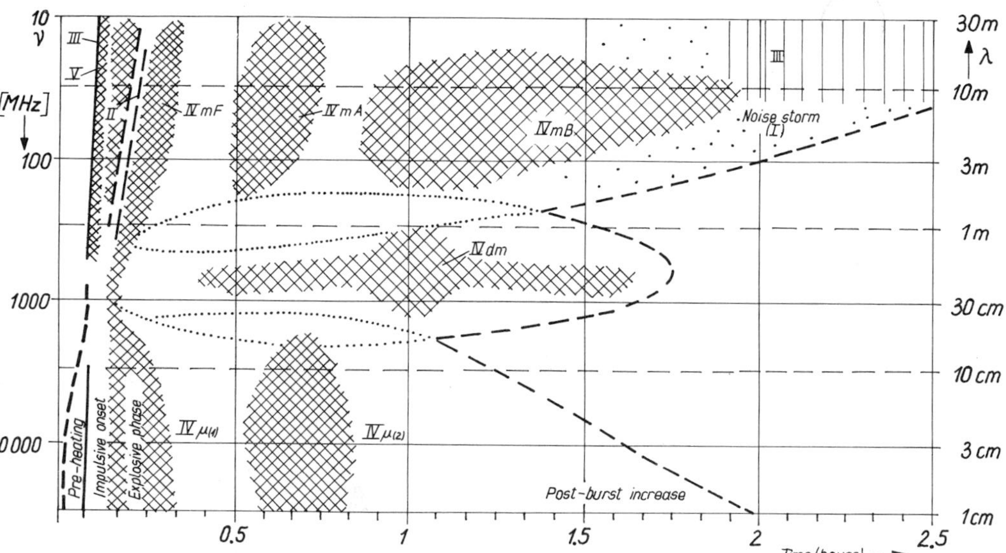

Fig. III.26. Schematic representation of different spectral components of solar radio bursts.

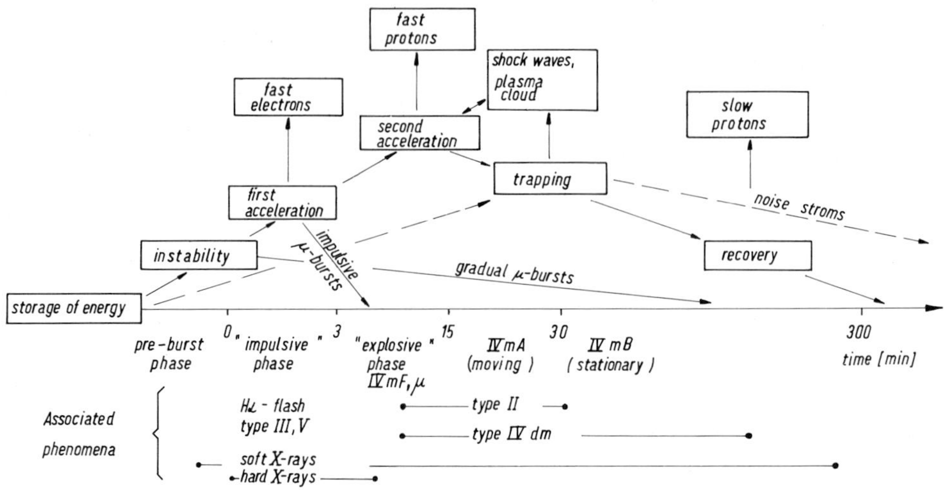

Fig. III.27. Idealized scheme of the development of type IV bursts.

(4) type IVmA bursts (moving type IV bursts), and
(5) type IVmB bursts (stationary type IV bursts)
(cf. Figure III.26).

d. *Outline of a Complete Picture*

The temporal succession of the different stages of an idealized type IV burst and possible ways of the development of solar continuum radio burst events are schematically sketched in Figure III.27.

Some confusion still exists in the literature in using the notations 'flash phase', 'impulsive phase', 'explosive phase', 'hard X-ray phase', 'type IV burst phase', etc. Following Křivský *et al.* (1976) we recommend distinguishing the *impulsive phase*, which corresponds to the onset of an impulsive microwave burst, from the *explosive phase*, which corresponds to the occurrence of a type IVμ burst. The terms flash phase, hard X-ray phase, and impulsive phase denote one physically connected complex. The terms explosive phase and type IV burst phase belonging together would denote another physically connected complex as indicated by Figure III.27. (Formerly, the denotations 'first' and 'second part events' were used in a similar sense.) Since in real solar events both complexes are often mixed and sharp boundaries are generally missing, the hitherto existing difficulties in developing a unique classification are not surprising.

3.6.2. The cm- and dm-Region

a. *Type IVμ Bursts*

The type IVμ burst component is characteristically separated from the other parts of a type IV burst in the dynamical spectrum by a gap of the emission in the dm-region (especially in those cases, where no type IVdm burst is present). Since the type IVμ bursts tend to occur a little more frequently than big type IVm bursts, the microwave region sometimes contains the sole manifestation of a type IV event. It must be noted, however, that the distinction to minor complex microwave bursts, which will not be counted as type IVμ bursts, appears somewhat arbitrary, because only quantitative morphological criteria are as yet adopted and qualitative differences are obviously missing to distinguish between complex bursts and type IVμ bursts (cf. Section 3.3.2).

Nevertheless, in contrast to some minor microwave bursts, the spectrum of the type IVμ bursts exhibits typically a broad-band structure extending up to about 70 GHz (and possibly beyond) for special cases. Often not sharply defined, the low-frequency boundary of type IVμ bursts lies typically in the region at about 1 GHz. The spectral maximum varies in a wide range around $\lesseqgtr$ 10 GHz. The peak fluxes at the spectral maximum can exceed 10^4 s.u. The duration of type IVμ bursts is typically of the order of 1 h.

The polarization of the type IVμ bursts often corresponds to the extraordinary wave mode of the magneto-ionic theory, but also complex polarization patterns can be observed (cf. e.g. Böhme *et al.*, 1974). There are indications that the source region of the type IVμ bursts is progressively broadening during its lifetime and moving upward in accordance with the development (expansion) of magnetic loops observed in double-ribbon flares in Hα (Křivsky and Krüger, 1973). Estimations of the magnetic field strength and electron density distribution were obtained from the burst decay times under the vague assumption, that the decay is completely due to kinetic energy losses by gyro-synchrotron radiation (Kai, 1968).

b. *Type IVdm Bursts*

Type IVdm bursts were proposed to form a distinct subtype (cf. Table III.4) with regard to their polarization, morphological, and spectral characteristics. The frequency range of the type IVdm bursts extends from below 200 to about 2000 MHz. The sense of circular polarization is often reversed to that of the type IVμ bursts. In the spectral diagram (cf. Figure III.28) the type IV dm component forms patches or bands of definite bandwidth sometimes revealing remarkable high fluxes ($\gtrsim 10^4$ s.u.).

The correspondence in time with other type IV burst components is often very poor. It must be supposed, that the type IV dm burst is not a frequency-shifted analog of any type IVμ or type IVm burst component. Differences in the source structure, energy distribution, and emission mechanism in comparison to the adjacent type IV burst components are very likely.

3.6.3. The m- and Dm-Region

a. *Flare Continua* (*Quasi-Stationary Type IVm Bursts*)

There is some evidence for the existence of a distinct type IV burst component in the meter region which neither belongs to the moving ('IVmA') nor to the typical, stationary ('IVmB') part of a type IV burst event. The emission starts in a wide frequency band almost simultaneously with the onset of a type IVμ burst and a series of type III bursts, and lasts for a period of the order of half an hour (cf. the spectral

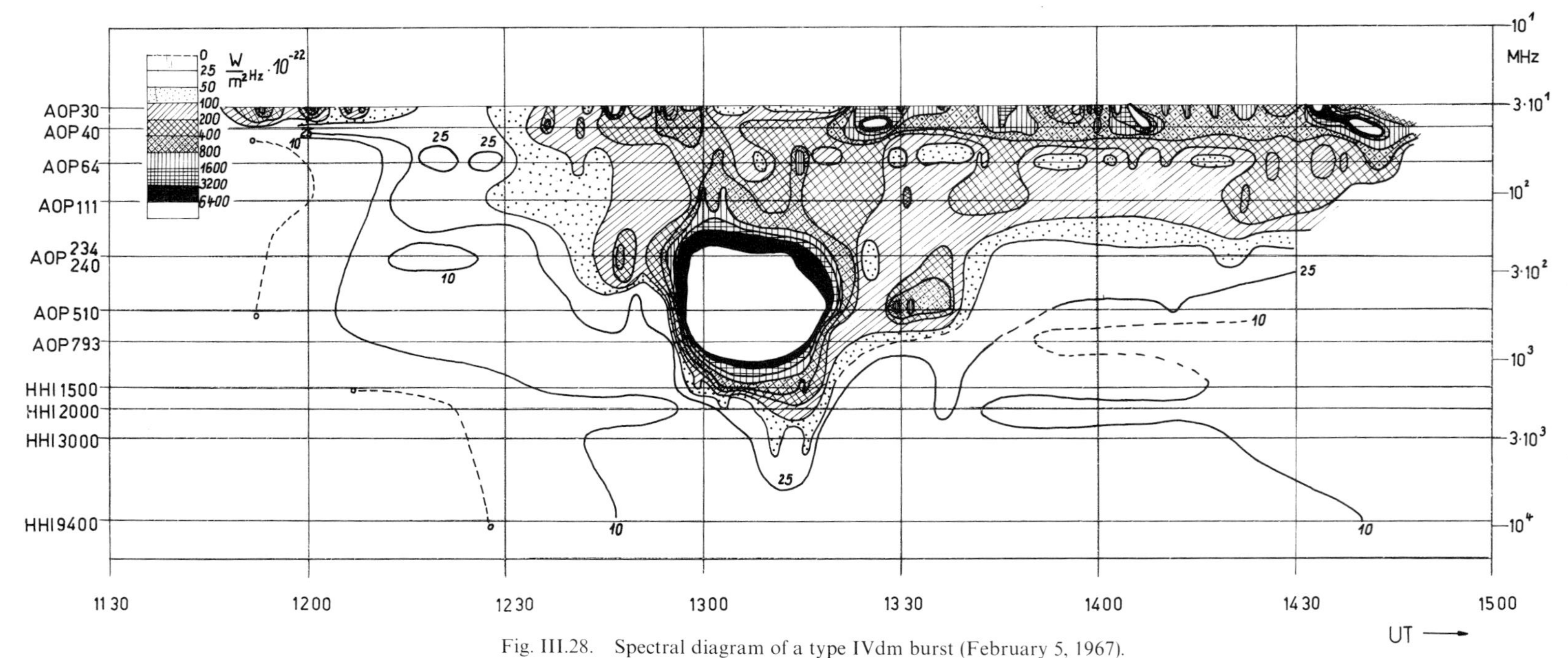

Fig. III.28. Spectral diagram of a type IVdm burst (February 5, 1967).

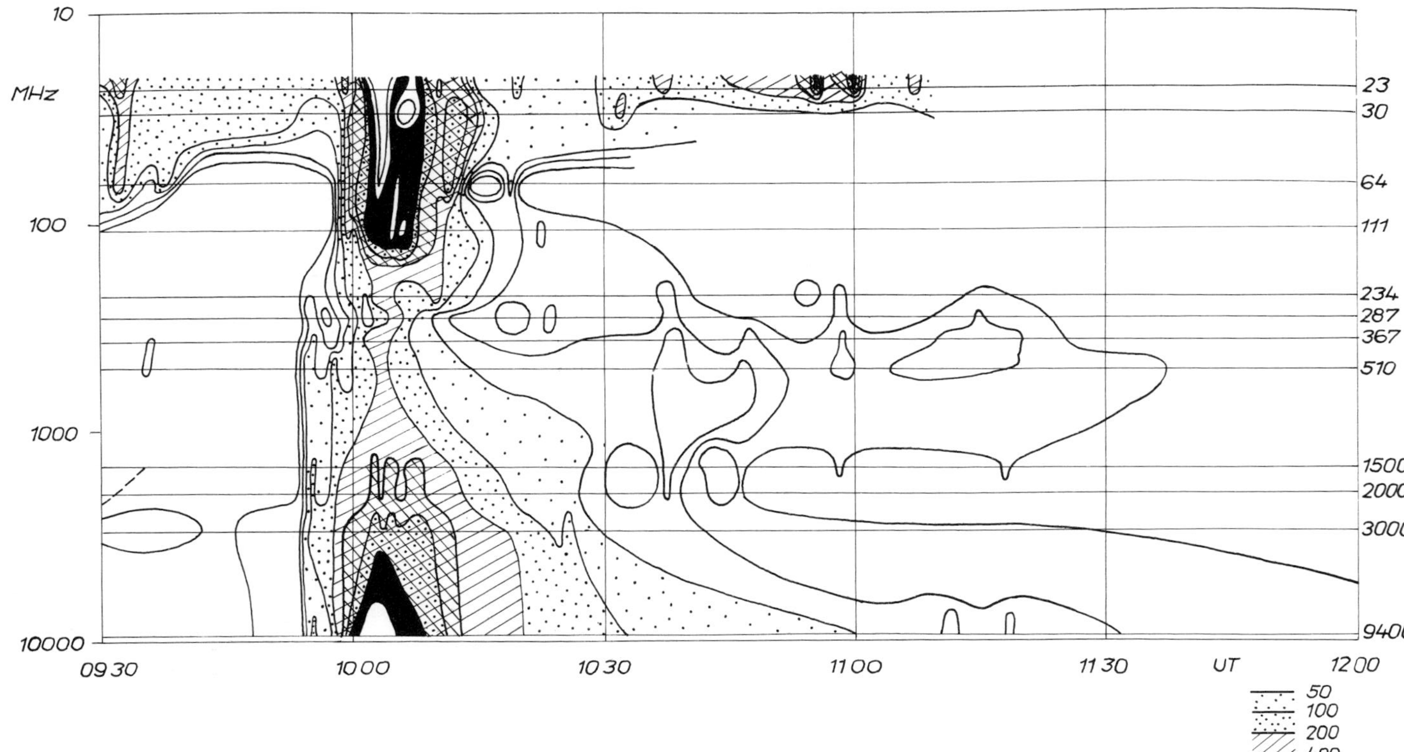

Fig. III.29. Spectral diagram of a type IV burst showing the typical distinction between the microwave and meterwave parts (June 5, 1969).

diagram of Figure III.29). In contrast to the 'moving' type IVm burst component the emission at any frequency starts clearly before the onset of a type II burst (if present) and exhibits the appearance of a 'stationary' source by heliographic observations. The phenomenon was named type $IVmA_1$ burst or *quasi-stationary type IVm* burst by Krüger (1968, 1972), Böhme (1971), and Akinyan *et al.* (1971), *flare continuum* by Wild (1970c), *broad-band continuum* by Böhme (1972), and *first stationary source* (S_1IV) by Boischot (1974a). Independently from these designations, as a matter of fact, the phenomenon can be recognized in numerous previous observations (cf. e.g. Thompson, 1961; Fokker *et al.*, 1966; Wild, 1968b).

For clarity we adopt here the term '*flare continuum*' and abbreviate it as type IVmF burst radiation.

Its most important characteristics can be summarized as follows:

(a) Typically same sense of (weak) circular polarization (extraordinary) and possibly the same emission mechanism (?) as for the type IVmA bursts;

(b) Broad-band emission spectrum ranging from decameter to decimeter waves;

(c) Association with type IVμ bursts and proton events; and

(d) Apparent stationary source position as observed at fixed frequencies at locations where later a type IVmB component may develop.

Sometimes the type IVmF continuum is merged with a strong type V burst emission, which, however, generally has a shorter duration (<10 min) than the former one. A complicated situation arises when the type IVmF component is masked by a subsequent type II burst or even type IVmA burst as discussed by Robinson and Smerd (1975). These authors propose a further subdivision of flare continua.

b. *Moving Type IV Bursts ('IVmA')*

Originally the labels 'moving' and 'stationary' type IV bursts were introduced by Weiss (1963b) as a result of an extended interferometric study of the positions of type IV burst sources in the range between 40 and 70 MHz. The moving type IV bursts can be distinguished from other burst types with certainty only by positional measurements well outside the center of the solar disk, thus limiting the chances of observing this burst type. The most valuable information about moving type IV bursts came recently from the Culgoora heliograph observations, where they provided well elucidated and the perhaps most spectacular results of that instrument on the whole (Wild, 1970c).

The observations delivered important information about the complexity of the nature of moving type IV bursts (Wild and Smerd, 1972; Smerd and Dulk, 1971; Schmahl, 1973) and much could be learned about the structure of coronal magnetic fields. An advantage over drift bursts is that the continuous burst types can be studied by single-frequency positional measurements during the whole lifetime of an event.

Although it is not always possible to classify each particular event with certainty, indications were obtained that the majority of the moving type IV bursts comprise four different categories (McLean, 1974), namely expanding magnetic arches, isolated sources (ejected plasma blobs), advancing shock fronts, and jets.

Now we briefly discuss these different forms of the appearance of moving type IVm bursts.

(1) *Magnetic arches*

A classical example of a magnetic arch was described by Wild (1969b). In the radio spectrum the arch is marked by three distinct sources. One source, widely unpolarized but showing polarized edges, is situated at the top of the arch and corresponds to the actual moving type IVm burst. The other two sources are strongly circularly polarized (in the ordinary mode) in opposite sense to each other thus marking the intersections of the arch with the corresponding plasma level. For the emission at the top of the arch synchrotron radiation was proposed. Perhaps there may be a link to the transient loops observed by satellite borne coronographs (Hildner *et al.*, 1976).

During the lifetime of the event the sources move apart consistent with an arch-expansion velocity of about 300 km s^{-1}, where a shock front (type II burst) is entirely missing. The velocity is remarkably smaller than the speed of advancing shock fronts ($\gtrsim 1000$ km s^{-1}), but also remarkably greater than the expansion velocities sometimes measured at lower heights in the Hα flare loops (~ 20 km s^{-1}).

(2) *Isolated sources*

Evidently a second variety of moving type IV bursts exists in a bounded source region, which exhibits a more or less uniform outward motion up to heights of several solar radii without any related foot point emission. A prototype of such a phenomenon was described by Riddle (1970a) (Figure III.30). The term 'isolated source' was introduced by Smerd and Dulk (1971).

The velocity of the movement of isolated sources is typically of the order of 300 km s^{-1} (local Alfvén velocity?) extending to heights beyond 5 $R_{\odot}$. Sometimes the source is spread into two or more distinct parts of different polarization. The polarization often increases with time reaching high degrees.

The early observations of backward motions (Boischot, 1958; Weiss, 1963b) were not confirmed by the modern heliographic observations. However, curved paths of the emission center and movements almost parallel to the solar limb have been detected (Smerd, 1971; Dulk *et al.*, 1971). The clearly observed termination of the emission was interpreted as an effect of an expansion of the plasma blob.

(3) *Advancing fronts*

Similar to the magnetic arches, but not anchored at the foot points, the obvious type IV burst emission in the wake of a shock wave occupies a larger region. An example of the appearance of such an advancing front was first reported by Kai (1970a) (Figure III.31).

The predominant features differing from the magnetic arches (and type II bursts) are the following:

- Greater expansion velocity ($\gtrsim 1000$ km s^{-1}) and association with a related type II burst;

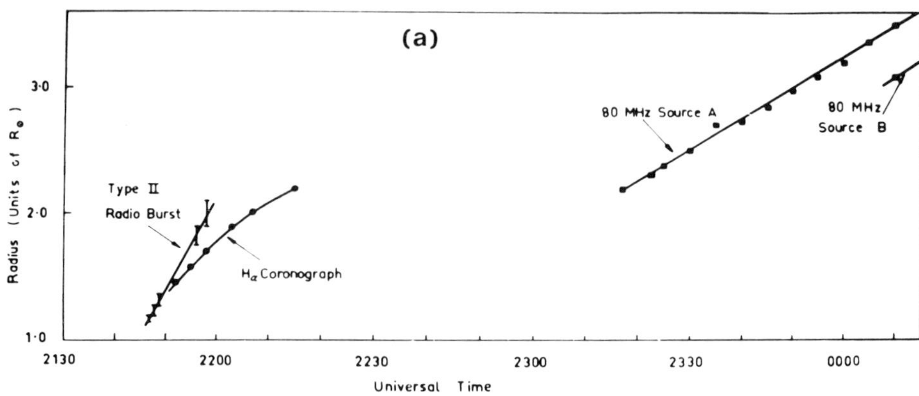

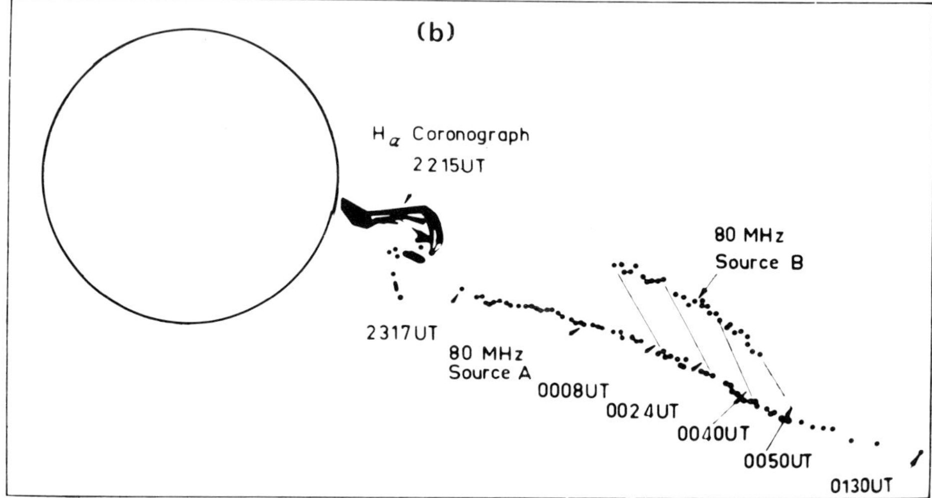

Fig. III.30 Temporal (a) and spatial (b) relationships between a type II burst, a rising Hα-prominence, and a moving type IV burst observed by the 80 MHz Culgoora heliograph (after Riddle, 1970a).

– Weak polarization throughout the whole source region; and

– Large extent of the source region perpendicular to the direction of motion.

Several models were proposed to explain the emission of moving type IV bursts invoking synchrotron radiation caused by fast electrons from behind the shock front (Warwick, 1965a, b; Boischot *et al.*, 1967; Lacombe and Mangeney, 1969).

(4) *Jets*

An outstanding observation of a jet-like source geometry and fast speeds was described by Riddle and Sheridan (1971). During a period shorter than 100 s a sequential brightening occurring suddenly during the late phase of a rather complex type

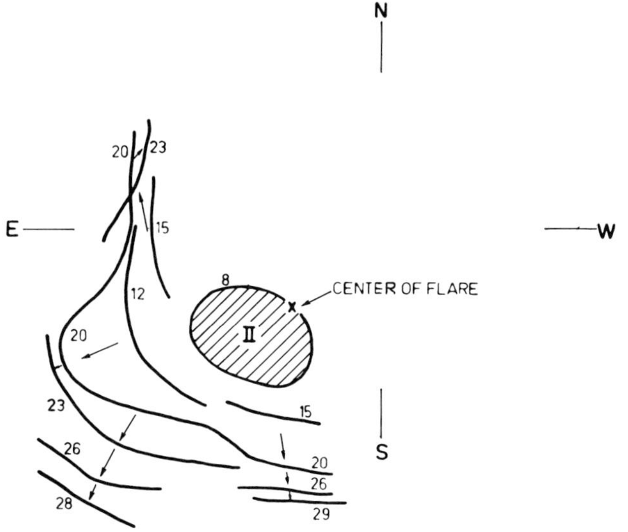

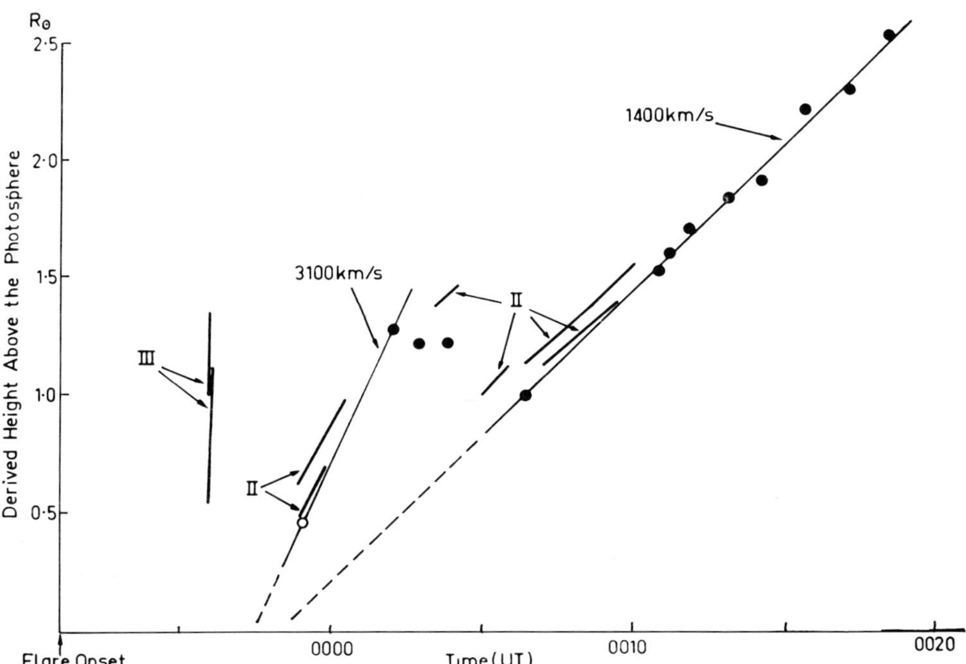

Fig. III.31. Developemtnt of a moving type IV burst (advancing shock front) as observed with the 80 MHz Culgoora heliograph (October 23/24, 1968; numbers indicate time in minutes after the flare onset at 23^h51^m). Bottom: Derived height–time plots (after Kai, 1970a).

IV burst was observed with the Culgoora radio heliograph indicating a disturbance traversing the corona with a speed of the order of 0.5*c*.

Evidently such fast moving ejecta form a special class at the boundary between continuum bursts and fast-drift bursts.

c. *Stationary Type IV Bursts ('IVmB')*

The stationary or type IVmB bursts differ in many respects from the other type IV burst components. The main characteristics are the following:

(a) Typical duration up to several hours; sometimes development into a noise storm;

(b) No systematic movements of the source region; the source heights are believed to be near the local plasma levels;

(c) Comparatively high degrees of polarization usually related to the ordinary wave mode of the magneto-ionic theory, increasing with time; and

(d) High directivity of the emission.

The type IVmB bursts are a typical phenomenon of the late phase of large solar flare events. The morphological connection to the type IVμ burst component and to electron events in the interplanetary space is rather bad, but a direct connection to the type IVmF (flare continuum) phase may exist, from which a type IVmB burst can develop (cf. Section 3.6.3.a).

The other important link exists to noise storm phenomena. No essential physical difference has yet been observed between the type IVmB burst continuum and a noise storm continuum. The development of a noise storm often passes the following stages:

$$\text{type IVmF} \rightarrow \text{type IVmB} \rightarrow \text{continuum storm} \rightarrow \text{type I noise storm} \\ \rightarrow \text{type III noise storm},$$

which are characterized by a subsequent increase of the degree of circular polarization and of the number of superimposed storm bursts.

Finally, little can be said about the emission mechanism and the origin of the type IVmB bursts, for which some kind of plasma radiation can be expected.

3.6.4. Type IV Burst Models

a. *Source Geometry*

The interpretation of type IV bursts contains three major tasks which are linked together:

(1) Exploration of the source-region characteristics (source geometry, distribution of the physical parameters in space and time);

(2) Determination of the physical processes taking place inside the source regions during the development of type IV bursts (emission processes, energy gain and loss processes, etc.);

(3) Investigation of the cause of the single type IV burst components and of a type IV event as a whole (questions of the flare origin, energy storage, plasma instabilities, trigger actions, etc.).

Dealing with the first task mentioned above, a number of geometrical considerations were made in the past ascribing the source locations of different type IV burst components roughly to different sites in an assumed structure of an active region. Different proposals referring to different stages of the flare-burst development are due to McLean (1959), Wild (1962), de Jager (1965a), Takakura and Kai (1966), Akinyan *et al.* (1971), Hundhausen (1972), Kane (1974), and others.

The problem of classification of the source structures is closely connected with the question of getting adequate information on the magnetic field distribution in the corona and the technical problem of obtaining high-angular-resolution radio observations in a wide spectral range. These tasks are still only beginning to be solved. In summary, a tentative schematical picture of a possible source geometry of a type IV burst is shown in Figure III.32.

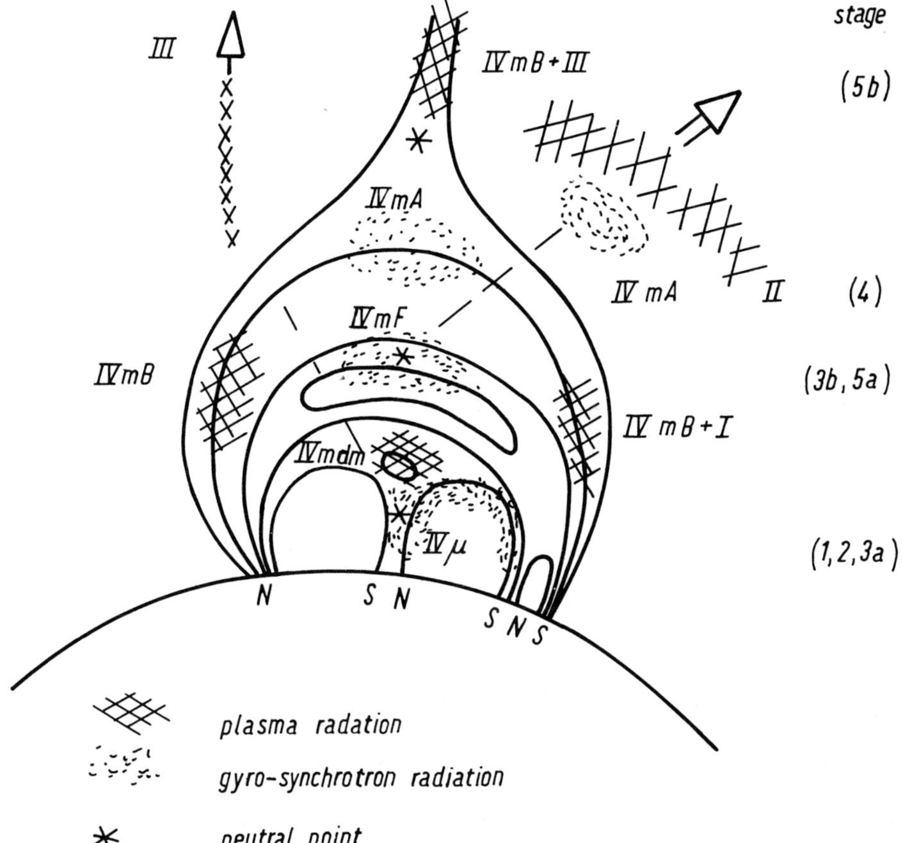

Fig. III.32. Tentative schematic geometric model of a type IV burst (numbers refer to Table III.5).

TABLE III.5

Possible origin and emission processes of different type IV burst components and other burst types

Origin	Location	'Gyromagnetic' radiation	'Plasma' radiation
(1) Heating	chromospheric-coronal interface	gradual μ-burst	–
(2) First acceleration (impulsive)	chromospheric-coronal interface	impulsive μ-burst	fast-drift bursts
(3) Second acceleration	chromospheric-coronal		intermediate-drift bursts
a) low energies	interface	type IVμ (?)	– (?)
b) high energies ('explosive')	(low) corona	type IVmF (?)	type IV dm (fast-drift bursts)
(4) Shock waves, traveling disturbances	corona	IVmA (?)	slow-drift bursts
(5) Third acceleration (quasi-continuous)	corona	– (?)	IVmB (?) (continuum and noise storms, type I)

b. *Emission Processes*

Concerning the interpretation of the physical processes taking place in type IV burst sources, the present (still incomplete) state of our knowledge of the emission processes will be briefly summarized. The main emission processes playing a role for the interpretation of type IV bursts (as well as for other burst types, too) are the gyro-synchrotron process and (analogous to the drift bursts) some kinds of plasma and MHD wave radiations, which must be transformed into electromagnetic radiation observable by radio methods. A compilation of the possible relevance of these two major groups of emission processes to the different stages and components of type IV bursts is tentatively outlined in Table III.5.

More detailed considerations on the related processes will follow in the next two chapters.

3.7. Noise Storms

3.7.1. Outline of the Phenomenon

a. *General Properties*

The noise storms are a very common expression of solar activity at metric and decametric wavelengths. They consist primarily of two different parts: a more or less slowly varying enhanced wide-band continuum radiation, and superimposed short-lived narrow-band bursts.

Though generally linked together, sometimes the one (continuum) or the other (burst) component can dominate over some periods. Further, noise storms can be (but don't have to be) flare-initiated. With respect to these properties the noise storm constitutes an intermediate phenomenon in between the classes of phenomena reflecting the long-period development of active regions (similar e.g. to the S-component), and of the short-lived radio bursts. Both aspects are joined together here.

The spectrum of the noise storms can be divided into two main regions, extending within the limits, roughly
(1) 60–600 MHz,
(2) $\lesssim$60 MHz.
These two spectral regions are characterized by both different peaks of the continuous spectrum and different forms of the appearance of the burst component. The latter forms the typical *storm bursts* or type I bursts predominant in the first region, while fast-drift bursts ('type III storms') prevail in the second. The low-frequency limit of the noise-storm spectrum appears not to be so well determined. An extension of the storm radiation up to hectometer and kilometer wavelengths has been established by satellite observations (Fainberg and Stone, 1971, 1974).

Physically, at least the type I storms can be regarded as manifestations of closed magnetic structures in the solar corona. Boischot *et al.* (1971) suggested that the transition from type I storms to type III storms corresponds to a transition of the source level from closed to open magnetic field structures. A full explanation of the noise storm complex is still missing in spite of a number already existing spirited attempts to solve the question.

There are few reviews on noise storms. The first comprehensive study is due to Fokker (1960), a general review is contained in the book of Kundu (1965), and a more recent summary was presented by Bougeret (1972). Some other extended studies were made by Malinge (1963), Clavelier (1967), and Gergely (1974). A monograph on noise storms was recently published by Elgarøy (1977).

b. *Related Phenomena*

There are many relationships between noise storms and other solar phenomena, which can be roughly grouped into the following categories:
(1) Large-scale magnetic fields on the Sun:
 interacting active regions,
 sector structures.
(2) Single active regions:
 sunspots,
 S-component.
(3) Flare-burst complex:
 initial flare phase, fast-drift bursts and type IVμ + IVmF/A bursts,
 type IVmB bursts.
(4) Particle radiation:
 low-energy particles, enhanced solar wind,
 magnetospheric disturbances, geomagnetic effects.

Some comments may illustrate the interrelations. Early in the study of solar radio astronomy the association between noise storms and large spot groups was detected (McCready *et al.*, 1947). Nonetheless, there are also many cases where large spot groups are found not to be related to any detectable noise storm activity. Inspite of this restriction, it appears justified to state that noise storms must have something to do with strong magnetic fields on the Sun.

Recent interferometric and heliographic studies show that noise storms, especially at longer wavelengths, are often related to not only one but several active regions on the Sun (Wild, 1968b; Kai, 1969a; Gergely, 1974). This demonstrates that the classical method of the investigation of solar activity in the form of a mere study of single independent (isolated) active regions is not sufficient for the exploration of noise storms. There is evidence that noise storms arise just from the interaction of different active regions. The centroids of the source regions of noise storms were found to be located near neutral sheets in the corona, which are marked by large filaments and sector boundaries. The magnetic structure associated with the storm regions consists most likely of arcs of quite different height and length connecting different magnetic polarities from one or more active regions.

Looking at single active regions, certain evolutionary relations exist unquestionably between the noise storms and the S-component of solar radio emission. A positive correlation of the intensities of both phenomena has been found with an average delay of 1 to 2 days of the peak of the noise storms behind the peak of the S-component (Kai and Sekiguchi, 1973).

It must be stated that the correlation between noise storms and the impulsive flare phase is rather poor. Extended statistical studies do not show a significant direct connection between the onsets or intensifications of noise storms and chromospheric flares (Gergely, 1974).

3.7.2. Storm Continua

a. *Phenomenological Characteristics*

The continuum of noise storms can be defined as an enhanced, smooth, long-period, and broad-band background radiation, on which, as usual, short-lived narrow-band bursts are superimposed. In practice a sharp distinction between the noise storm continua and either type IVmB bursts or merging, unresolved type I and type III bursts is generally difficult to make.

On the average, noise storms last several days; the duration of individual storms ranges from about half an hour for sporadic events up to perhaps half a solar rotation and more. It must be stated that observation is seriously hindered by the effect of the high directivity of the noise storm radiation in connection with the solar rotation. The directivity becomes greater at longer wavelengths.

The noise storm spectrum commonly has a relative maximum in the range of 100 to 200 MHz and a relative minimum around 40 to 60 MHz. The starting frequencies statistically show a peak near about 300 MHz. Below the drop at 40 to 60 MHz there often follows a flux increase towards lower frequencies. The intensities can vary in a wide range. The frequency of the spectral maxima can also vary with time and depends on the individual event. There exists a connection between the spectrum of noise storms and flare actions: Flare-associated noise storms tend to extend their spectral range toward higher frequencies. Systematic variations of the noise storm spectra with the solar cycle are also indicated (Böhme, 1971). A storm at decametric or longer wavelengths is almost always connected with a storm at metric wavelengths.

The development of the storm centers as seen from the Earth at a fixed frequency is illustrated by the example of Figure III.33 taken from the regularly published maps of E–W-scans observed with the 169 MHz interferometer at Nançay.

The sources of the storm continua exhibit comparatively stable positions, which often depart from a location on a straight line radially above a related spot group. In this way, a one-to-one correspondence between a noise storm and a single active region can be missed, especially if decametric and even longer wavelengths are

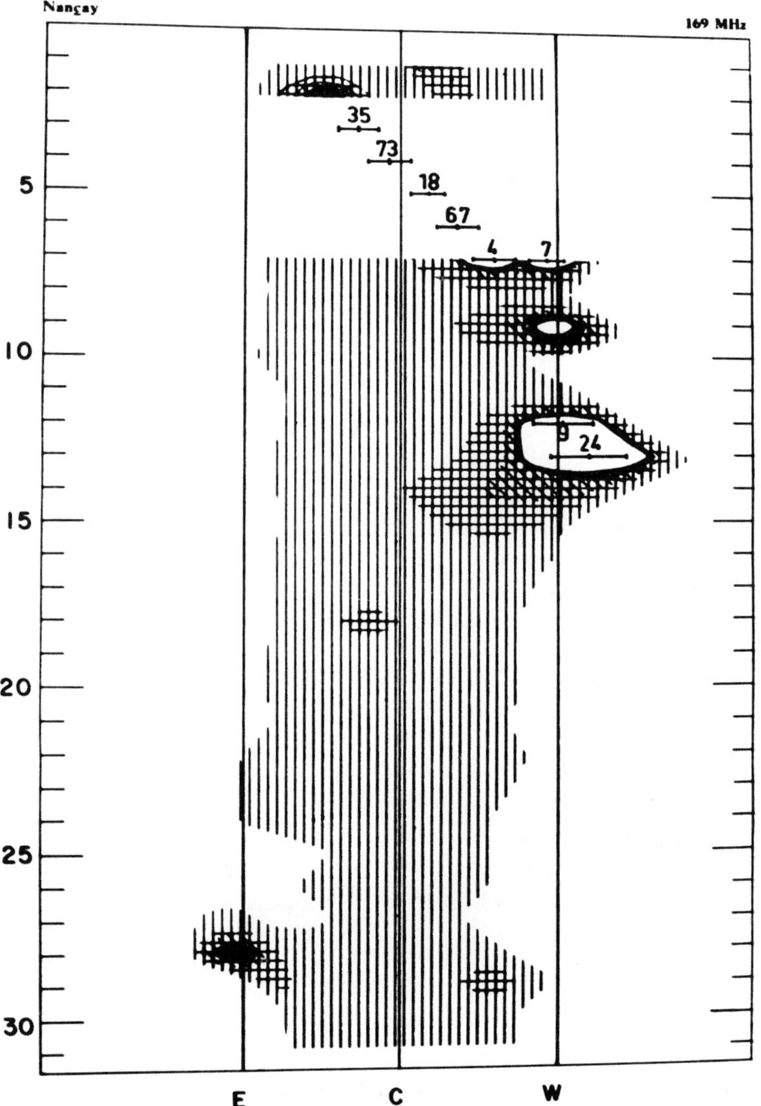

Fig. III.33. Evolution of the noise-storm radiation as observed from the E–W scans at 169 MHz at Nançay for a selected monthly period (August 1972).

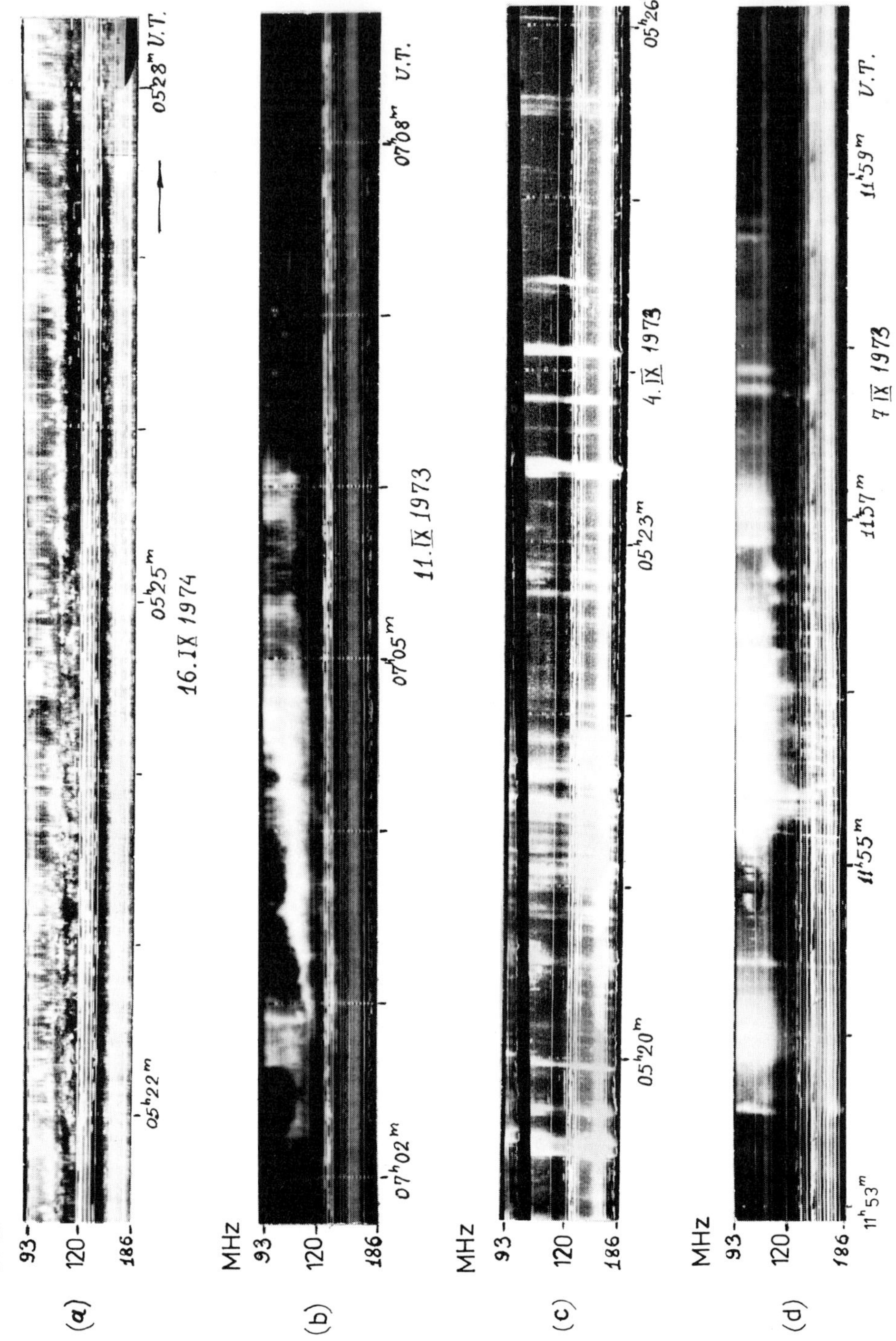

Fig. III.34. Dynamic spectra of different burst spectral types: a – noise storm with chains of type I bursts at the boundary to a type III storm, b – type II burst, c – groups of type III bursts, and d – type III + V bursts. (Courtesy of V. V. Fomichev).

considered. Further, remarkable height variations of the source regions for a fixed frequency are observed for different events. As an orientation, the heights of the decimetric/metric component lie between about 0.1 and more than 1 $R_\odot$ above the photosphere (Clavelier, 1967; Malinge, 1963). These levels are attributed to the local plasma levels for a given fixed frequency, so that considerable variations of the density scale must be assumed in the corona.

The observed source diameters of noise storms (1–10′ in the meter region, increasing with wavelength) are probably influenced by scattering, and complex sources are likely to exist (Daigne, 1968; Bougeret, 1973). The radiation is strongly circularly polarized favoring an interpretation in terms of the ordinary wave mode of the magneto-ionic theory, but the assignment of the needed magnetic polarity is not always very certain.

Effects of an E–W asymmetry can be ascribed to a systematic tilt of the axis of the storm regions to the east (Malinge, 1963).

A characteristic dynamic spectrum of a part of a noise storm together with those of basic types of drift burst is shown in Figure III.34.

b. *Relation to Small-Band Features*

One important question is, whether the continuum is a mere superposition of many unresolved small bursts (and only the largest bursts are being observed as distinct phenomena), or not. The question was treated by means of models, which artificially synthesize the time profile of a noise storm by a statistical summarizing of a large number of elementary burst pips (Takakura, 1959; Fokker, 1960; Titulaer, 1967). The idea of a superposition was supported by the facts that a positive correlation exists between the fluxes of the continuum and burst component, and that the sense of the polarization is usually the same for both components.

Nevertheless, there is a number of convincing arguments against the interpretation of the storm continuum as a consequence of a merging of an appreciable number of unresolved storm bursts (Fokker, 1960; Daigne, 1968; Bougeret, 1972), namely:

(a) Examples show, that a reduced number or even the entire absence of type I bursts must not necessarily influence the continuum level of noise storms.

(b) The spectral distribution of type I bursts for a given time interval must not be related to that of the storm continuum.

(c) The spatial extension of the source region of the storm continuum may differ from that of the sum of the storm bursts, which appears to be smaller.

(d) Storm bursts do not appear to be distributed randomly as is postulated in the synthetic models.

3.7.3. Type I bursts

a. *Basic Characteristics*

The storm bursts in the metric region (type I bursts) can be roughly characterized by the following properties (cf. Figure III.35):

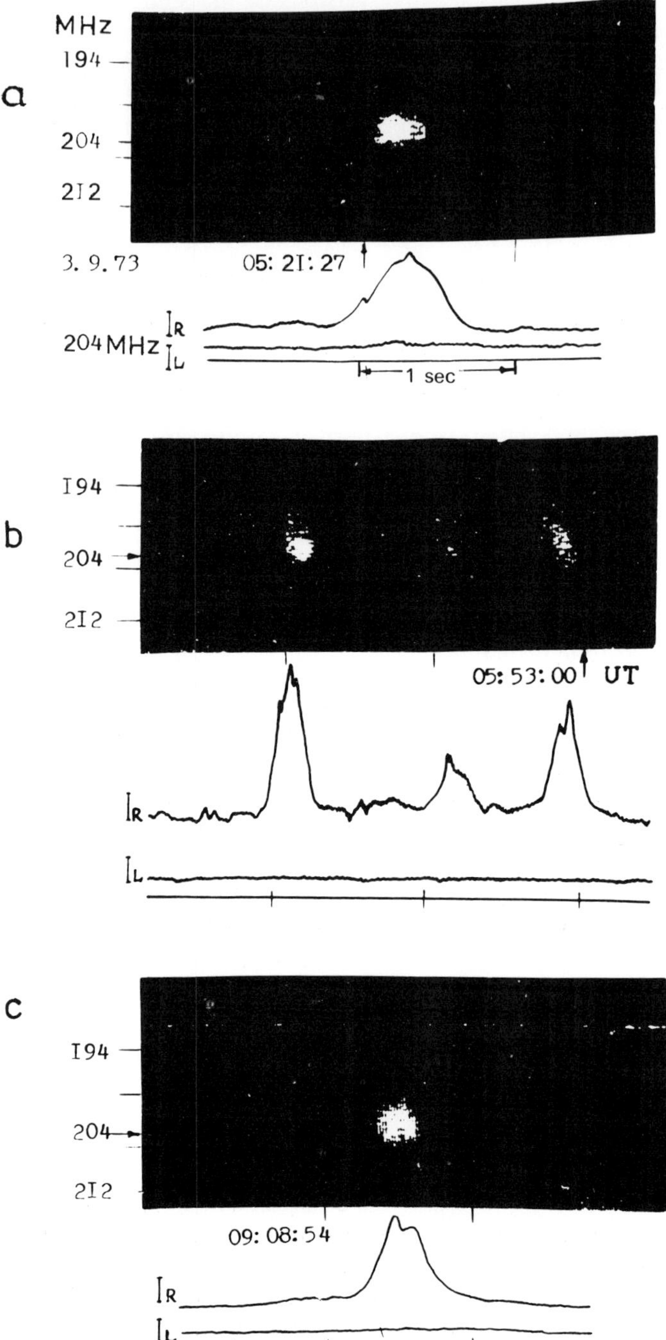

Fig. III.35. Dynamic spectra and single-frequency records of type I bursts showing fine structure; observation of September 3, 1973. (Courtesy of V. V. Fomichev, cf. also Chernov, 1976b).

- Duration: 0.1 to 2 s, increasing with wavelength;
- Bandwidth: 2 to 10 MHz, decreasing with wavelength;
- Angular size: $\lesssim 1$–5′, increasing with wavelength;
- Emitted energy: 10^{11} to 10^{14} erg, various dependencies on wavelength; and
- High degree of circular polarization, independent of wavelength (?).

(The degree of polarization approaches 100%, except for the initial stage of a storm, a center-to-limb effect, and possible irregularities.)

High frequency- and time-resolving radio spectrographs reveal the following subclasses of type I bursts (Elgarøy, 1961; Bougeret, 1972):

(1) *Stationary bursts* – the maximum of the emission remains stable with frequency;

(2) (Common) *drifting bursts* – the emission maximum moves with time with various rates (1–20 MHz s^{-1}) towards higher or lower frequencies;

(3) *Split bursts* – the emission maximum is separated into two or more parts in frequency and time; and

(4) *Spike bursts* – fast drifting bursts of very short duration at a fixed frequency (De Groot, 1966; Eckhoff, 1966; Markeev and Chernov, 1970).

All these variants of type I bursts are expected to originate in highly disturbed coronal regions close to the corresponding local plasma level. For illustration cf. Figures III.36, III.37, and III.38.

b. *Chains of Type I Bursts*

Dynamic radio spectra show that the storm bursts do not always occur in a random distribution during a noise storm. Sometimes they exhibit a tendency to cluster together and to form narrow-band *chains* consisting of tens or hundreds of type I bursts (cf. Figure III.34). This phenomenon is of interest with respect to a possible explanation of the physical nature of the type I bursts. The existence of the chains was first reported by Wild (1957) and Elgarøy (1961) as well as by Vitkevich and Gorelova (1960). Later on the phenomenon was considered in more detail by Wild and Tlamicha (1965), Hanasz (1966), and Elgarøy and Ugland (1970).

The chains show a negative or positive frequency drift, which corresponds to radial velocities ranging between about ± 800 km s^{-1} with a high peak for outward motions at velocities of the order 200 km s^{-1}. These velocities may be attributed to magnetohydrodynamic waves. Although overlapping, the velocity distribution function is distinct from that of the type II bursts. Moreover the drift rates of individual drifting bursts in a chain can be quite different, and differ also from that of the whole chain.

The distribution of the chains on the observing frequencies appears to be similar to the distribution of single type I bursts reflecting roughly the average spectrum of a noise storm as a whole. However, chains could not be detected at frequencies higher than 200 MHz.

No essential distinction has been found between chains and single type I bursts regarding the position of the sources on the solar disk, the intensity, and the polarization characteristics. The lifetimes of the chains follow approximately an exponential law with some scatter. In particular cases a systematic association with weak type

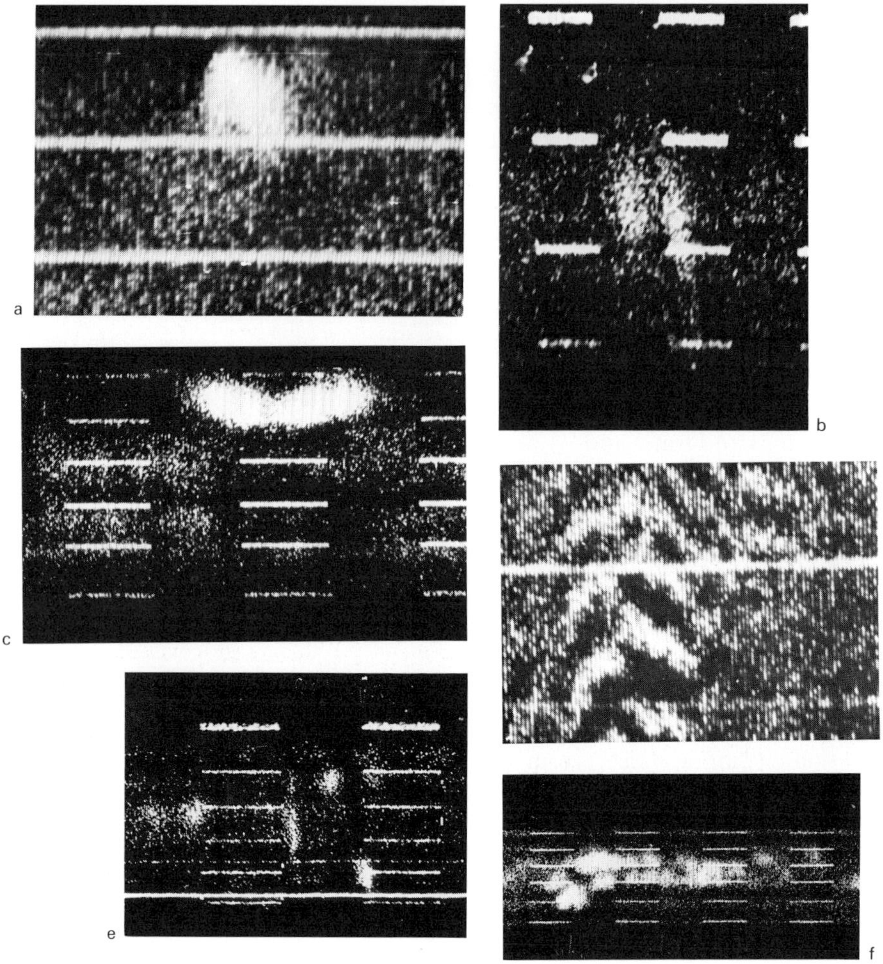

Fig. III.36. Examples of dynamic spectra of different kinds of storm-burst activity in the 190–215 MHz region: a – type I burst (U-shaped), b – reverse-drift pair, c – oppositely drifting type I bursts, d – band structure, e – spike burst and stable type I bursts, and f – split-band bursts. (Courtesy of Ø. Elgarøy).

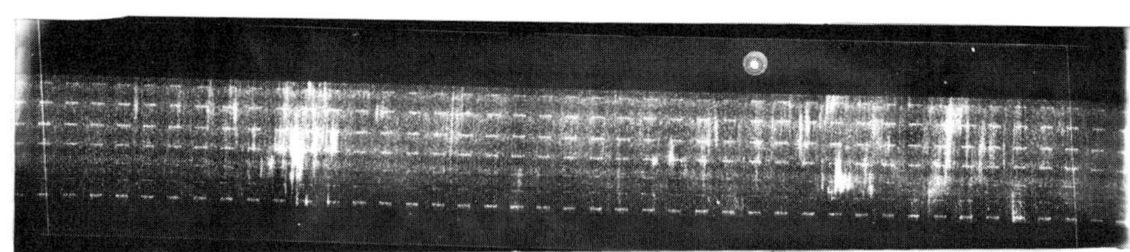

Fig. III.37. Spike bursts in the 310–340 MHz region with relatively large band widths (September 6, 1976, around 11.41 UT). (Courtesy of Ø. Elgarøy).

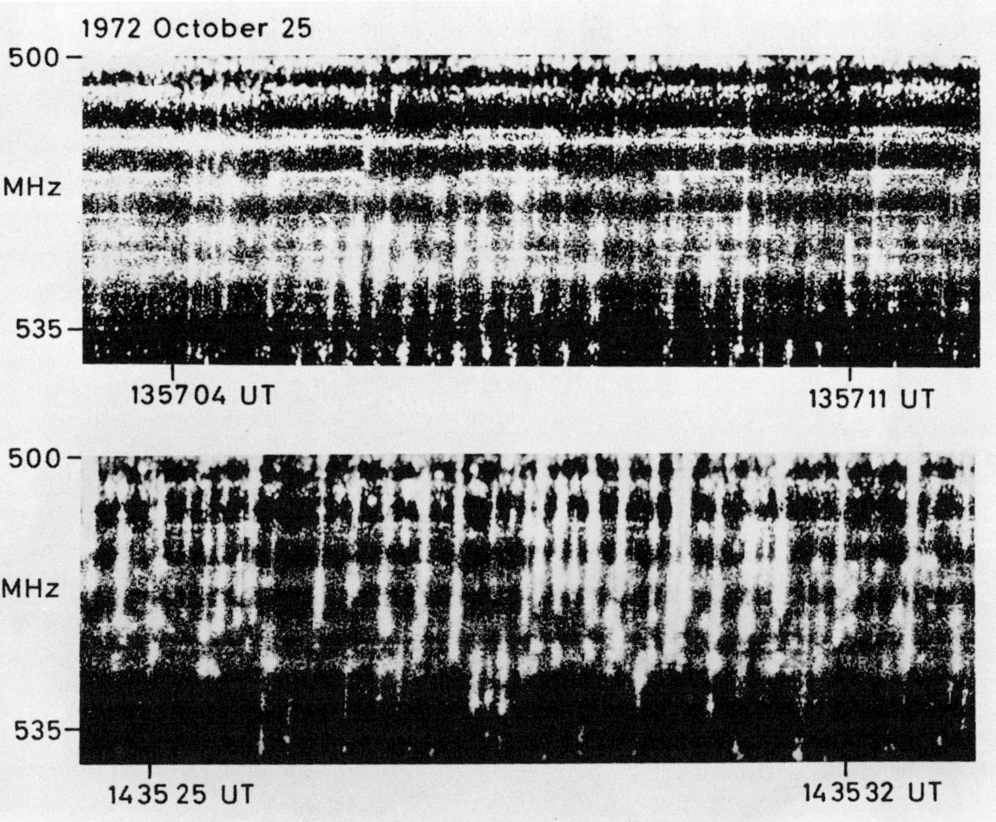

Fig. III.38. Fine-structure emission pattern in the 500–535 MHz region. (Courtesy of Ø. Elgarøy).

III bursts, giving rise to the impression that the type III bursts are growing out of the chains, is an interesting feature (Hanasz, 1966).

A center-to-limb effect in the lifetimes of the chains seems to be indicated. Furthermore, there is an indication for an enhanced background emission in chains (Elgarøy and Ugland, 1970).

c. *Attempts at Interpretation*

At present the knowledge of the mechanism and origin of noise storms in general and of storm bursts in particular is still limited. Only nonthermal processes can account for the observed brightness temperatures of 10^9 to 10^{11} K of a storm region, and for the short duration of the storm bursts. The present theories on type I bursts invoke either (coherent) gyro-synchrotron radiation at low harmonics of the electron gyrofrequency or plasma radiation processes.

The early theories based on incoherent or coherent gyro-synchrotron emission (Kiepenheuer, 1946; Kruse *et al.*, 1956; Takakura, 1956; Twiss, 1958) have been overcome by more recent approaches taking into account the effect of the ambient plasma on the radiation processes (cf. Fung and Yip, 1966). However, the principal difficulties in explaining the type I bursts by gyromagnetic emission still exist, which mainly consist in the interpretation of the high directivity, the high degree of polarization corresponding to the ordinary wave mode, the dispersion in the source heights with frequency, the bandwidth, and the lack of harmonic radiation.

For all these reasons plasma waves seem more likely to account for the type I burst radiation. The present theories invoking plasma radiation have their forerunners in the ideas of Ginzburg and Zheleznyakov (1959) and Denisse (1960). In essence a stream of accelerated particles is invoked to excite plasma waves, which, in turn, are partly converted into electromagnetic radiation. Beyond others, a special model for the origin of the type I bursts was proposed by Takakura (1963b) and later improved by Trakhtengerts (1966).

A beam of accelerated electrons is assumed to be produced at a collision of two packets of Alfvén waves propagating in opposite directions along the magnetic field lines. The acceleration takes place in an electric field, which results from a charge separation at the wave front. Then, if the conditions for an instability exist, coherent plasma waves are excited. These plasma waves according to the early terminology of Ginzburg and Zheleznyakov give rise to Rayleigh scattering at thermal density fluctuations causing a transformation into radio waves. The theory seems capable of satisfactorily explaining some important features of the type I bursts, e.g. the bandwidth, duration, fine structures, and the chain phenomenon. Other features, however, are left unexplained, e.g. the relationship to the continuum, the spectral range, and the transition to type III storms at longer wavelengths.

The polarization is explained, as in most approaches to a type I burst theory, by the supposition that the emission occurs inside that region where the escape condition for the ordinary wave mode is fulfilled, i.e. in the simplest case

$$1 > X > 1 - Y. \qquad \text{(III.32)}$$

(An alternative possibility to obtain an ordinary-mode radiation invoked by the gyro-synchrotron radiation theories would be the gyro-resonance absorption of the extraordinary wave mode at low harmonics of the electron gyrofrequency.)

A modification of the above sketched theory was made by Sy (1973), who criticized the necessity of assuming high magnetic field strengths ($\omega_H \approx \omega_p$) in the former explanations, which may not be fulfilled especially at longer wavelengths. He invoked an induced-scattering mechanism leading to an amplified radiation at frequencies less than the average frequency of the electron plasma frequency waves. Under the condition $\omega_{He} \ll \omega_{pe}$ it was shown that the scattering process is more efficient for the ordinary mode, explaining directly the observed sense of the polarization.

A combined theory considering both the continuum and the burst component was suggested by Mollwo (1970). The proposed emission mechanism consists of radiation at the fundamental gyrofrequency caused by an ensemble of electrons gyrating in a magnetic field in nearly circular orbits. On the basis of the effects of wave propagation in a warm plasma near $Y \lesssim 1$, $X \approx 1$, the wave propagation was found to be possible only for very small angles of θ with a sufficiently small cyclotron damping and a not too high refractive index. A turbulent structure of coronal

TABLE III.6
Attempts at an interpretation of some type I burst properties

Property	Interpretation	References
Duration	Collisional damping,	de Jager and van't Veer (1958)
	beam lifetime,	Takakura (1963b)
	scattering,	Steinberg *et al.* (1971)
	density variations and beam-velocity dispersion	Elgarøy and Eckhoff (1966)
Bandwidth	Local plasma frequency,	Takakura (1963b)
	local gyrofrequency,	Fung and Yip (1966)
	Doppler broadening	de Jager and van't Veer (1958)
Frequency drift	Dispersion relation of plasma waves	Elgarøy and Eckhoff (1966)
Polarization	Wave-escape condition,	Denisse (1960)
	resonance absorption at low gyro-harmonics,	Twiss (1958), Fung and Yip (1966)
	induced scattering	Sy (1973)
Frequency splitting	Magnetic splitting (ω_p and ω_{uh}),	Ellis (1969)
	nonmagnetic splitting	Melrose and Sy (1971)
Time splitting	Echo phenomenon,	Jaeger and Westfold (1950)
	Echo by different modes	Elgarøy (1968)
Directivity	Coupling at $X \approx 1$, $Y \gtrsim 1$ for small angles θ	Mollwo (1970)

densities provides a crossing of the coupling point $X = 1$ by the waves, so that the ordinary mode is generated which can propagate to the observer thus showing a high directivity effect due to the coupling process at small values of θ.

By the same author the type I burst radiation was proposed to be generated by Čerenkov plasma radiation (inverse Landau damping) from that part of the electron population which exerts strong velocity components in the direction of the magnetic field within the same region $Y \lesssim 1$, $X \approx 1$, as discussed above. Though not avoiding relatively high magnetic fields and small electron energies (it was suggested $\beta \approx 0.5$), this approach tries to explain the relation between the storm continua and the type I bursts, the emission at the plasma frequency, the high directivity and polarization, as well as the frequency extent by means of the conditions to X and Y. The transition to type III bursts at longer wavelengths was not considered explicitly there, but it seems reasonable to connect it with a transition between closed and open magnetic field configurations.

In conclusion, in Table III.6 the earlier type I burst theories are summarized with some proposals for the interpretation of the different characteristics of the storm bursts; however, they may soon be obsolete if newer concepts, e.g., of solitions are taken into account.

3.7.4. Type III storms

a. *The Phenomenon*

Observations indicate that at frequencies below, say, 40 to 60 MHz the properties of noise storms appear drastically altered. Superimposed on the background continuum, the number of type I bursts is reduced. Instead of this often a large number of type III bursts is observed forming a type III noise storm. The rate of occurrence of these type III bursts and their spectral extent vary from case to case. Also during a single storm the characteristics of the type III bursts such as the starting frequency, intensity, etc. can be changed from time to time. The position of some type III bursts can be displaced from that of the continuum. In this respect, two groups of type III bursts were observed interferometrically during noise storms:

(1) Type III bursts coinciding in location with the continuum source ('on-fringe' type III bursts), and

(2) Type III bursts displaced from the source of the background continuum ('off-fringe' type III bursts) (Gergely, 1974).

To explain the physical nature of type III storms, it may be important to stress that a decametric or hectometric storm always appears as an extension of a metric storm. Underlying the metric storm regions, in turn, a source of the S-component is always present.

Type III bursts are often observed to start near the low-frequency limit of the type I bursts (Malville, 1962; Hanasz, 1966; Boischot *et al.*, 1971). This feature suggests a connection between both types of burst components. A proposal to explain this feature was made by Gordon (1971) invoking a nonthermal plasma process in which the comparatively low-phase-velocity plasma waves are scattered to high velocities causing a rapid acceleration of electrons up to the type III burst exciter velocities.

Such a process would imply nearby locations of the type I and type III burst sources. In order to explain the often observed displacements between both sources, Stewart and Labrum (1972) proposed a trigger action of a magnetic instability caused by a MHD disturbance similar to those needed to interpret the chains of the type I bursts, traveling along the magnetic field lines diverging from a type I storm region. Arriving at the top of a closed magnetic field structure in the vicinity of a T-shaped neutral point a pinch-like instability is expected to arise (Sturrock, 1966). In this way, at the interference between closed and open (helmet) magnetic structures, the acceleration of the type III burst exciters is assumed to be efficient (cf. Figure III.39). The proposed structure would easily explain that the type I bursts are generated in regions of comparatively high magnetic fields and trapped electrons, whereas the type III bursts refer to regions of comparatively low magnetic fields. The role of the storm continuum is not so well represented in this picture.

Satellite measurements of hectometer bursts reveal type III storm-burst positions which are remarkably dislocated from the roots of the type III storm (Fainberg and Stone, 1970), which in special cases may correspond to the 80 MHz level observed by the Culgoora radioheliograph. The observed difference in the positions implies the existence of a deflection of the magnetic field direction from the radial direction (Stewart and Labrum, 1972). Evidence for the occurrence of such tilted structures is also indicated principally by the optical eclipse observations of nonradial streamers.

The directivity of decameter storms indicates a cutoff of the visibility of the sources at distances corresponding to about ± 5 days from the central meridian (ca. $\pm 70°$

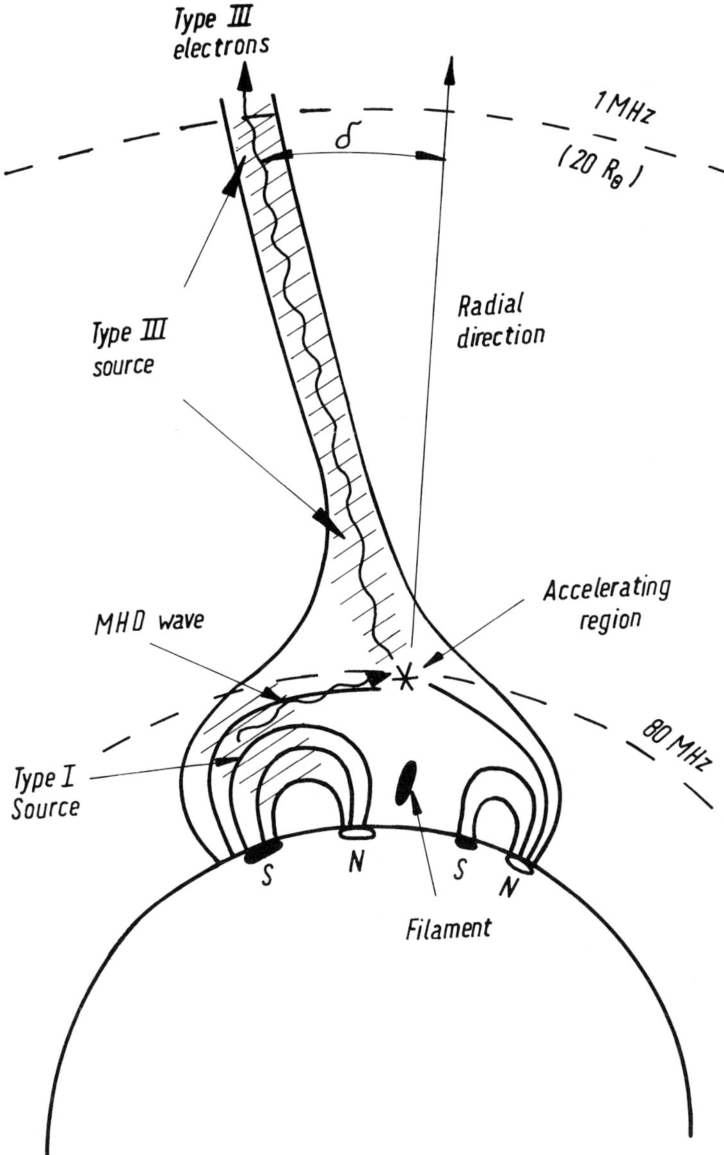

Fig. III.39. The model proposed by Stewart and Labrum (1972) for the positions of the type I and type III storm sources deduced from meter-wavelength observations. The direction of the deflection of the streamer from the radial direction (angle δ) demonstrates only a particular case.

in longitude). No variation of the directivity with the frequency inside the decameter range could be stated. The east–west asymmetry (western hemisphere favored) found at meter wavelengths (Malinge, 1963) also appears to persist at decameter wavelengths (Gergely, 1974).

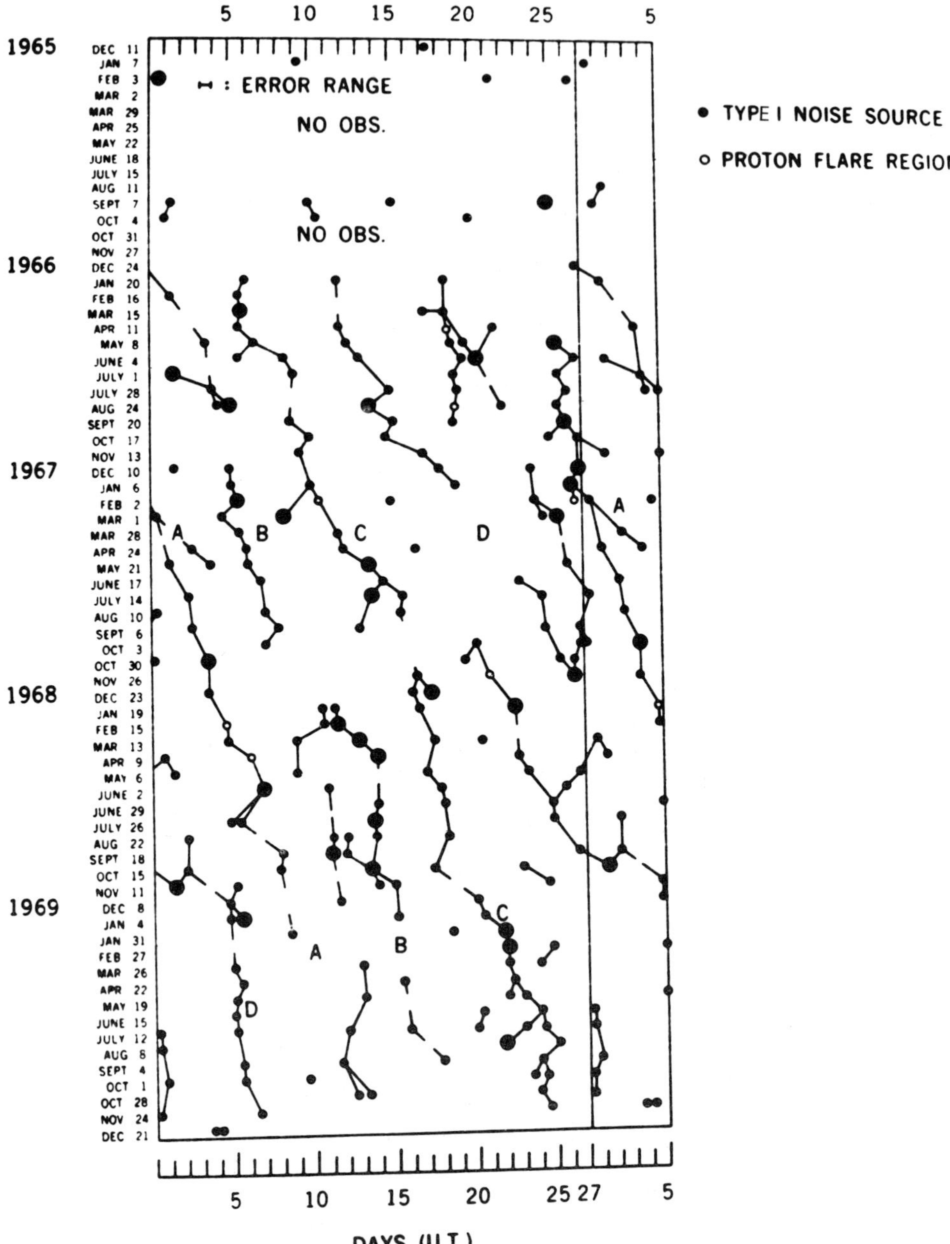

Fig. III.40. A 27-day recurrent diagram of type I noise-storm sources (the black and white circles indicate the times of the central-meridian passage of the type I noise-storm sources and active regions which produced proton flares, respectively). Four main trends are indicated by A, B, C, and D (after Sakurai, 1972b).

b. *Implications*

The type III storm phenomenon poses a number of important questions, e.g., what is the origin of the energy transformed in noise storm regions in general, for which the type III storms may give a clue. From the output of about 10^4 type III bursts per day during a storm a necessary energy supply about 10^{31} erg results, if each burst requires an average energy of about 10^{27} erg. This value, which only refers to the burst component, is unlikely to be attributed to an origin in single flares, but requires a quasi-continuous energy conversion extending over the whole lifetime of a storm, as has been stated e.g. by Boischot (1974a).

The energy stored in the magnetic field at the heights of the source region also seems insufficient to explain the whole type III storm phenomenon. On the other hand, the required amount of energy is comparable with the instantaneous energy content of an underlying source of the S-component, which can be estimated to be about 10^{31}–10^{32} erg. The problem of the coronal energy transfer, however, is not yet satisfactorily solved.

Another interesting aspect arises from the relationship between the noise storm centers and the large-scale structures of the Sun or sector boundaries in interplane-

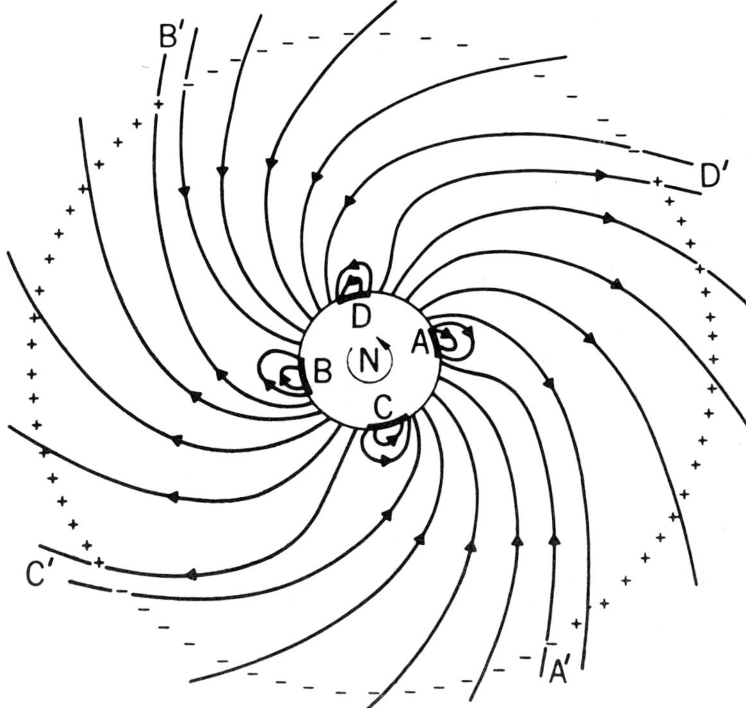

Fig. III.41. A geometrical model of the relationship between the type I noise-storm regions (A, B, C, and D) and the sector structure of the interplanetary magnetic field (A', B', C', D' – sector boundaries) (after Sakurai and Stone, 1971).

tary space. A recurrent nature of the type I storms over many solar rotations including small time shifts was established by Sakurai (1972b) from an analysis of the Nançay interferometer observations which is shown in Figure III.40. There seems to exist a certain correspondence between the interplanetary sector structures and the location of noise storm regions. Usually a noise storm is located just westward of a corresponding sector boundary (Sakurai and Stone, 1971). A model sketch of this relationship is reproduced in Figure III.41. Closely connected with the appearance of a storm region a tendency of a release of energetic electrons was found (cf. e.g. Sakurai, 1971a).

3.8. Solar Radio Pulsations

3.8.1. General features

The knowledge of the solar radio emission is by no means exhausted with the establishment of a large-scale classification of the radio phenomena distinguishing between the quiet Sun, the S-component, the noise storms, and the major burst types, as discussed in the foregoing sections. Rather there are many more phenomena and a wealth of structural details opposing any single classification scheme. It seems that fine-structure effects of the solar radiation can open the entry to new vistas in solar physics. The study of solar radio pulsations may be considered as a concrete example of these matters.

Regular fluctuations of the solar radio emission occur at different frequency bands within a wide range of frequencies (cf. also Section 3.2.5). In the following we are concerned with *pulsating patterns* superimposed on continuum bursts exhibiting characteristic periods of the order of about 1 s and some minutes.

Such more or less regularly fluctuating (pulsating) patterns are detectable by means of high-time-resolution dynamic spectra and single-frequency records. They occur preferably during special, well defined phases of particular type IV bursts almost simultaneously over an extended spectral range. We adopt here the most commonly used term 'pulsation' (Maxwell, 1965) to denote a regular emission pattern in the above described manner. (It should be remarked, that this term is not always uniformly used. Other labels are e.g. fluctuations, quasi-periodic modulations, oscillations, etc.) A sharp definition seems only fruitful if the physical process could be precisely determined and could thus be attributed to the suitable label.

In principle, pulsations are oscillatory in nature and are to be distinguished from a sequence of single pulses without an oscillatory character (Dröge, 1967). Considering complex events, this distinction appears often difficult to achieve in practice.

There are few reviews about radio pulsations. Preliminary summaries were given by Abrami (1972b) and Wild and Smerd (1972). Considerable progress in the field was achieved by the Utrecht multi-channel radio spectrograph operating at Dwingeloo (Slottje, 1972; Rosenberg, 1970). A brief compilation of the most prominent features observed with this instrument and other equipment is given in Table III.7. As is shown below, the amount of information provided by radio pulsations and fine structure should provide an opportunity for sensitive tests of theoretical models exposing the related physical processes.

TABLE III.7
Main features of solar radio pulsations and fine structures in type IV bursts

Frequency range	Pattern form	Period	Structure elements	Special features
cm–dm	Decimetric fluctuations, microwave pulsations	0.5–50 s	wide-band emission ridges	sometimes associated intensity modulations of X-rays
dm–m	Pulsating structures, harmonic pattern, rain-type bursts	1–5 s	broad-band emission ridges almost without definite drift rate	broad-band absorptions, wedge-shaped absorptions
	Parallel drifting bands, spaghetti structure, zebra patterns	./.	fiber bursts, normal and reverse intermediate-drift bursts, tadpole bursts	medium-band absorptions
Dm	Medium period fluctuations	2–5 min	quasiperiodic modulation of the background continuum	occurring in type IVmB bursts and noise storms

3.8.2. Periodicities in the cm- and dm-burst radiation

Pulsations of the solar radio burst emission were first detected at decimetric wavelengths. There are many early descriptions of the phenomenon given e.g. by Boischot and Fokker (1959), Elgarøy (1959, 1961), and Young *et al.* (1961) (cf. also Thompson and Maxwell, 1962; Kundu *et al.*, 1961; Kundu and Spencer, 1963; Hughes and Harkness, 1963, who sometimes interpret the pulsation patterns as a sequence of fast-drift bursts). After the first recognitions more systematic studies followed, based on both spectrographic records (de Groot, 1966, 1970) and single-frequency observations with enhanced time resolution (Dröge, 1967; Abrami, 1968, 1970a, b, 1972a, b).

New attention was drawn to the phenomenon when it became apparent that sometimes a close association exists between fast pulsations of hard X-ray bursts, EUV flares, and microwave-burst emissions (Parks and Winckler, 1969, 1971; Maxwell and Fitzwilliam, 1973). The pulsations analyzed in the microwave region exhibit various periods ranging from 0.5 to 200 s. But it seems likely that even shorter periods may be observed with higher time resolution. The amplitudes are often very faint and do not clearly depend on the intensity of the continuum. The pulsations were observed spectrographically at frequencies up to 2000 MHz (Gotswols, 1972; Maxwell and Fitzwilliam, 1973). Systematic studies at higher frequencies are still lacking.

The low-frequency cutoff of the pulsation structure occurs usually, more or less sharply marked, in the region between about 200 and 250 MHz. Sometimes this boundary coincides with the high-frequency cutoff of a noise storm. Inside the pulsating structure the roots of type III bursts can often be observed. The 'drift rates' of the pulsations themselves, however, appear often infinite, and deviations in

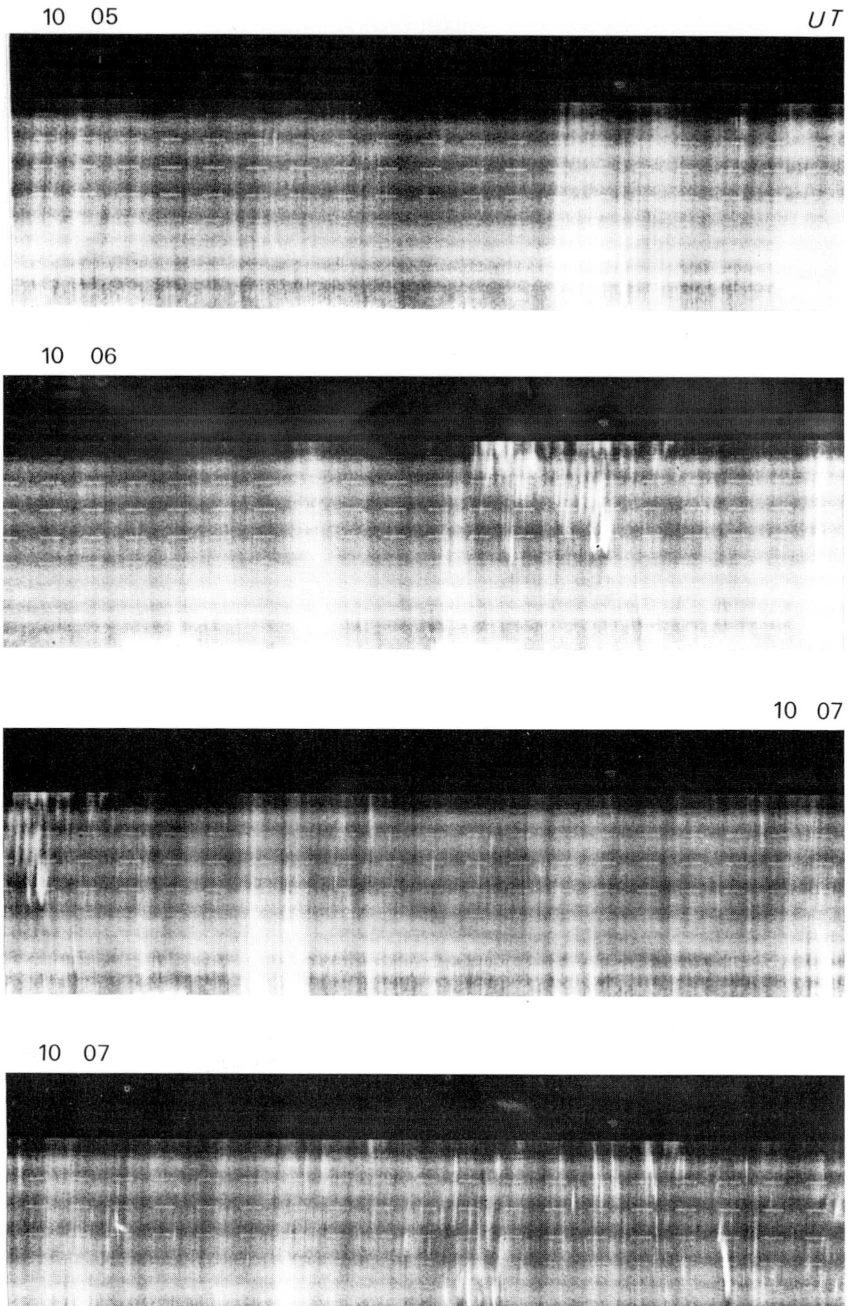

Fig. III.42a. Examples of dynamic spectra showing pulsations and various kinds of superimposed disturbances (event of October 4, 1973): 310–340 MHz range. (Courtesy of Ø. Elgarøy).

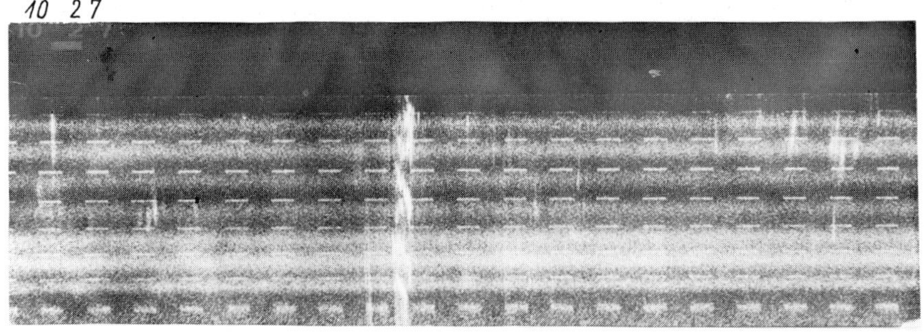

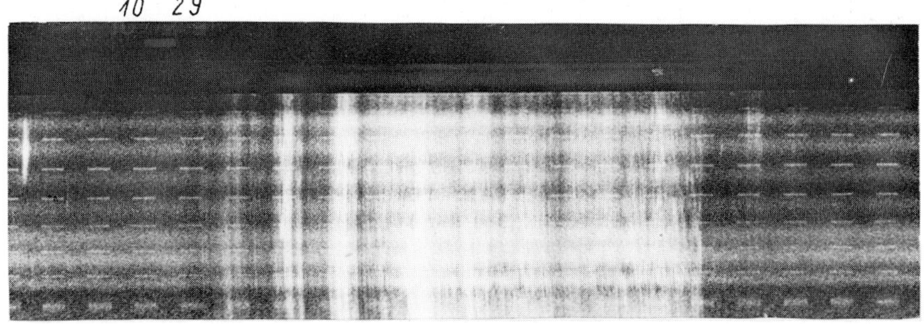

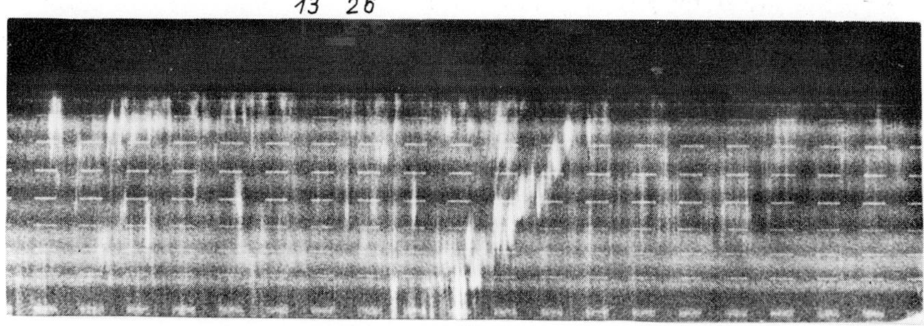

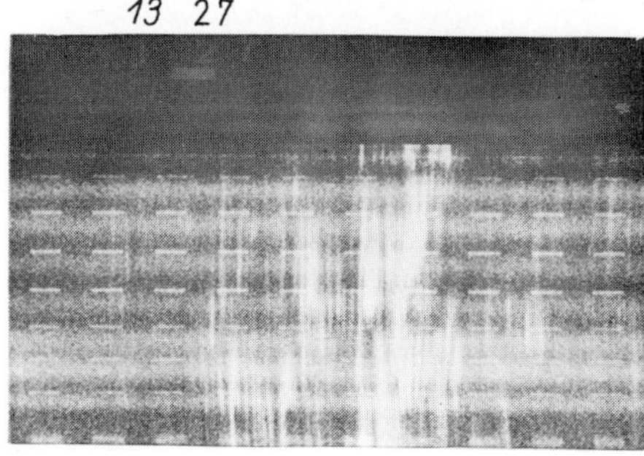

Fig. III.42b. (Continuation of Figure III.42a).

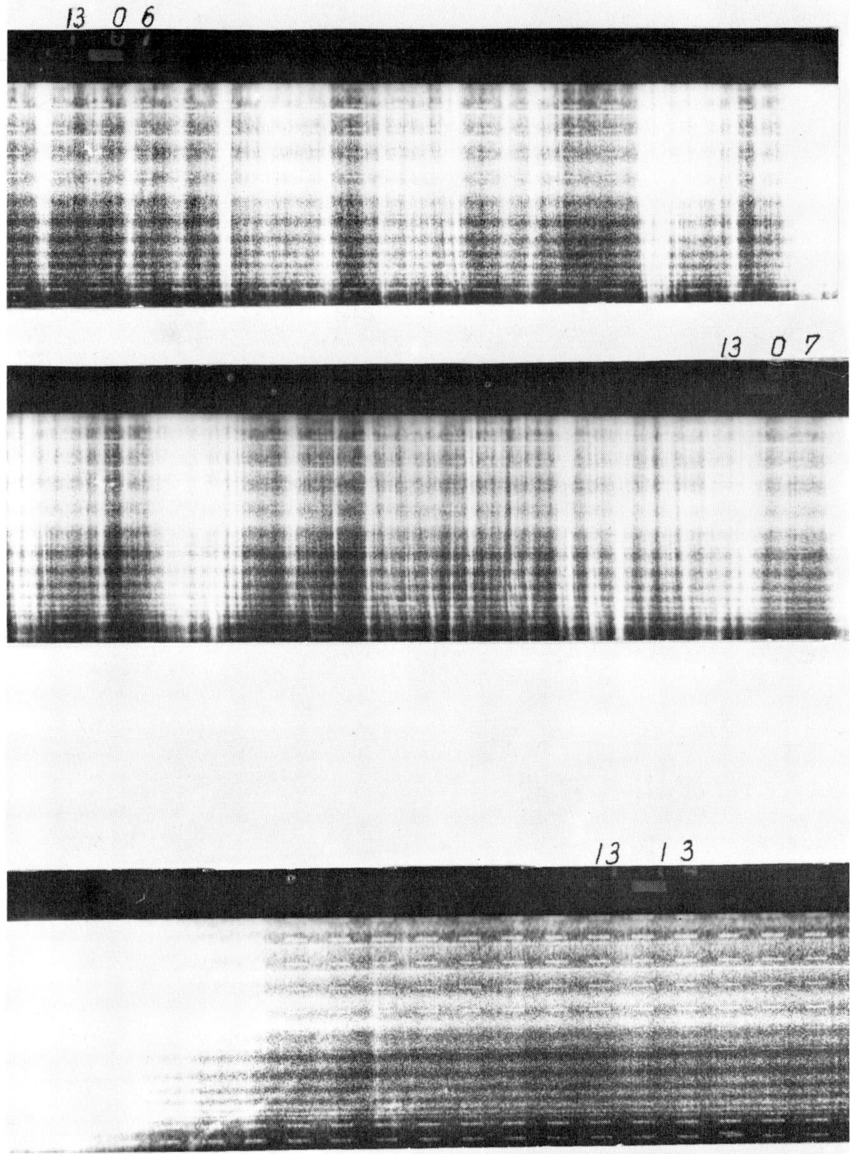

Fig. III.42c. Same as Figure III.42a,b, but 510–545 MHz range.

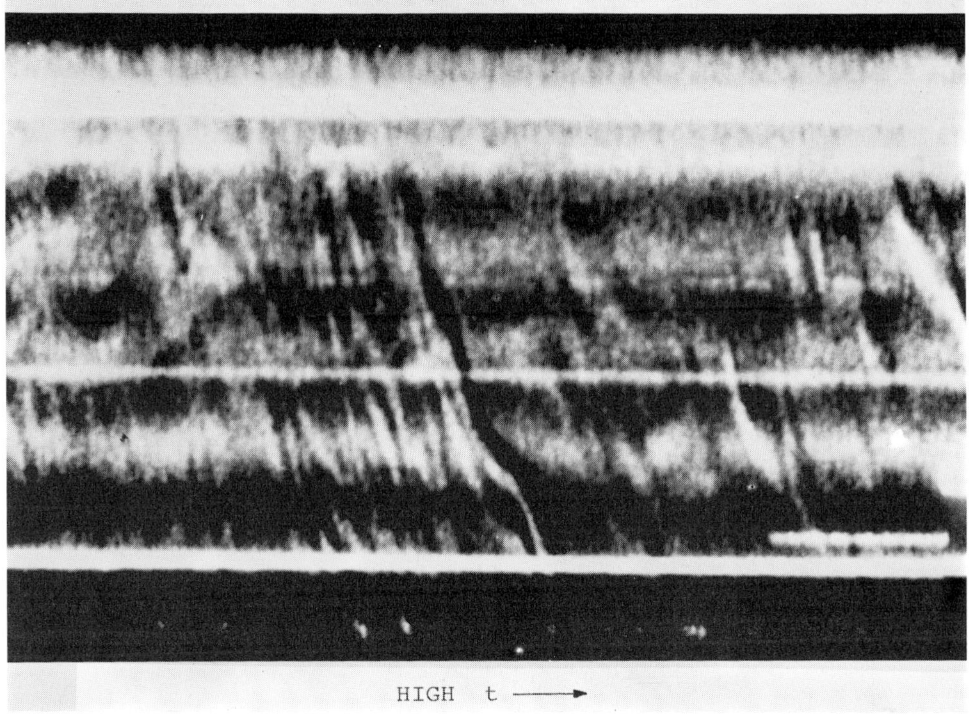

Fig. III.43. Example of details of zebra-like fine structures showing reverse drifting emission bands and shadow-like absorption features in the 195–220 MHz range. (Courtesy of Ø. Elgarøy).

both positive and negative directions occur. This feature strongly supports the assumption that the pulsations, in contrast to type III bursts, are a stationary phenomenon.

An illustration of various pulsating phenomena is given by the dynamic spectra of Figures III.42a, b, c and III.43.

3.8.3. Fine Structure and Short-Period Fluctuations in Type IV Bursts

a. *Pulsating Structures in the m-Region*

Pulsating structures in the dm/m-region were best displayed by the Utrecht/Dwingeloo radio stpectrograph (Rosenberg, 1970; Slottje, 1972). The pulsations are practically identical with the decimetric part of the pulsations described in the foregoing section. Nevertheless, they also occur in the meter region at frequencies below 200 MHz (McLean *et al.*, 1971).

By no means all pulsations exhibit an obvious periodic nature. In most cases the single broad-band elements with characteristic time scales of the order of 0.1 to 0.3 s appear to be distributed without any clear periodicity, often merging together (those structures were called also 'rain-type' bursts).

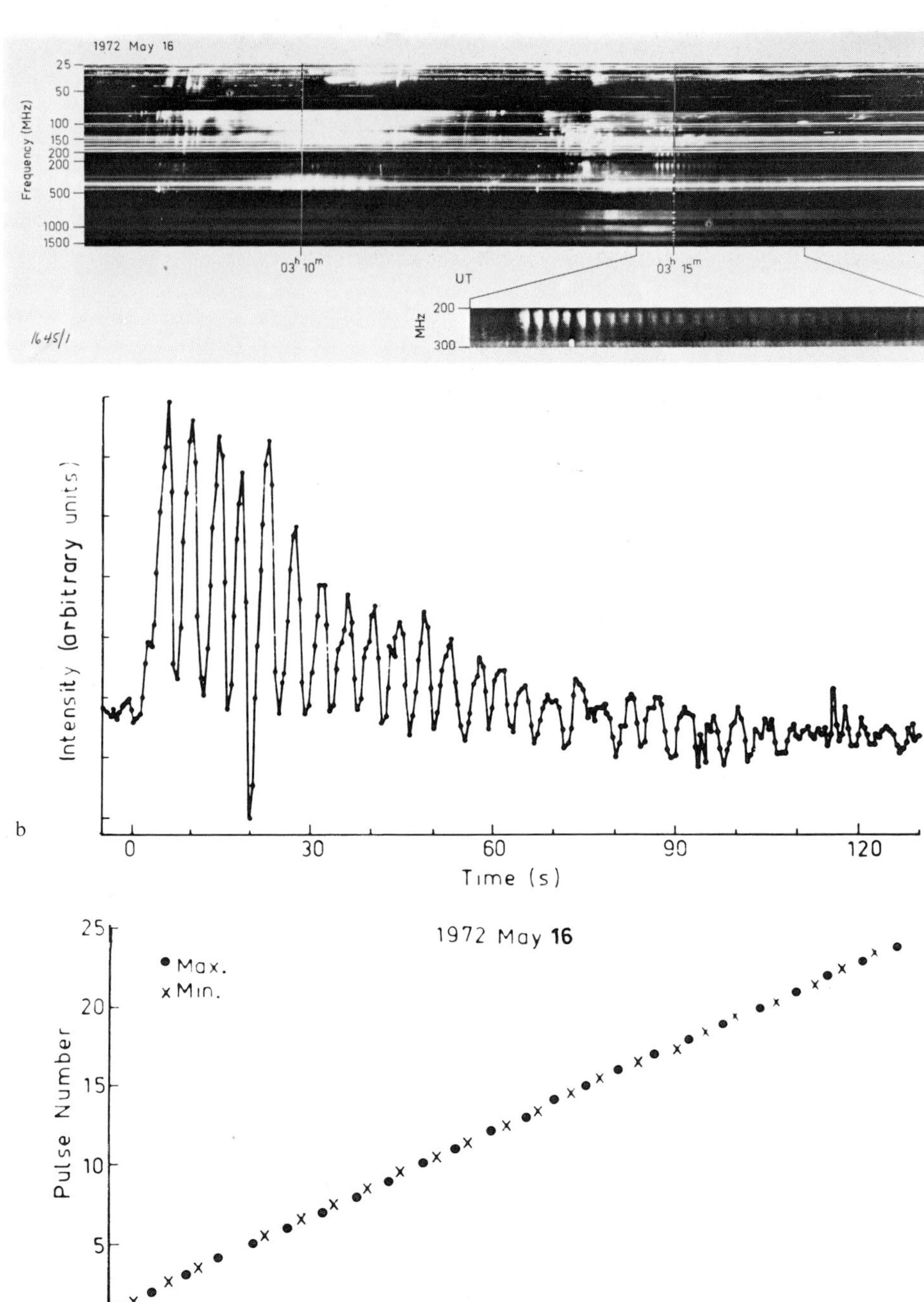

Fig. III.44. Example of regular radio pulsations following a type II/IVdm burst: a – dynamic spectrum, b – microphotometer tracing at 230 MHz, c – plot of the times of individual maxima and minima against pulse number (after McLean and Sheridan, 1973).

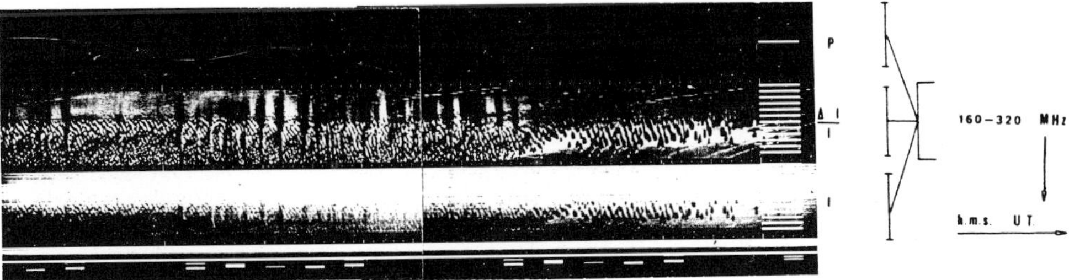

14.18.44- 14.20.05

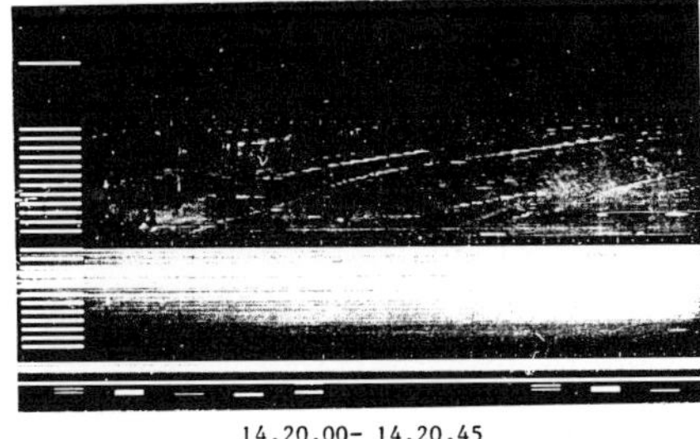

02-03-1970 14.20.00- 14.20.45

Associated events:
Type IV dm, m : 13.48 - 14.40 U.T.
H_α : 1N; N06 E33: 14.20 - 15.16 U.T.
1N; N05 E35: 13.28 - 14.05 U.T.
-- F; S10 E65: 14.12 - 14.23 U.T.
X-ray flare : 13.35 - 14.15 U.T.
S.I.D. : 13.36 - 15.38 U.T.

Type IV : above 220 MHz : tadpole - zebra pattern evolving into separate tadpole bursts.
wedge-shaped broadband absorptions
fiberbursts

Fig. III.45. Example for broad-band wedge-shaped absorptions, tadpole bursts, and zebra structures. (Courtesy of C. Slottje).

On the other hand, though infrequent, there are observations of extremely regularly distributed pulsations showing a clear periodicity at dm- or m-waves. An example of such a damped train of regular pulses is shown in Figure III.44 (McLean and Sheridan, 1973).

A very remarkable fact is that pulsations do not only appear in emission but also in absorption. As yet, three morphological types of short-lived absorption phenomena have been distinguished (Slottje, 1972):

(1) *Medium-band absorptions* (bandwidth about 40 to 70 MHz) related to tadpole bursts (cf. next subsection);

(2) *Broad-band absorptions* (bandwidth about 200 MHz), related to pulsations; and

(3) *Wedge-shaped absorptions* (broad-band absorptions with a typical increase in duration towards lower frequencies).

Some examples for these features are shown in Figure III.45. Interpretations are given in Section 3.8.5.

b. *Parallel Drifting Bands (Zebra Patterns) and Groups of Tadpole Bursts*

Although not belonging directly to the pulsations, some other spectral patterns of burst fine structures are to be briefly added for completeness.

TABLE III.8

Characteristic features of tadpole bursts (Slottje, 1972)

Property	Emission eye	Absorption body	Emission tail
Frequency range [MHz]	>320–240	300–220	270–220
Band width [MHz]	1–2	10–50	0–25
Duration [s]	≤0.1	0.2–0.3	0.2
Intensity [dB]	>3	< −3	0–2
Frequency drift (tilt) [MHz/s]	—	−150– +100	−150– +100

The first pattern to be mentioned has been known for a long time and was called 'spaghetti' structure, *intermediate drift bursts*, or more recently, harmonic patterns (Rosenberg, 1971a) and bunches of fiber bursts (Slottje, 1972). The frequency range of these *fiber bursts* lies roughly between 500 and 200 MHz. The characteristic intrinsic bandwidths and durations of the fiber bursts are about 1.5 MHz and 0.3 s, respectively. At the low-frequency side sometimes *absorption edges* of about the same duration and bandwidth are present. Both normal and reverse drifting fiber bursts are to be observed. The drift rates range between those of the type II and of the type III bursts.

A second kind of pattern is constituted in connection with the phenomenon of the so-called '*tadpole bursts*' (Slottje, 1972). These strange and comparatively rare bursts consist mainly of an absorption 'body'; sometimes also emission 'eyes' and emission 'tails' exist embedded in a continuum (Figure III.45). Special characteristics are reproduced in Table III.8 and Figure III.46.

Sometimes the tadpole bursts lose their individual identity and form an organized pattern of dark and bright stripes in the dynamic spectrum which is called the *zebra pattern*. In such a pattern the absorption is a dominant feature together with numerous dots of emission (Figure III.45). The slopes of the zebra stripes are often found to be arranged with time or frequency.

It should be noted that zebra-like patterns also occur without any association of tadpole bursts as a mere modulation of the background continuum having some resemblance with the phenomenon of the fiber bursts (Slottje, 1972). Recently zebra structures were also detected in conjunction with type III bursts (Fomichev and Chertok, 1977).

3.8.4. Medium-period Fluctuations in Type IVmB Bursts and Noise Storms

Besides the short-period pulsations, which are characteristic for the dm/m- range as mentioned above, sometimes peculiar modulations of the radio flux at longer wave-

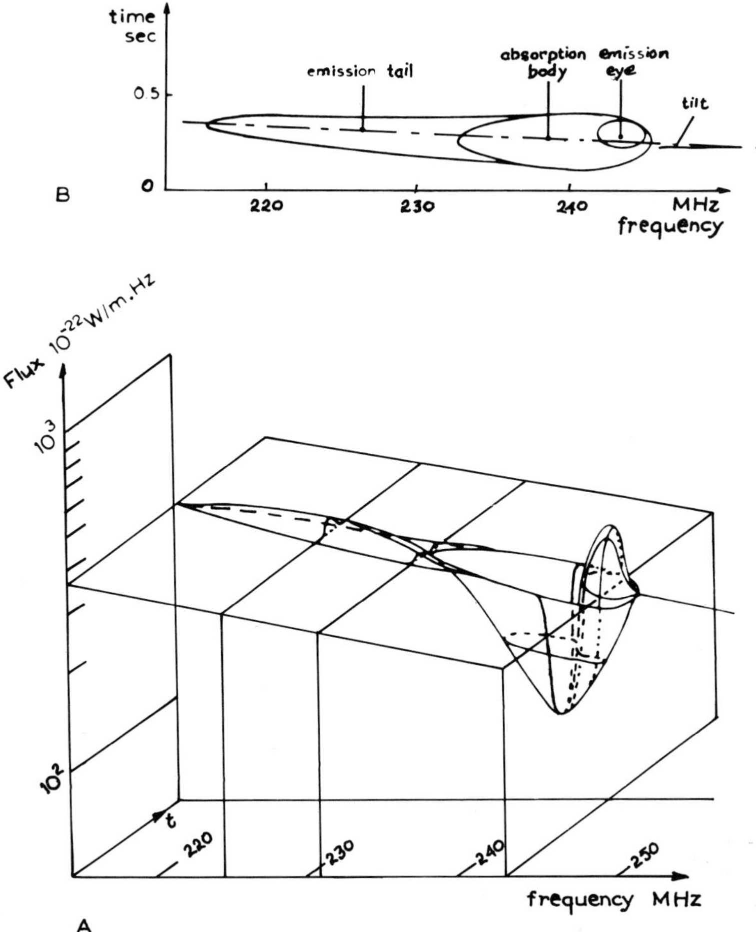

Fig. III.46. Intensity–time–frequency structure of a tadpole burst (after Slottje, 1972).

lengths occur with periods of the order of a few minutes (Böhme and Krüger, 1973). The modulations appear superimposed on the continuum of a type IVmB burst or noise storm. A number of strange properties have been stated favoring an interpretation by standing-wave phenomena in different coronal magnetic flux tubes (Aurass *et al.*, 1976), but more systematic investigations are still necessary.

3.8.5. Interpretations

The richness of detail delivered by solar fine-structure features is challenging refined physical explanations which are subjugated to precise opportunities for checking. In this way the fine-structure radio phenomena can allow a completely new entry to the hitherto insufficiently explored basic processes present in solar flare regions. The efforts hitherto undertaken to explain particularly the pulsating radio structures are by far not exhausted and can be divided into two main groups. These groups consider either a modulation of some important plasma parameters (**H**, **E**, T, N) influencing the emission, or the ejection of particle streams in analogy to the generation of the fast-drift bursts as basic mechanisms.

a. *Modulation of Plasma Parameters*

There are several attempts to propose a modulation particularly of the magnetic field, which then via gyro-synchrotron emission would result in a related modulation of the generated radio continuum. A special model was suggested by Rosenberg (1970, 1972) and elaborated by McLean *et al.* (1971) and Gotwols (1973). In this model a periodic variation of the magnetic field in a flux tube is assumed to be attributed to a standing magnetohydrodynamic wave, which is triggered by the passage of a shock wave. Assuming radial fluctuations of the magnetic field strength propagating perpendicularly to the magnetic field direction and postulating synchrotron emission, a relation between the radius of the flux tube r, the Alfvén velocity v_A, and the period T of the pulsations can be derived, viz.

$$2\pi r/(Tv_A) = \text{const}\,(\approx 1.8). \qquad \text{(III.33)}$$

Though the model looks quite promising in various respects, a number of features are left unexplained (e.g. the abrupt ending of the pulsations, the relationship to the intensity of the continuum, and the asymmetric (saw-tooth) wave form) demanding future work (McLean and Sheridan, 1973).

Another approach concerning the microwave part of pulsations in particular was made by Chiu (1970), who assumed a modulation by whistler waves in magnetic loop structures generated by a pitch-angle instability. The bounce motion of these whistlers should lead to enhanced modulated microwave and X-ray emission by pitch-angle scattering (as well as modulated type III burst emission), which cause an enhanced precipitation of relativistic electrons. The adopted emission process again is synchrotron emission and bremsstrahlung for the microwave and X-ray burst emission, respectively. The application of this hypothesis, which appears attractive in many respects, depends, among other things, severely on the assumed existence of relativistic electrons trapped in coronal magnetic fields, which is, however, subject to some doubt.

Another modulation effect concerning the plasma temperature was proposed by Abrami (1972a, b). In this case the modulated thermal absorption within localized regions would directly influence the radio emission.

b. *Particle Ejections*

The resemblance of the pulsations with the appearance of fast-drift bursts as well as occasional observations of groups of type III bursts with quasi-periodic spacings (Wild, 1963; Maxwell, 1965) stimulated the assumption of repeated streams of particles as a possible cause of the pulsation phenomena. This hypothesis was worked out in more detail by Zaitsev (1970, 1971), who suggested that pulsations may be formed by large-amplitude plasma waves due to a stream instability in a stabilized jet of ions or protons trapped in a bipolar magnetic structure. According to Smith and Fung (1971) and Zaitsev (1971) the stabilization of the stream instability may result in a pulsating region under certain conditions. Some deficiencies of the model became apparent by the work of Heyvaerts and de Genduillac (1974).

A mechanism for the generation of broad-band fast-drift absorption features was proposed by Zaitsev and Stepanov (1974, 1975). In this model the continuum emission is assumed to be caused by a plasma-wave mechanism due to a loss-cone instability near the upper hybrid frequency

$$\omega_{uh} = (\omega_p^2 + \omega_H^2)^{1/2} \tag{III.34}$$

set up in a magnetic trap. The absorption effect is suggested as a result of a switching off the plasma waves by the action of streams of fast electrons injected into the trap filling the loss-cone distribution and quenching the instability. In this theory the emission fringes of the absorption pattern can also be interpreted as a result of a stream expansion. Wedge-shaped absorptions are also explainable.

Finally it should be mentioned that loss-cone distributions also give rise to the generation of Bernstein modes. Interactions between these modes or with plasma waves at the upper hybrid frequency were treated in order to explain the zebra patterns (Rosenberg, 1972; Zaitsev and Stepanov, 1975; Zheleznyakov and Zlotnik, 1975).

3.9. Summary

Table III.9 gives a gross recapitulation of the different phenomenological properties of solar radio-burst radiations described in Chapter III.

TABLE III.9

Compilation of mean burst characteristics

Type	'Microwave bursts' Impulsive	Gradual	Type IV μ	'Meter bursts' Type IV dm	Type IV mF/mA	Type IV mB	Noise storms continuum	Noise storms Type I	Type II	Type III	Type V
Frequency range	$\gtrsim$1000 MHz			~150–1500 MHz	$\lesssim$200 MHz		$\lesssim$500 MHz		$\lesssim$200 MHz	$\lesssim$500 MHz	$\lesssim$200 MHz
Emission	continuous							'noncontinuous'			quasi-continuous
Harmonics	—							two harmonics of ν_p		(three?)	—
Variability	weak			fine structures	weak	slowly varying		strong	complex	group structure	
Number in solar maximum	$\lesssim$1 per hour	$\gtrsim$1 per day	~1–10 per month, very strong events avoiding the solar maximum				~1–3 per week	$\gtrsim 10^3$ per hour	~0.5 per day	$\lesssim$3 per hour (groups)	$\lesssim$1 per hour
Band width [MHz]	~500–50000		~5000–$\gtrsim$50000	~1000	$\gtrsim$200	$\lesssim$200	<500	$\gtrsim$4	~5	50–100	
Frequency drift	(no drift)				sometimes similar type II		not typical	sometimes fast drifting	~1 MHz $\times s^{-1}$	~20 MHz $\times s^{-1}$	following type III
Duration	$\lesssim$10 min	$\gtrsim$10 min	10 min–several hours			up to several hours	hours, days	<1 sec (groups ~0.5 min)	~5–30 min	$\lesssim$10 s (groups ~1 min)	$\lesssim$1 min
Max. flux [s.u.]	$\lesssim 10^3$	$< 10^2$	$\lesssim 10^4$	$\lesssim 10^5$	$< 10^5$	$\lesssim 10^4$		$\gtrsim 10^4$	$\lesssim 10^5$		
Equivalent temperature [K]	$\sim 10^6$–10^{10}	$\lesssim 10^6$	$\lesssim 10^9$		10^7–10^8	$\lesssim 10^{10}$		$\lesssim 10^{10}$	$\sim 10^8$–10^{11}		

Circular polarization	no or partial		partial	partial, strong	weak		very strong			very weak	partial	
Mode	e			o	e		o				o	
Linear polarization	weak	?	partial ?		≲10%			?	very weak		questionable	
Directivity	moderate		weak	moderate			strong			weak		
Height of the source	$\sim 0.05R_\odot$			$0.1–0.4R_\odot$	$0.5–5R_\odot$		$0.2–1R_\odot$			$0.2–200R_\odot$	$0.2–200R_\odot$	$0.2–2R_\odot$
Source diameter	1–4′	≲1–2′	2–4′	3–5′	6–12′		2–6′	2–10′	1–6′	6–≥30′	3–12′	
Source movement	stationary				quasi-stat.	ascending	stationary			ascending $\sim 10^3$ km s^{-1}	ascending (descending) ($\sim 0.3c$)	
Emission mechanism	bremsstrahlung and gyro-synchrotron emission			induced scattering	gyro-synchrotron emission (?)		mode coupling		solitions ?	plasma waves, mode transformations		
Exciter	thermal and super-thermal trapped electrons			plasma waves	superthermal electrons		?		?	MHD shock waves	electron streams	
Associated phenomena	flares, X-ray bursts		flares, relativistic particles					sometimes flares		type IV	often flares (flash phase)	type III
Pecularities	'spike' bursts, fluctuations	(absorptions)	different stages	fluctuations; absorption bands, intermediate drift bursts	fiber bursts, tadpole bursts, zebra patterns		medium-band fluctuations		reversed-drift pairs, split pairs, spike bursts	frequency splitting, herring-bone structure	U-type and J-type bursts, type III b	attached and merging subtypes

CHAPTER IV

THEORY OF SOLAR RADIO EMISSION

4.1. Basic Properties of the Solar Atmosphere as a Plasma Medium

4.1.1. Compilation of Important Plasma Parameters

Any theory and interpretation of solar radio waves is inevitably connected with certain parameters characterizing the physical conditions of the plasma medium of the solar atmosphere, in which the relevant processes of the wave origin and propagation take place.

Basic plasma parameters attributed to solar radio physics are the electron density N_e, the (external) magnetic field $\mathbf{H}$, and the (electron) temperature T_e or the electron energy distribution function $f(E)$.

These quantities are related to three characteristic frequencies

(a) the electron plasma frequency

$$\nu_{pe} \equiv \frac{\omega_{pe}}{2\pi} = \left[\frac{e^2 N_e}{\pi m_e}\right]^{1/2} = 8.98 \times 10^3 \sqrt{N_{e_{[\mathrm{cm}^{-3}]}}}\ [\mathrm{MHz}]\ , \qquad \text{(IV.1)}$$

(b) the (nonrelativistic) electron gyrofrequency

$$\nu_{He} = \frac{\omega_{He}}{2\pi} = \frac{eH}{2\pi m_e c} = 2.8 H_{[\mathrm{G}]}\ [\mathrm{MHz}] \qquad \text{(IV.2)}$$

(valid for $\beta < 0.3$, i.e. $E_{\mathrm{kin}} < 24$ keV),

(c) the electron–ion collision frequency, among different approximations we quote

$$\nu_{\mathrm{coll}} = 6.6 \times 10^{-6} Z N_e / T^{-3/2} g \qquad \text{(Bekefi, 1966)}$$

$$\approx 50 N_{e[\mathrm{cm}^{-3}]} T_{[\mathrm{K}]}^{-3/2}\ [\mathrm{s}^{-1}] \qquad \text{(Wild } et\ al.\text{, 1963)} \qquad \text{(IV.3)}$$

(e – electron charge, m_e – electron mass, c – vacuum velocity of light, Z – charge number, g – Gaunt factor).

The thermodynamic temperature is related to the following velocities characterizing a Maxwell distribution:

$$v_{th}^{[\text{root-mean-square}]} = (3\ KT/m)^{1/2} \qquad \text{(IV.4)}$$

$$v_{th}^{[\text{average}]} = (8\ KT/(\pi m))^{1/2}$$

$$\approx (2.5\ KT/m)^{1/2} \qquad \text{(IV.5)}$$

$$v_{th}^{[\text{most probable}]} = (2\ KT/m)^{1/2} \qquad \text{(IV.6)}$$

(cf. Sears, 1953). Sometimes also the expression $v_T = (KT/m)^{1/2}$ is used as thermal velocity (referring to one degree of freedom) (cf. e.g. Kaplan and Tsytovich, 1972).

From the basic parameters some other important quantities can be derived, e.g. the *mean free path* of electrons

$$\lambda_f = \frac{1.3 \times 10^5 T^2}{\ln \Lambda N_e} \text{ [cm]} \tag{IV.7}$$

(ln Λ – Coulomb logarithm for electron-ion collisons),

the *Larmor radius* of electrons

$$r_H = \frac{m v_\perp c}{eB}, \tag{IV.8}$$

the *degree of ionization r*

$$r = N_e / N_{\text{neutral}}, \tag{IV.9}$$

the *Debye length* l_D

$$l_D = \left[\frac{KT}{4\pi N_e e^2} \right]^{1/2}, \tag{IV.10}$$

which gives the dimension below which the fluid description of a plasma cannot be used.

A further basic parameter is given by a *characteristic length L* of a considered plasma region; this can be, e.g., a representative diameter of an emission volume, the distance between magnetic mirror points, scale heights, etc.

4.1.2. Density and Temperature

a. *Electron Density*

Numerical values of the electron density distribution of the solar atmosphere derived by optical and radio methods were given by numerous authors. Representative summaries of older results are due to van de Hulst (1953) and de Jager (1959). These data are commonly cited in reviews on solar radio astronomy (e.g. Minnaert, 1959; Kundu, 1965; Zheleznyakov, 1964, 1970). A compilation of the more recently achieved observational results was made by Newkirk (1967) and by Athay (1976). For illustration, Figure IV.1 shows a distribution of N_e according to Newkirk (1967) and others representing a reference level for quasi-quiet conditions in the equatorial (ecliptic) plane. The data have been supplemented by spaceborne observations extending the resulting distribution to distances beyond 1 AU. Closer to the Sun, for comparison, a density profie obtained from solar eclipse observations (on March 7, 1970, in Mexico) for a particular streamer was included in Figure IV.1 (Saito, 1972).

Since the chromosphere and corona undergo remarkable temporal and spatial changes, it is not astonishing that there is an appreciable range of scattering in the results of different individual observations. It was usually assumed in constructing coronal density models that the average corona is spherically symmetric although it has considerable structure.

Tables of the numerical values of various physical quantities on the Sun are contained in Allen (1973) and *Landoldt-Börnstein* (Hellwege, 1965).

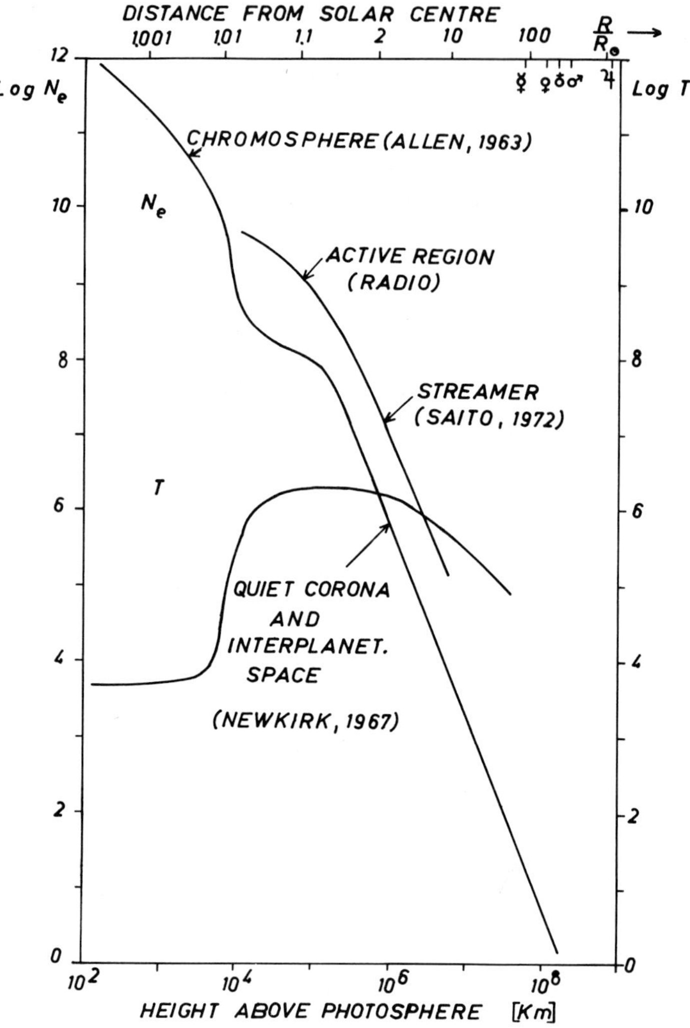

Fig. IV.1. Distribution of electron density and temperature in the solar atmosphere and interplanetary space.

b. *Thermal Equilibrium*

Though the local thermodynamic equilibrium (LTE) is only an idealization for certain real conditions in the solar atmosphere, the concept of temperature is often applied because of its simplicity and utility as a standard reference quantity. A rough orientation about the shape of the mean temperature distribution over the height above the photosphere is contained in Figure IV.1. However, a number of problems arise, especially when the results of different methods of a temperature estimation are compared in detail. A careful distinction between excitation, ionization, and kinetic temperatures becomes necessary (in practice, kinetic electron temperatures are preferably used). Moreover, temporal and spatial inhomogeneities, the height variation of the optical depth, and other influences must be taken into account. Therefore the often made assumption of an isothermal corona is only a first approximation. A determination of the energy balance by means of reliable models is inseparably connected with the knowledge of the detailed temperature or energy structure. A satisfying solution of this task, which also involves the exact determination of the coronal temperature maximum (which probably occurs at about 1.1 $R_{\odot}$) is still missing. For a review of these problems the reader is referred to the monographs of Billings (1966), Zirin (1966), and Athay (1976).

Among the empirical methods the measurements of the brightness temperatures at the center of the solar disk by radio observations provide the most direct means of a determination of the temperatures in the corona (cf. Section 3.1.2). However, the resulting brightness temperatures are influenced by the reduced optical thickness in the corona.

c. *Nonthermal Energy Distributions*

In many cases, especially in flares, departures from LTE become significant, so that they cannot be longer ignored. The most commonly used forms of nonthermal energy distributions, apart from the Delta function, are the power-law distribution

$$N(E) \sim E^{-\alpha} \tag{IV.11}$$

and, more rarely used, the exponential distribution

$$N(E) \sim \exp(-E/E_c) \tag{IV.12}$$

(cf. e.g. Takakura and Kai, 1966).

Solar X-ray bursts provide a more useful tool for the determination of the energy distribution of the radiating electrons than radio bursts. Whereas the soft X-rays can be interpreted preferably by thermal bremsstrahlung determining a Maxwell distribution, the hard X-ray burst spectra are often successfully approximated by a power law

$$I(\varepsilon) = a\varepsilon^{-\lambda} \tag{IV.13}$$

(ε – photon energy), which can be converted into a power-law electron-energy distribution of the form of Equation (IV.11), or

$$f(v) = A(v_s/v)^{\gamma} \tag{IV.14}$$

where $\gamma = 2\lambda$.

In the case of a combined spectrum consisting of a Maxwellian part and a power-law tail with a high-energy cutoff, the following normalized spectrum can be obtained (Böhme *et al.*, 1976):

$$f_e(v) = A_e f_M(v) \times \theta(v_s - v) + (1 - A_e - C_e) f_P(v) \times \theta(v - v_s) \times \\ \times \theta(v_c - v) + C_e f_C(v) \times \theta(v - v_c) \times \theta(v_c + \delta v - v) \tag{IV.15}$$

where

$$f_M(v) = A_M \exp(-v^2/v_{th}^2)$$

(Maxwell distribution),

$$f_P(v) = A_P(v_s/v)$$

(power-law tail),

$$f_C(v) = A_C(1 - v/(v_c + \delta v)]^2$$

(sharply decreasing function simulating a cutoff).

θ is the step function:

$$\theta(a) = \begin{cases} 1 & \text{if} \quad a > 0 \\ 0 & \text{if} \quad a < 0. \end{cases} \tag{IV.16}$$

As illustrated in Figure IV.2, the following velocities determine the distribution:

$v_{th} = (2KT/m)^{1/2}$ (thermal velocity),

v_s – velocity, denoting the joining point between $f_M(v)$ and $f_P(v)$, besides

$$v_0 = (2\varepsilon_0/m)^{1/2} \geq v_s$$

denotes the energy, above which the X-ray spectrum can be well represented by a power law, and

v_c = cutoff velocity of the power-law distribution ($\delta v \ll v_c$).

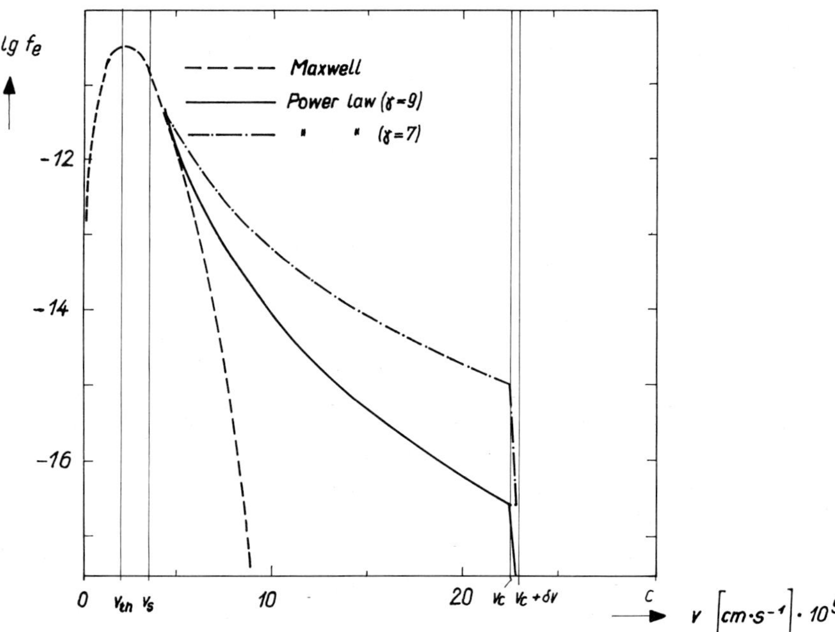

Fig. IV.2. Examples of energy distributions consisting of a Maxwellian and different power-law tails with an energy cutoff (after Böhme *et al.*, 1977).

In general one finds

$$v_{th} < v_s \leq v_0 < v_c \lesssim v_c + \delta v \leq c.$$

The quantities A_M, A_P, and A_C are the normalization constants for each part of the distribution function (normalization is made according to the density N_e). A_e and C_e are the constants which result from the continuity condition at v_s and v_c and the normalization of the whole distribution to N_e.

4.1.3. Magnetic Field

a. *Empirical Distributions*

The structure and dynamics of the solar atmosphere are widely controlled by magnetic fields. The dominant role of the magnetic fields is particularly evident in many phenomena of solar activity. Therefore it seems inconsistent to state that we still have insufficient information about the magnetic fields at higher levels than the photosphere: It is no exaggeration to say that the main lack of information in solar physics is due to inadequate methods of measurements of magnetic fields in the chromosphere and the corona and the future progress in the field depends just on the solution of this question.

The most reliable measurements of solar magnetic fields come from optical observations of the Zeeman splitting of photospheric Fraunhofer lines. In comparison to other methods the application of the Zeeman effect is free from uncertain hypotheses. The absolute value of the magnetic field strength is directly proportional to the distances between the single Zeeman components of a line. The line intensity and polarization determine the magnetic field direction. For weaker fields outside the sunspots longitudinal field components $B_{\parallel}$ can be measured with a sufficient accuracy; information on the transverse field is more difficult to obtain by a second order effect. The angular resolution of ground-based optical observations is limited by the influences of the terrestrial atmosphere (scattered light, atmospheric blurring) and reaches at best 1 s of arc ($\approx$725 km on the Sun).

In the chromosphere the observations of the Zeeman effect are less conclusive, since even the strongest Fraunhofer lines are not free from unwanted photospheric light, particularly in the line wings. Other complications arise from the necessity to take into account complex processes of line formation (non-LTE conditions). It is hoped that observations from space platforms will result in higher resolution measurement of magnetic fields in the chromosphere and corona.

A promising approach to a measurement of coronal magnetic fields consists in the utilization of the linear polarization caused by the resonant scattering in emission lines (interference of Zeeman sub-levels, Hanle-effect) (Lamb, 1970; House, 1972). But the evaluation of the magnetic fields depends strongly on, more or less, complicated model assumptions.

Since in wide regions of the solar atmosphere the magnetic pressure $p_m = B^2/(8\pi)$ is not small in comparison with the gas pressure p_g and the kinetic pressure $p_k = \rho v^2/2$, the structure of the solar atmosphere observed in different spectral regions (X, UV, radio) contains implicitly information about the magnetic structure. Plasma motions appear guided by magnetic fields, so that e.g. the Hα-filtergrams can be used as indicators for the magnetic field structure of the middle chromosphere (Zirin, 1972), but special problems of the interpretation of such observations do not appear as fully solved.

Beside optical and UV observations, radio observations are potential candidates for magnetic field estimations. To derive information about the magnetic fields almost all kinds of solar radio emission have been exploited, e.g. by spectral and polarization analyses of the S-component and microwave bursts, by means of bursts of all spectral types including the chains of type I bursts, reversing heights of U- and J-type bursts, adopting the Razin-effect at type IV burst emission bands, etc. Unfortunately, the reliability of the obtained information is reduced by the vagueness of some (necessarily) postulated physical mechanisms and/or the lack of sufficient spatial resolution. Among several discussions of the different ways of a determination of solar magnetic fields by radio methods we quote the papers of Takakura (1966), Newkirk (1967, 1971), and Gelfreikh (1972). General accounts on the problems of solar magnetic fields were edited by Lüst (1965) and Howard (1971).

Summarizing the hitherto achieved results, a rough sketch of an averaged magnetic field is shown in Figure IV.3. This figure is based mainly on the presently most probable estimates compiled by Newkirk (1971). Because all hitherto existing estimations of the magnetic field above the photosphere must be considered with some caution, Figure IV.3 contains a probable range in which the chromospheric and

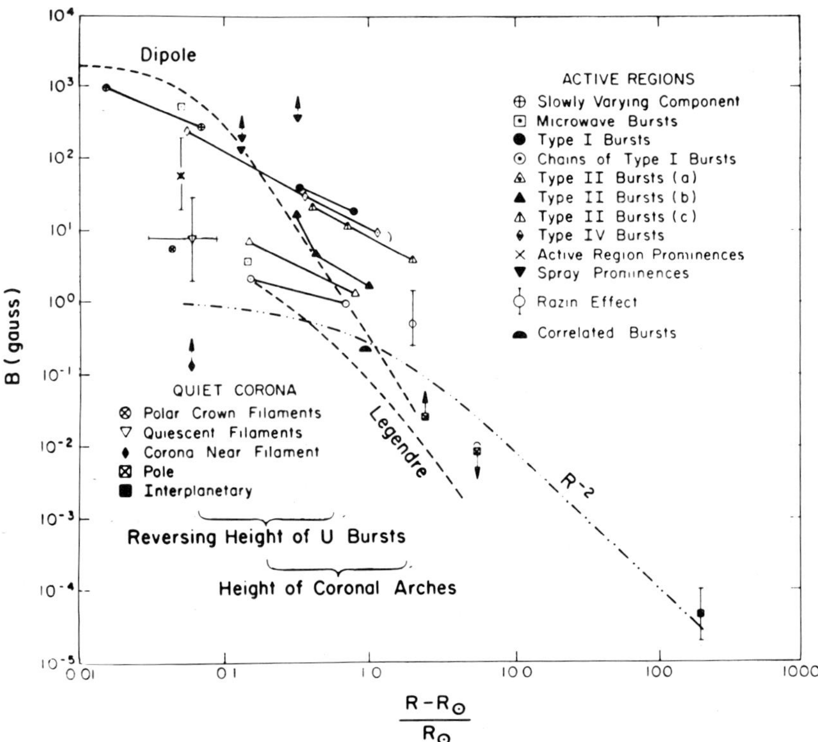

Fig. IV.3. Summary of magnetic field estimations in the solar corona after Newkirk (1971).

coronal magnetic fields are expected to vary. A support of the magnetic field distribution in greater distances from the Sun is given by direct magnetometric measurements from space vehicles in the vicinity of the Earth.

b. *Models*

Modeling the spatial magnetic field distribution in the solar atmosphere, two main forms can be distinguished:

(a) Theoretical extrapolation of measured magnetic field strengths, and

(b) pure analytical magnetic field models (e.g. dipole distributions).

Concerning the first group it is remarked that, since the photospheric magnetic fields are the most surely measured at present, attempts have been made to extrapolate the known field values into greater heights. To do this, special physical conditions must be assumed. Choosing such a condition, the field in the chromosphere and lower corona is usually assumed to be force-free:

$$\operatorname{curl}\mathbf{B} = \alpha\mathbf{B}, \qquad \operatorname{div}\mathbf{B} = 0. \tag{IV.17}$$

The scalar α is a measure for the torsion of adjacent field lines, α is proportional to the electric current flowing in the field direction.

In the case $\alpha \equiv 0$ a current-free magnetic field (potential field) is obtained. Such a field can be uniquely computed, if the distribution of the normal component B_z is known on a basic surface. Magnetic field computations were carried out either for restricted areas on the Sun (Schmidt, 1964, 1966) assuming a plane basic surface or with the whole-Sun photosphere as basic surface (Schatten *et al.*, 1969; Altschuler and Newkirk, 1969) thus describing the fine structure or the gross structure of the magnetic field distribution, respectively. An atlas of large-scale magnetic fields in the corona was published by Newkirk *et al.* (1972). An example of a map from this publication is shown in Figure IV.4.

In spite of all the success of the above mentioned magnetic field extrapolations it must be taken into account that the use of potential fields is only a first approximation. The next step of an approximation would include fields with a constant $\alpha \neq 0$ (Trkal fields) or even fields with a variable α (Nakagawa *et al.* 1973; Seehafer, 1975). A generalization of the method of Altschuler and Newkirk (1969) by the application of a Trkal field was due to Nakagawa (1973). In order to obtain higher degrees of approximation all these calculations become progressively more complicated, and, moreover, require additional information (or assumptions) for the estimation of the value α.

A second group of magnetic field models consists of more theoretical developments and is based, more or less, on artificial assumptions, e.g., cylindric symmetry. These models refer to idealized local regions and may be constructed to serve as an input parameter configuration for other model calculations (e.g. in gyro-synchroton emission models). For such purposes sometimes the dipole approximation is used e.g. in the form

$$B = \frac{B_m R_m^3}{2R^3}(1 + 3\sin^2\vartheta)^{1/2} \tag{IV.18}$$

$$\theta = \arccos\{[\sin\psi\cos(\vartheta - \lambda)\cos\varphi + \cos\psi\sin(\vartheta - \lambda)],$$

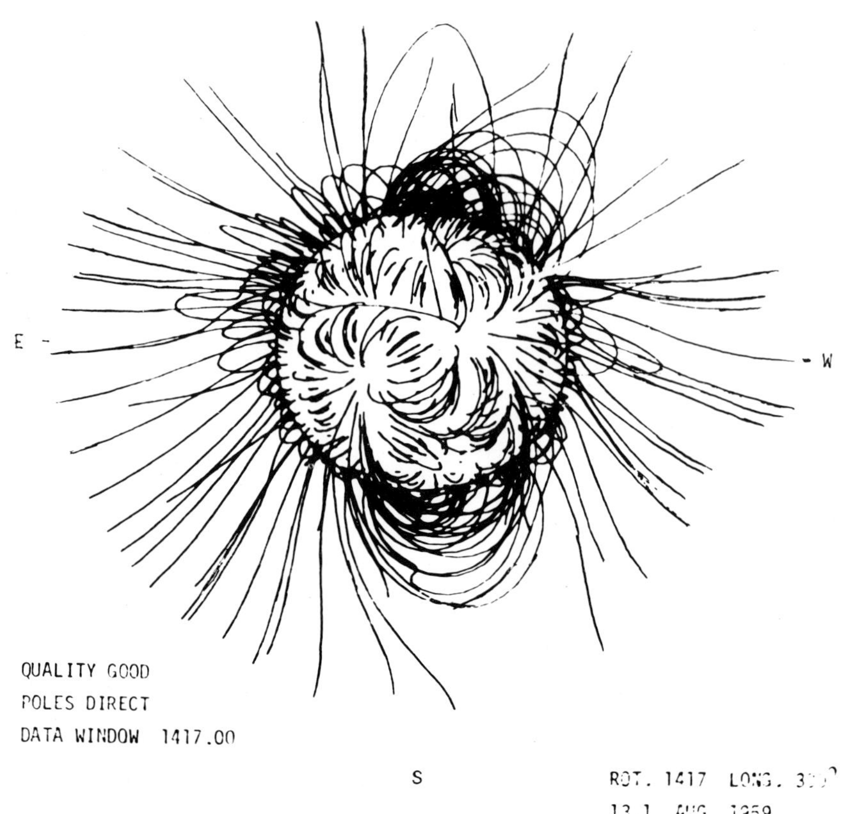

Fig. IV.4. Example of current-free magnetic-field extrapolations of the solar corona according to Newkirk *et al.* (1972).

where $\theta = \sphericalangle(\mathbf{B}, \mathbf{k})$, R_m, B_m – free model parameters, $\mathbf{k}$ – wave vector, R, φ, ϑ – spherical coordinates, $\psi = \sphericalangle(\mathbf{k}, \vartheta = 90^\circ, \lambda = \sphericalangle(R, \mathbf{B}))$ (cf. Takakura and Scalise, 1970; Alfvén and Fälthammar, 1963).

Another more artificial distribution with a R^{-2} space dependence is given by

$$B = \frac{B_m R_m^2}{\zeta R^2 \cos\vartheta + (R \sin\vartheta + R_m)^2}, \tag{IV.19}$$

$$\beta = \operatorname{arccot} \frac{1}{\zeta} \frac{R \sin\vartheta + R_m}{R \cos\vartheta}$$

where $\beta = \sphericalangle(\mathbf{B}, \vartheta = 90^\circ)$, ζ, B_m, R_m – free model parameters (cf. Böhme *et al.*, 1976).

Finally, concluding our short excursion on magnetic field models, the extrapolation of the magnetic field into the interplanetary space up to 1 AU taking into account models of the solar wind should be mentioned. A special model of this kind was proposed e.g. by Stenflo (1972).

c. *Alfvén-Wave Velocity*

The influence of the magnetic field basically determines the dynamics of a plasma. Magnetic fields are also responsible for the existence or modifications of several wave types different from radio waves, but nonetheless being of high interest for solar radio astronomy. As a basic example we quote the MHD waves, which in the simplest case of an incompressible, ideally conducting MHD-plasma propagate with the well known Alfvén velocity

$$v_A = B/(4\pi\rho)^{1/2} \tag{IV.20}$$

(ρ – mass density).

As an important feature, above a spot group the Alfvén velocity varies with height exhibiting a maximum at a certain altitude (above ca. 10^4 km) reaching typical values of the order of 10^4–10^5 cm s^{-1}. In weaker magnetic fields outside active regions the corresponding Alfvén velocity is much smaller and essentially determined by the mass density (cf. also Section 3.5.3.b).

4.1.4. Characteristic Time Scales and Transport Quantities

a. *Redistribution and Diffusion Times*

When nonstationary processes are considered, relaxation times are frequently used to denote the time required for bringing the system back into some state of equilibrium. Following the classic treatise of Spitzer (1962), a number of relaxation times describing the influence of distant encounters between charged particles can be defined, which well describe the state of a plasma. The first quantity to be quoted is the *deflection time* t_D

$$t_D = w^2/\langle(\Delta w_\perp)^2\rangle, \tag{IV.21}$$

which describes the time between collisions expressed by that time, in which the test particles are gradually deflected by an angle of 90°. w is the particle velocity and $\langle(\Delta w_\perp)^2\rangle$ denotes the diffusion coefficient in the velocity space perpendicular to the initial direction of w of the test particles.

Analogously an *energy exchange time* t_E is defined by

$$t_E = E^2/\langle(\Delta E)^2\rangle. \tag{IV.22}$$

The rate at which the mean velocity of some test particles is decreased by encounters is described by the *slowing-down time* t_s:

$$t_s = -w/\langle\Delta w_\parallel\rangle. \tag{IV.23}$$

Furthermore sometimes an *equipartition time* t_{eq} is used considering the rate at which equipartition of energy is established between two groups of particles.

Finally, a special case of the deflection time t_D defined by Equation (IV.21) is given by a group of particles which interact with themselves yielding the *self-collision time* t_c (Spitzer, 1962; Alfvén and Fälthammar, 1963).

b. *Electric and Heat Conductivity*

Concerning steady nonequilibrium phenomena the flow of energy expressed in terms of different transport coefficients is to be considered. Beside other forms of energy exchange (e.g. mass and momentum transfer, radiation) in particular the charge and heat transport are of importance for many questions of solar physics. The corresponding transport coefficients are the electric conductivity σ and the thermal conductivity κ, which are closely related to each other.

In a uniform isotropic medium the d-c conductivity $\sigma = 1/\eta$ (η – resistivity) is given by

$$\sigma = e^2 N/(m_e \nu_{\text{coll}}) \tag{IV.24}$$

and hence depends on the temperature and density. The main difficulty in evaluating σ arises from the necessity of the determination of a representative value for the collision frequency. A practicable treatment was proposed e.g. by Spitzer and Härm (1953). Numerical values of σ are not very sure and cover a wide range of magnitudes. So σ is believed to vary between about 10^{10} e.s.u. in the photosphere and 10^{18} e.s.u. in coronal flare plasmas (cf. e.g. Kopecký, 1971; Alfvén and Fälthammar, 1963).

In the case of the presence of magnetic fields it is convenient to distinguish three components of the conductivity:

$$\begin{aligned} \sigma_{\parallel} &= \sigma \\ \sigma_P &= \frac{\sigma}{1 + \left(\dfrac{\omega_H}{\nu_{\text{coll}}}\right)^2} \\ \sigma_H &= \frac{\sigma}{\dfrac{\nu_{\text{coll}}}{\omega_H} + \dfrac{\omega_H}{\nu_{\text{coll}}}} \end{aligned} \tag{IV.25}$$

The quantity σ_P is called the *Pedersen conductivity* or cross-conductivity, and σ_H is called the *Hall conductivity*. They are related by the following equations

$$\begin{aligned} j_{\parallel} &= \sigma_{\parallel} E_{\parallel} \\ j_{\perp} &= \sigma_P E_{\perp} + \sigma_H (\mathbf{E} \times \mathbf{B})/E. \end{aligned} \tag{IV.26}$$

It can be mentioned, that basically similar methods as used for the evaluation of the electrical conductivity are also applicable for the calculation of the thermal conductivity (as well as for other transport quantities, e.g. the viscosity).

Explicitly the coefficient of heat conduction is written approximately

$$\kappa = 5 \sqrt{\frac{\pi}{6} \frac{NKT}{m_e \nu_{\text{coll}}}}. \tag{IV.27}$$

It should be noted that a temperature gradient not only produces a flow of heat but also influences the velocity distribution of the plasma particles giving rise to an electric current and vice versa.

4.2. Fundamentals of the Emission and Propagation of Radio Waves

4.2.1. Elementary Processes of the Generation of Radio Waves

a. *General Remarks and Definitions*

The question of the emission mechanisms plays a basic role in solar radio astronomy. Perhaps more pronounced than in other spectral ranges the detailed knowledge of the emission and absorption processes is important for both the study of the relevant physical processes and the diagnostics of the physical conditions (i.e. plasma parameter configurations) on the Sun. To demonstrate this feature, the run of information from the Sun to the Earth is illustrated in a simplified way in Figure IV.5. By some physical process a source of energy released in a certain plasma volume represented by special distributions of the plasma parameters causes radiation by the action of one or more particular emission mechanisms. If the radiation can propagate to the observer, information about the interesting physical processes, the energy exchange, and plasma parameters (e.g. the magnetic field) is obtained in a coded way and can be decoded by running back the scheme of Figure IV.5. Evidently this is only possible by taking into account known propagation and emission (absorption) mechanisms.

In contrast to most other spectral regions the radio emission exhibits the following peculiarities:

(a) From an energetic point of view radio waves are completely negligible; nevertheless, they are best suited as indicators for important plasma processes, which, as in the case of the type IV bursts and proton flares, are sometimes connected with just the most violent excessive energy exchange present in the whole solar system.

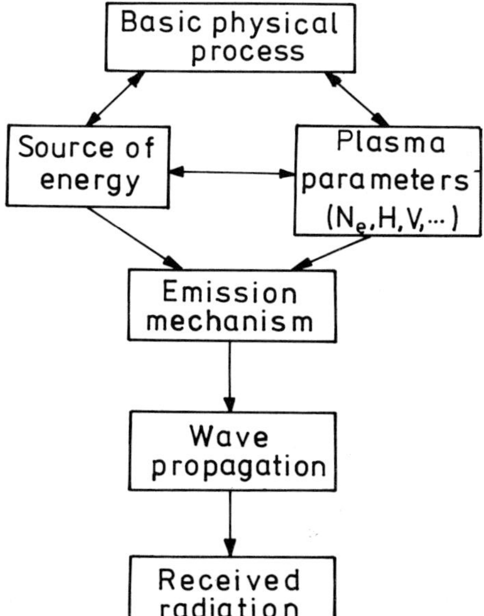

Fig. IV.5. General scheme of the run of information.

(b) The radio waves are sensitively influenced by magnetic fields and by particle-drift processes, which make them predestined as attractive means for the diagnostics of especially such regions, to which an approach by other methods is lacking or more difficult.
(c) The investigation of the radio emission mechanisms is principally complicated by the necessary inclusion of the physics of magnetized plasmas, and also by the fact that the number of unknown variables often exceeds the number of the observed independent parameters.
(d) Up to now, line emissions do not play any appreciable role in solar radio astronomy.

Now, before going into more detail, some further introductory remarks about definitions and classifications will be given.

(1) *Continuous and noncontinuous radio emission*

As continuous or *continuum radiation* at radio frequencies a broad-band radiation has been empirically defined, opposing the more narrow-band emissions of the drift bursts.

Owing to a certain correlation between the spectral and temporal extent of different solar radio emissions, the defined continuum radiation refers mostly to almost stationary or long-period emissions, i.e. to the quiet Sun, the S-component, microwave bursts, several kinds of type IV bursts, and noise storm continua. Since line emissions are of minor importance in solar radio astronomy, it is evident that the present use of the term 'continuum' differs remarkably from that defined in optics according to which simply all nonline emissions would be called 'continuous' (cf. Section 3.3.1).

(2) *Particle-generated and wave-generated radio emissions*

Theoretically a division of the radio emission mechanisms can be made between those which are primarily based on accelerated particles and those based on wave-mode transformations. These mechanisms can also be named *direct* (primary) and *indirect* (secondary) *emission processes*, respectively (cf. Krüger, 1972). For the most part this division will correspond to the above given empirical classification (1), since it is supposed that continuum bursts refer preferably to particle-generated (i.e. direct) and drift bursts to wave-generated (i.e. indirect) emission processes. However, the reality may be somewhat more complicated, and a total coverage of both schemes cannot be *a priori* expected (cf. next subsection).

(3) *Incoherent and coherent emission mechanisms and radiation fields*

A further distinction can be made between *incoherent* and *coherent emission mechanisms* in a plasma. An *incoherent* radiation mechanism denotes an origin in a great number of uncorrelated elementary emission processes (e.g. from single electrons) and can be statistically interpreted in terms of a radiation (electron) temperature. The related absorption is always positive or at least zero. Sometimes incoherent wave phenomena in a plasma are identified with turbulence (Davidson, 1972; Tsytovich, 1971). On the other hand, *coherent* emission processes were defined in the literature in different ways, namely either on the basis of correlated (collective) processes involving a large number of particles (cf. Heald and Wharton, 1965), or by the condition of negative absorption (occurrence of instabilities, amplification of radiation) as has been proposed by Zheleznyakov (1964, 1970). Since both concepts are not necessarily identical, we shall adopt the latter definition here.

From the above concept of coherent and incoherent *emission processes* the coherency of a *radiation field* as defined in optics is principally to be distinguished. In optics the term coherent radiation is used to denote the equality of the phase relations in two or more considered waves. The latter concept is applied to derive coherence times and coherence lengths (and also coherence areas, radii, volumes, etc.) of the waves.

(4) *Method of Einstein coefficients*

Applying the quantum theory, Einstein (1917) showed that principally three types of radiative processes may exist by which photons can be emitted or absorbed in interaction with matter: *spontaneous emission*, *stimulated* (or induced) *emission*, and *absorption* (attenuation). The first two terms refer to the already mentioned mechanisms of incoherent and coherent emission, respectively.

Though $h\nu \ll KT$ for radio waves, the quantum methods were sometimes also applied in radio physics. This leads to a calculation of the Einstein coefficients $A_{\varepsilon\to\varepsilon-h\nu}$, $B_{\varepsilon\to\varepsilon-h\nu}$, and $B_{\varepsilon-h\nu\to\varepsilon}$ in isotropic media under the assumption of the validity of the principle of detailed balance (Smerd and Westfold, 1949). A description of the method was given e.g. by Bekefi (1966) and Wild *et al.* (1963).

(5) *Resonant and nonresonant emission processes*

A further classification of emission processes is concerning *resonant* and *nonresonant radiations*. For resonant processes mostly a special frequency is marked. This can be e.g. the electron gyrofrequency or the upper hybrid frequency. Another class of resonant interactions comes into play, if the particle speed equals the phase velocity of the waves in a medium (Čerenkov effect). Resonant processes are also relevant for wave–wave interactions.

(6) *Thermal and nonthermal emissions.*

Finally, the division between *thermal* and *nonthermal* emission should be briefly noted without touching the problem of giving quantitative criteria for the concrete realization of the former or latter emission. The limiting case of thermal radiation for high optical depths is the blackbody radiation.

Since under real conditions the thermal equilibrium is often insufficiently realized, the concept of local thermodynamical equilibrium (LTE) was introduced in astrophysics by Schwarzschild (1914) and Milne (1921): Instead of an isothermal temperature T a local temperature $T(x, y, z)$ is defined and the formalism of thermal radiation employed. It is to be noted, that this procedure does not necessarily imply radiative equilibrium. Radiation which originates from regions with apparent deviations from the LTE, but which none the less is interpreted thermally, is sometimes called 'quasi-thermal'. But in spite of the mathematical difficulties the real nonthermal emission processes are increasing in importance in current solar radio physics.

b. *Primary Emission and Absorption Processes*

In the following we distinguish primary emission processes and secondary emission processes, where the term *primary emission* refers to particle-generated wave generation processes and the term *secondary emission* to wave–wave interactions and scattering (mode transformations). The terms wave–wave and particle–wave interactions are widely used in plasma physics also in connection with processes involving nonelectromagnetic radiations. Having in mind the final generation of radio waves we propose the following scheme:

A. Primary emission processes
(generation of radiation by accelerated particles – classical and relativistic emission mechanisms)
   (a) generation of electromagnetic radiation
   ('direct' emission processes of radio waves)
   (b) generation of plasma radiations, etc.

B. Secondary emission processes
(wave-mode transformations)
   (a) generation of electromagnetic radiation
   ('indirect' generation of radio waves)
   (b) generation of other wave modes.

The direct emission processes of radio waves constitute the first large group of radiation mechanisms, which have been known for a longer time and are extensively treated in the literature. Nevertheless, although the basic concepts are well developed, for a number of interesting conditions (e.g. nonequilibrium radiation in a hot magnetoplasma, strong absorption, etc.) much work still remains to be done.

Considering applications to solar radio waves, we are dealing with radiation processes connected with changes of the external motions of electrons and ions. Thus we have mainly to do with *free–free transitions* of electrons in the Coulomb field of protons or positive ions (Coulomb bremsstrahlung) and with the gyromagnetic radiation caused by electrons spiralling in a magnetic field (magneto-bremsstrahlung, cyclotron or gyro-synchrotron emission). Free–bound or bound–bound (line) radiations have comparatively little importance for solar radio astronomy. Also transitions associated with the internal structure of the ions and atoms will be omitted in our discussion.

A schematic compilation of the most important direct elementary emission processes (primary emission processes), is given in Table IV.1, where the mechanisms are plotted over the energy range and magnetic field range as basic characteristics. It is indicated in Table IV.1, that the Coulomb bremsstrahlung is basically treated for the absence of external magnetic fields and lower energies, but it may also exist to a certain extent in a magnetized plasma causing a modification of the radiation. Then there is a continuous transition to the cyclotron radiation depending on the conditions, whether the Coulomb collisions or the gyrations dominate in the motion of the radiating electrons (cf. Section 3.2.3.b). Depending on the energy distribution, the bremsstrahlung may be thermal or nonthermal; in the latter case a transition to the Čerenkov effect comes into play. However, on the Sun nonthermal processes occur

TABLE IV.1
Primary (direct) emission processes

	$\beta \ll 1$ ('quasi-thermal')	$1 - \beta \ll 1$ ('nonthermal')
$H \approx 0$	Bremsstrahlung (nonresonant generation of radiation)	Čerenkov radiation (resonant at ω_p)
$H > 0$	Cyclotron radiation (gyro-synchrotron emission) (resonant at $s\omega_H$)	

often in connection with strong magnetic fields, so that in such cases gyro-synchrotron radiation is expected to dominate.

The Čerenkov radiation is obtained as a particular case of 'particle-wave' interactions, when the particle velocity is close to the phase velocity of the waves in the medium. It is interesting, that the Čerenkov radiation can be also regarded as a limiting case of the synchrotron radiation, viz. for $\omega \gg \omega_H$. It must be also remarked that in a cold infinite isotropic plasma no Čerenkov radiation of transverse electromagnetic waves is possible, because the phase velocity there cannot be smaller than the speed of light. In this way, apart from particular cases in magnetized plasmas, the Čerenkov effect is more important for the generation of longitudinal plasma waves than for the direct emission of radio waves. The inverse process of the generation of Čerenkov waves is the *Landau damping*.

It is to be noted also that the other quoted primary emission mechanisms, i.e. bremsstrahlung and cyclotron emission, are not restricted to the direct generation of radio waves, but can also be efficient in generating plasma waves etc. Possible other primary emission processes, e.g. transition radiation (Ginzburg and Frank, 1946; Zheleznyakov, 1964, 1970), have not yet attained greater attention in solar radio astronomy.

c. *Indirect Emission and Absorption Processes*

Knowing the relevant primary emission processes (and in particular the direct emission mechanisms for radio waves) in a plasma, only half the truth about the origin of solar radio waves is known. The other half comes from the secondary (and tertiary, ...) emissions leading to an indirect generation of radio waves according to our nomenclature. These processes consist in a transformation (mode coupling) of waves of different types into electromagnetic (radio) waves.

The recently achieved progress in plasma physics is due to a systematic consideration of the different types of plasma waves and oscillations as characteristic features of departures from the thermal equilibrium, which was widely neglected before (Tsytovich, 1967, 1971; Akhiezer *et al.*, 1974; Kadomtsev, 1976).

For the purpose of solar radio-wave physics the task of evaluation of the indirect (or more generally, secondary) emission processes can be divided into the following different steps:

(a) Survey of the possible and most preferable wave modes in the solar atmosphere (after a consideration of their primary generation, cf. foregoing subsection).
(b) Investigation of the possible and most probable transformation processes of the different wave modes.
(c) Calculation of the efficiency and coupling parameters especially for the transformation into radio waves.

Turning to the first question, it is quite evident that the plasma state shows a multitude of waves and oscillations. They are a consequence of the combined actions of mechanical, thermo- and hydrodynamical quantities (e.g. pressure, particle speed, density, temperature, etc.) with the electromagnetic forces, which otherwise are widely absent in neutral media.

For the above reason it is here not possible to obtain such a simple picture as could be given for the primary emission processes (direct generation) of the radio waves (cf. Table IV.1). In the following we give a short orientation about what basic wave modes may be generally possible in a plasma medium.

It is well known that in the relatively simple case of a homogeneous plasma without external magnetic fields three basic types of waves are possible, namely: *electromagnetic waves*, *plasma waves*, and *(ion-) sound waves*.

For an explanation, a compilation of some properties of these wave types is given in Table IV.2.

In the more interesting case of a plasma in the presence of an external magnetic field the abovementioned types are branched out and a greater variety of waves results. This feature is schematically shown in Figure IV.6, where in a dispersion diagram for a particular homogeneous plasma the different branches of single wave modes are marked. In an inhomogeneous plasma additionally the existence of drift waves leads to a further branching of the low-frequency parts of the dispersion diagram. Some of these aspects relevant to solar radio astronomy will be discussed in Section 4.5.

We briefly note also a slightly differing classification of waves in a plasma distinguishing

TABLE IV.2

Waves in a plasma without external magnetic field

No.	Name	Type	Dispersion relation	v_k
1	Electromagnetic waves	transv.	$\omega^2 = \omega_{pe}^2 + c^2k^2$	$v_k \geq c$
2	Langmuir (electron plasma or space-charge) waves	long.	$\omega^2 = \omega_{pe}^2 + 3k^2v_{Te}^2$	$v_k \geq v_{Te}$
3	Ion-sound waves	long.	$\omega^2 = \dfrac{k^2v_s^2}{1 + k^2v_{Te}^2/\omega_{pe}^2}$	$v_k \leq v_s \ll v_{Te}$ $v_s \gg v_{Ti}$
	(a) Sound waves		$\omega = kv_s$	$v_k = v_s$
	(b) Ion-plasma waves		$\omega = \omega_{pi}$	$v_k = v_s(1 + k^2\lambda_0^2)^{-1/2}$

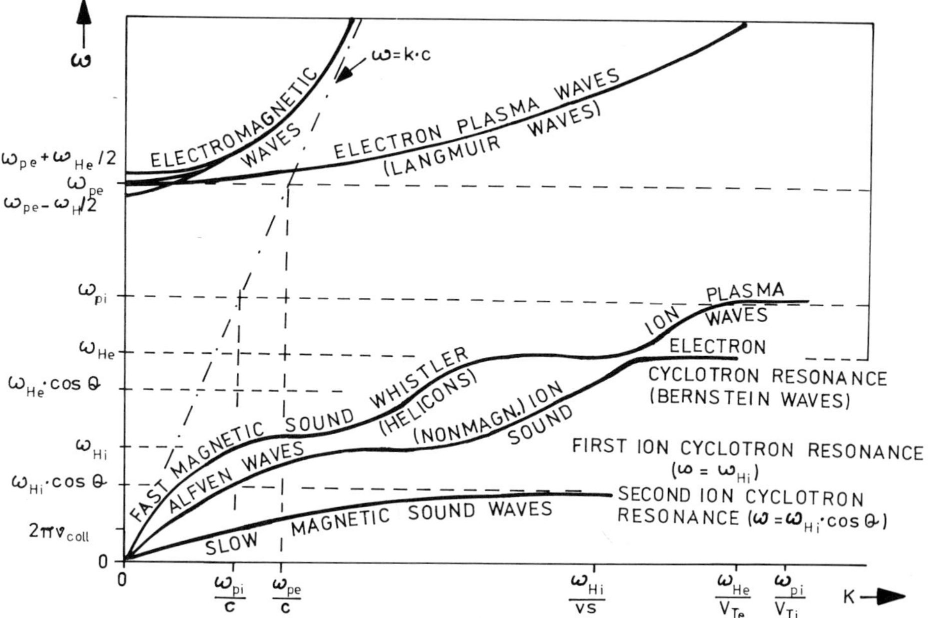

Fig. IV.6. Dispersion diagram demonstrating the possible main wave modes in a homogeneous warm magnetoplasma.

(a) electromagnetic waves $(\mathbf{E}, \mathbf{j} \perp \mathbf{v}_{kj})$,
(b) electrostatic waves $(\mathbf{E}, \mathbf{j} \parallel \mathbf{v}_{kj}, \mathbf{E} = -\nabla\phi, B_{\parallel} = 0)$,
(c) hydromagnetic waves (restoring forces: $\mathbf{j} \times \mathbf{B}, \omega < \omega_{pi}$),

where the presence of an external magnetic field is already taken into account.

Concerning the task of the examination of possible and important wave transformation processes some progress has been achieved in recent time, but it still remains difficult to obtain a comprehensive picture. For a detailed study of the fundamentals and special aspects of wave–wave interactions in a plasma the reader is referred to the surveys by, e.g., Akhiezer *et al.* (1974), Kaplan and Tsytovich (1972), and Sauer (1972). Here we merely consider some basic questions on these matters.

According to the linear dispersion theory the different wave modes in a plasma propagate independently (uncoupled) of each other. When nonlinearities in the basic equations (of motion) are taken into account, a transformation of the different wave modes arises, which is connected with the generation of combination frequencies. The wave transformation can be described by the introduction of a nonlinear conductivity tensor formalism, which is used e.g. in the approximation of weak nonlinearity (cf. below). The general task comprises a solution of the system of Maxwell equations together with an equation (nonlinear in general) of motion in order to obtain the functional dependence between the current $\mathbf{j}$ (or polarization) and the electric field strength $\mathbf{E}$. This task can be solved only approximately.

The kind of approximation can be characterized by the following steps: *linear theory*, *quasi-linear theory*, *weak turbulence* (weak nonlinearity), *moderate turbulence* (moderate nonlinearity), and *strong turbulence* (strong nonlinearity). In the first approximation (linear dependence) one can write for the case of spatial homogeneity

$$\mathbf{j}^{(1)}(x, t) = \iint \mathrm{d}x' \,\mathrm{d}t' \,\boldsymbol{\sigma}^{(2)}(x', t') \cdot \mathbf{E}^{(1)}(x - x', t - t'). \tag{IV.28}$$

Here the conductivity tensor $\boldsymbol{\sigma}^{(2)}(x, t)$ is of second rank describing the linear macroscopic connection between $\mathbf{j}$ and $\mathbf{E}$ in the plasma and hence determining the dispersion characteristics of the different wave modes.

A quadratic dependence between $\mathbf{j}$ and $\mathbf{E}$ is realized by a 'nonlinear' conductivity tensor $\boldsymbol{\sigma}^{(3)}$ of third rank according to the relation

$$\mathbf{j}^{(2)}(x, t) = \iiiint \mathrm{d}x' \,\mathrm{d}t' \,\mathrm{d}x'' \,\mathrm{d}t'' \,\boldsymbol{\sigma}^{(3)}(x', t', x'', t'') : \mathbf{E}^{(1)}(x - x', t - t') \times \\ \times \mathbf{E}^{(1)}(x - x' - x'', t - t' - t''). \tag{IV.29}$$

The quadratic relationship hinders the principle of superposition. It also leads to the occurrence of combination frequencies and particularly to second-harmonic emissions. (The latter can be easily demonstrated, if two waves are assumed:

$$E_1 \sim E \cos\omega_1 t, \qquad E_2 \sim E \cos\omega_2 t.$$

The resulting current varies according to

$$j \sim \cos\omega_1 t + \cos\omega_2 t + 1/2[\cos(\omega_1 + \omega_2)t + \cos(\omega_1 - \omega_2)t].$$

Hence for $\omega_1 = \omega_2$ the second harmonic results.)

As the next step of approximation the third-order current $\mathbf{j}^{(3)}$ is given by

$$\mathbf{j}^{(3)}(x, t) = \iiiiiint \mathrm{d}x' \,\mathrm{d}t' \,\mathrm{d}x'' \,\mathrm{d}t'' \,\mathrm{d}x''' \,\mathrm{d}t''' \boldsymbol{\sigma}^{(4)}(x', t', x'', t'', x''', t''') : \\ : \mathbf{E}^{(1)}(x - x', t - t') \mathbf{E}^{(1)}(x - x' - x'', t - t' - t'') \times \\ \times \mathbf{E}^{(1)}(x - x' - x'' - x''', t - t' - t'' - t'''), \tag{IV.30}$$

where $\boldsymbol{\sigma}^{(4)}$ is the conductivity tensor of fourth rank. In general, the current can be represented by an expansion

$$\mathbf{j}(x, t) = \sum_n \mathbf{j}^{(n)}(x, t). \tag{IV.31}$$

Here often the condition of 'weak nonlinearity' is used:

$$\mathbf{j}^{(n)} \gg \mathbf{j}^{(n+1)} \tag{IV.32}$$

Considering harmonic wave phenomena, the Fourier components of the electric field and the current are of interest

$$\mathbf{E}(\mathbf{k}, \omega) = \iint \mathrm{d}\mathbf{x} \,\mathrm{d}t \,\mathbf{E}(\mathbf{x}, t) \exp[i(\omega t - \mathbf{k} \cdot \mathbf{x})] \tag{IV.33}$$

$$\mathbf{j}(\mathbf{k}, \omega) = \iint \mathrm{d}\mathbf{x} \,\mathrm{d}t \,\mathbf{j}(\mathbf{x}, t) \exp[i(\omega t - \mathbf{k} \cdot \mathbf{x})] \tag{IV.34}$$

leading to the following relations for the first three orders:

$$\mathbf{j}^{(1)}(\mathbf{k}, \omega) = \boldsymbol{\sigma}^{(2)}(\mathbf{k}, \omega) \cdot \mathbf{E}^{(1)}(\mathbf{k}, \omega)$$

$$\mathbf{j}^{(2)}(\mathbf{k}, \omega) = \iint \mathrm{d}\mathbf{k}' \, \mathrm{d}\omega' (2\pi)^{-4} \boldsymbol{\sigma}^{(3)}(\mathbf{k}, \omega, \mathbf{k}', \omega') : \\ : \mathbf{E}^{(1)}(\mathbf{k}', \omega') \, \mathbf{E}^{(1)}(\mathbf{k} - \mathbf{k}', \omega - \omega')$$

$$\mathbf{j}^{(3)}(\mathbf{k}, \omega) = \iiiint \mathrm{d}\mathbf{k}' \, \mathrm{d}\omega' \, \mathrm{d}\mathbf{k}'' \, \mathrm{d}\omega'' (2\pi)^{-8} \boldsymbol{\sigma}^{(4)}(\mathbf{k}, \omega, \mathbf{k}', \omega', \mathbf{k}'', \omega'') : \\ : \mathbf{E}^{(1)}(\mathbf{k}'', \omega'') \, \mathbf{E}^{(1)}(\mathbf{k}' - \mathbf{k}'', \omega' - \omega'') \, \mathbf{E}^{(1)}(\mathbf{k} - \mathbf{k}', \omega - \omega') \tag{IV.35}$$

with

$$\boldsymbol{\sigma}^{(2)}(\mathbf{k}, \omega) = \iint \mathrm{d}\mathbf{x} \, \mathrm{d}t \, \boldsymbol{\sigma}^{(2)}(\mathbf{x}, t) \exp[i(\omega t - \mathbf{k} \cdot \mathbf{x})],$$

$$\boldsymbol{\sigma}^{(3)}(\mathbf{k}, \omega, \mathbf{k}', \omega') = \iiiint \mathrm{d}\mathbf{x} \, \mathrm{d}t \, \mathrm{d}\mathbf{x}' \, \mathrm{d}t' \, \boldsymbol{\sigma}^{(3)}(\mathbf{x}, t, \mathbf{x}', t') \times \\ \times \exp[i(\omega t - \mathbf{k} \cdot \mathbf{x}) + i(\omega' t' - \mathbf{k}' \cdot \mathbf{x}')],$$

$$\boldsymbol{\sigma}^{(4)}(\mathbf{k}, \omega, \mathbf{k}', \omega', \mathbf{k}'', \omega'') = \iiiiiint \mathrm{d}\mathbf{x} \, \mathrm{d}t \, \mathrm{d}\mathbf{x}' \, \mathrm{d}t' \, \mathrm{d}\mathbf{x}'' \, \mathrm{d}t'' \times \\ \times \boldsymbol{\sigma}^{(4)}(\mathbf{x}, t, \mathbf{x}', t', \mathbf{x}'', t'') \exp[i(\omega t - \mathbf{k} \cdot \mathbf{x}) + \\ + i(\omega' t' - \mathbf{k}' \cdot \mathbf{x}') + i(\omega'' t'' - \mathbf{k}'' \cdot \mathbf{x}'')]. \tag{IV.36}$$

These equations contain the conservation of energy and momentum of the interacting waves expressed by

$$\omega = \omega' + \omega'', \qquad \omega = \omega' + \omega'' + \omega''', \qquad \ldots, \tag{IV.37}$$

and

$$\mathbf{k} = \mathbf{k}' + \mathbf{k}'', \qquad \mathbf{k} = \mathbf{k}' + \mathbf{k}'' + \mathbf{k}''', \qquad \ldots, \tag{IV.38}$$

respectively, following from the invariance of the equations in local space against spatial and temporal translations.

The required explicit calculation of the conductivity tensor is often complicated. Such calculations were carried out for different cases e.g. by Sauer (1972) and Mollwo and Sauer (1977). The full solution also contains an estimation of the efficiency of the wave transformation process, which is needed for quantitative computations of the intensity of the radiation.

An approximate estimation of a number of wave–wave as well as wave–particle processes by means of transition and emission probabilities was undertaken by Kaplan and Tsytovich (1972). The formulae given by these authors may be useful for an orientation about the order of magnitude of the radiation originating by different processes. A further discussion of these topics will be given in the Sections 4.5 to 4.7.

4.2.2. Basic Theoretical Developments

The basic equations from which the wave emission and propagation characteristics are derived, in general, are

(1) Maxwell's field and/or charge conservation equations and

(2) equations describing the particle motion or momentum transfer.

Adapted to the special purpose and the allowed degree of approximation these equations may be represented in different forms. Particularly for the particle motion different classes of theories are to be distinguished, using

(a) the equation of motion of a single particle (electron),
(b) the magnetohydrodynamic (generalized Navier–Stokes) equations of motion, and
(c) the Boltzmann or Vlasov equations.

These theories are related to the approximations of the *vacuum*, *cold plasma*, and *warm plasma*, respectively, as is shown by Table IV.3.
The basic equations are nonlinear in general, but a linearization is often made (in order to avoid mathematical difficulties) by the assumption of small perturbations, i.e. the assumption of small amplitudes of the **E** and **H** fields. In the past the single-particle and hydrodynamic theories have been widely applied for the treatment of the direct emission processes of radio waves. An extended survey of these matters was given by Zheleznyakov (1964, 1970, 1977) mainly in the framework of the cold plasma. The warm plasma theories are to be applied for the investigation of resonant and large-amplitude phenomena and different kinds of wave–wave interactions.

A plasma is called 'cold', if the thermal particle velocities can be neglected and accordingly the gas pressure p_g vanishes:

$$8\pi p_g/H^2 \ll 1 \tag{IV.39}$$

TABLE IV.3
Basic theories in plasma physics

Theory		Application
Single-particle theory	⟶	Vacuum approximation
	↘	
Hydrodynamic theory (Magneto-ionic theory)	⟶	Cold plasma
	↘	
Statistical theory (kinetic theory)	⟶	Warm plasma

(cf. e.g. Frank and Kamenetskij, 1964). This condition leads to the absence of all types of elastic waves. On the other hand, if thermal velocities and the gas pressure must be taken into account, the plasma is called 'warm' and elastic waves become possible.

If the electric conductivity is assumed to be infinite, the plasma is called 'ideal' in the hydrodynamic sense. In such a case of very high conductivity electromagnetic waves cannot propagate.

4.2.3. Polarized Radiation in a Plasma

a. *Basic Definitions*

The state of polarization is a characteristic property of transverse waves. It was already outlined in Section 2.4.1, that the intensity (or flux density) of such radiation is generally composed of a polarized and a randomly polarized part, whereas the degree of polarization is defined by the ratio of the polarized part to the whole intensity. The polarization is in general elliptical; linear and circular polarization are the limiting cases. The representation of the polarization by means of the Stokes parameters I, Q, U, and V was given by Equations (II.43) and (II.44). An alternative representation is given e.g. by the use of the intensity I_ν, the degree of polarization ρ, the axial ratio p, and an orientation angle χ of the polarization ellipse. The axial ratio p (defined as the ratio of the small to the large half main axes of the ellipse) ranges between 0 and 1 corresponding to linear and circular polarization, respectively (cf. Section 2.4.1), and is given by

$$p = \tan((1/2)\arcsin \rho_c/\rho) \tag{IV.40}$$

(ρ_c – degree of circular polarization).

As a basic property, elliptical polarization can be right or left handed (defined either with regard to the direction of the wave propagation or with regard to the magnetic field vector) as illustrated by Table II.2.

b. *Polarization of Radio Waves*

Considering radio waves, the axial ratio of the polarization ellipse is described by the ratios of the components of the wave vectors in Cartesian coordinates E_y/E_x, E_z/E_x, and H_y/H_x, H_z/H_x where z is the direction of the wave propagation (Ratcliffe, 1959).

Solving the dispersion relation two characteristic waves are marked for which these ratios are constant in the direction of propagation, satisfying the condition

$$P_x/E_x = P_y/E_y \tag{IV.41}$$

(P_x, P_y – components of the dielectric polarization vector). For these waves the axial ratio of the polarization ellipse is determined by the quantity

$$R = i/tg\sigma = E_x/E_y = -H_y/H_x = P_x/P_y\,. \tag{IV.42}$$

Using the expressions of the magneto-ionic theory

$$X = (\omega_p/\omega)^2, \qquad Y = \omega_H/\omega, \qquad Z = \nu_{\text{coll}}/\omega \tag{IV.43}$$

for both waves ($j = 1, 2$) it can be written

$$R_j = -\frac{i}{Y\cos\theta}\left\{\frac{Y^2\sin^2\theta}{2(1-X-iZ)} \mp \left[\frac{Y^4\sin^4\theta}{4(1-X-iZ)^2} + Y^2\cos^2\theta\right]^{1/2}\right\}$$

$$= -iT_j^{-1} \tag{IV.44}$$

where $\theta = \sphericalangle(\mathbf{k}, \mathbf{H})$.

Since $R_1 \times R_2 = 1$, both polarization ellipses have the same form but inverse orientations. Both ellipses are placed symmetrically to a straight line of 45° between the x- and y-axes. For $Z = 0$, R_j becomes imaginary. At a critical collision frequency corresponding to

$$Z_0 = Y^2\sin^2\theta/(2Y\cos\theta) \qquad \text{and } X = 1 \tag{IV.45}$$

both ellipses degenerate into straight lines. This is related to a critical angle θ_0 given by

$$\sin^2\theta_0/\cos\theta_0 = 2\nu_{\text{coll}}/\omega_H. \tag{IV.46}$$

More detailed discussions of special topics of polarized waves were given e.g. by Ratcliffe (1959), Heald and Wharton (1965), Krüger (1972), and many others.

The designation of the two electromagnetic wave modes is not always uniform in the literature. In Table IV.4 the wave modes are listed in dependence of the propagation angle θ and the related properties of R_j according to the notation of Stix (1962), which is widely used in plasma physics. In magneto-ionic theory and radio astronomy, however, often the notations extraordinary and ordinary waves are used throughout the whole range of angles θ. Here the sense of the circular polarization corresponds to that of the electronic gyration (cf. Figure II.7). In the case of linear polarization the vector of the **E**-field of the extraordinary wave is perpendicular to the external magnetic field direction, and the **E**-field vector of the ordinary wave is parallel to the external magnetic field.

TABLE IV.4

Designation of electromagnetic waves modes in a cold plasma according to Stix (1962)

Propagation angle	Modes	Notation	R_j	Polarization
$\theta = 0$	r, l	right handed, left handed	imaginary ($\pm i$)	circular
$0 < \theta < \frac{\pi}{2}$	s, f	slow, fast	complex	elliptical
$\theta = \frac{\pi}{2}$	$e(x), o$	extraordinary, ordinary	real ($\infty, 0$)	linear

Wave propagation in an inhomogeneous medium with varying parameters X, Y, and Z in general causes a change of the state of polarization. This can be illustrated by a representation of R_j in the complex half plane by conformal mapping of

$$\operatorname{sign}(\cos\theta)\, R_j = \frac{-iZ}{1-X-iZ}\left[\frac{Y\sin^2\theta}{2Z\,|\cos\theta|} \mp \operatorname{sign}(\cos\theta)\left\{\left(\frac{Y\sin^2\theta}{2Z\,|\cos\theta|}\right)^2 + \left(\frac{1-X-iZ}{Z}\right)\right\}^{1/2}\right] \tag{IV.47}$$

into the unit circle as proposed by Mollwo (1969). The real part of $[\operatorname{sign}(\cos\theta)\, R_j]$ is always positive. The mapping of the positive half plane of a rectangular coordinate system into the unit circle is realized by the complex function

$$w = \frac{\operatorname{sign}(\cos\theta)\, R_j - 1}{\operatorname{sign}(\cos\theta)\, R_j + 1} \tag{IV.48}$$

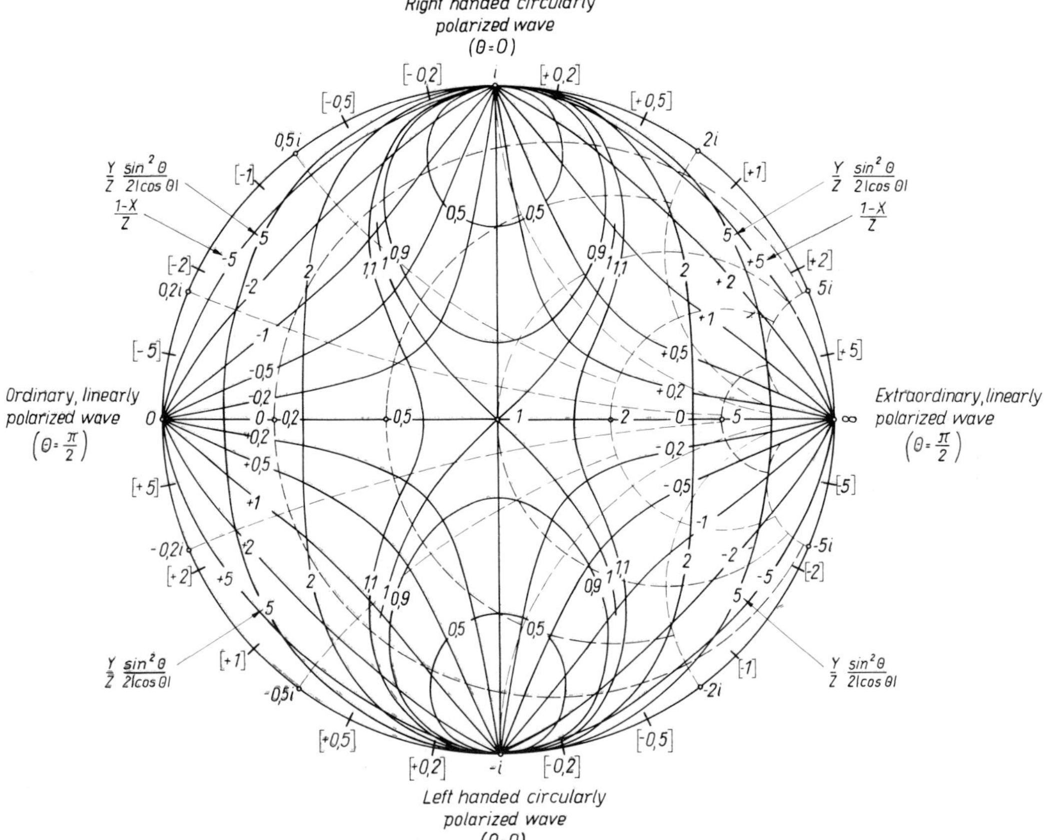

Fig. IV.7. Representation of the wave polarization R_j in the complex half plane by conformal mapping of sign $(\cos\theta) \times R_j$ into the unit circle after Mollwo (1969).

which is shown in Figure IV.7. There are represented curves of $(Y/Z) \times \sin^2\theta/(2\cos\theta) = \text{const}$ and $(1 - X)/Z = \text{const}$. The critical point (condition (IV.45)) corresponds to the center of the diagram. For every other combination of the plasma parameters X, Y, Z a set of two points exists, which are located symmetrically to the center according to $R_1 \times R_2 = 1$. The case $Z = 0$ is represented by the periphery of the unit circle and here one parameter $[Y/(1 - X)] \times \sin^2\theta/(2\cos\theta)$ suffices to describe the polarization. If $Z \neq 0$, this parameter and the quantity $(Y/Z) \times \sin^2\theta/\cos\theta$ define a series of orthogonal curves in the complex R-plane, which can be used for an alternative representation (Snyder and Helliwell, 1952). In Figure IV.7 R_j is <1 in the left half circle and >1 in the right half circle corresponding to ordinary and extraordinary waves, respectively. The upper and lower half circles correspond to right- and left-handed polarized waves, respectively, according to the definition 'c' of Table II.2.

In cases where the longitudinal component E_z of the wave vector is to be taken into account, a longitudinal polarization coefficient

$$G_j = E_z/E_x \equiv iK_j \tag{IV.49}$$

can also be defined, which reads explicitly

$$G_j = \\ = \frac{iXY\sin\theta}{(1 - X - iZ)\left\{1 - iZ - \dfrac{Y^2\sin^2\theta}{2(1 - X - iZ)} \pm \left[\dfrac{Y^4\sin^4\theta}{4(1 - X - iZ)^2} + Y^2\cos^2\theta\right]^{1/2}\right\}} \tag{IV.50}$$

c. *Faraday Effect*

If a linearly polarized plane wave is passing a magnetized, i.e. anisotropic plasma in the field direction, the plane of the linear polarization rotates (Faraday effect). The linearly polarized wave can be thought to consist of two counterrotating circularly polarized waves of equal amplitude but different velocities. The phase angle between these waves is changed by

$$\Delta\varphi = \frac{\omega}{c}\int_0^s (n_2 - n_1)\,\mathrm{d}s \tag{IV.51}$$

(n_j – refractive index, cf. next section). From this the rotation angle of a linear (or elliptical) wave can be calculated directly. For instance, in the case of the quasi-longitudinal approximation (cf. Section 4.2.4.b) one obtains

$$\Delta\chi = \frac{\Delta\varphi}{2} = \frac{1}{\omega^2 c}\int_0^s \frac{\omega_p^2\omega_H^2\cos\theta}{n_1 + n_2}\,\mathrm{d}s. \tag{IV.52}$$

Hence a large influence of the Faraday rotation on the propagation of radio waves in the solar atmosphere can be deduced (Hatanaka, 1956). The rotation is independent of the sense of the magnetic field direction (parallel or antiparallel to the wave-propagation direction) and the propagation is nonreciprocal.

If the two counterrotating components constituting a linearly polarized wave are differently attenuated, the polarization becomes elliptical. If one of the waves is fully absorbed, the remaining wave is circularly polarized and a Faraday rotation can no longer be observed. Since the position angle χ and its variation depend on the wave frequency in general, χ is dispersed throughout the spectrum. As a consequence, due to the finite bandwidth of the receiving equipment, the measured polarization is different from that at the source ('instrumental depolarization'). The dispersion angle ϑ' is a measure for the range in which the position angles χ are spread over the frequency band $\Delta\nu$ around a center frequency ν_0 ($\Delta\nu \ll \nu_0$):

$$\vartheta' = \frac{2\Delta\nu\chi}{\nu_0}. \tag{IV.53}$$

Another effect of the Faraday dispersion is the change of the shape of the polarization ellipse during the wave propagation through an anisotropic medium. An elliptically polarized wave tends to be changed into circular plus random polarization (cf. Akabane and Cohen, 1961).

4.2.4. Propagation of Radio Waves

a. *Ray Trajectories – Refraction and Scattering*

Within the range of the validity of the approximation of geometrical optics the calculation of ray trajectories has been proven to be a useful tool. Applications are made e.g. in ray-tracing programs computing the emerging radiation from a given emitting and propagating plasma volume. The main effects influencing the geometry of a ray path for solar radio waves are the refraction in a regular medium and the scattering.

The refraction is determinated by Snellius' law, which can be written in Cartesian coordinates as follows

$$n_j(z)\sin\varphi(z) = \text{const}, \tag{IV.54}$$

where $n_j = c/v_{kj}$ – refractive index of the jth wave mode, φ – angle of incidence between $\mathbf{k}_j$ and grad n_j ($= \sphericalangle(\mathbf{k}_j, z)$ for a plane stratified medium).

If spherical symmetry is present, Snellius' law is conveniently written in spherical coordinates

$$n_j(r)\, r \sin\varphi(r) = \text{const} \tag{IV.55}$$

($\mathbf{r}$ – radius vector, $r = |\mathbf{r}|$, $\varphi = \sphericalangle(\mathbf{k}_j, \mathbf{r})$). For $n_j \to 1$ and $r\sin\varphi \to p^*$ Equation (IV.55) becomes

$$n_j(r)\, r \sin\varphi(r) = p^*$$

(p^* – ray parameter, cf. Figure IV.8).

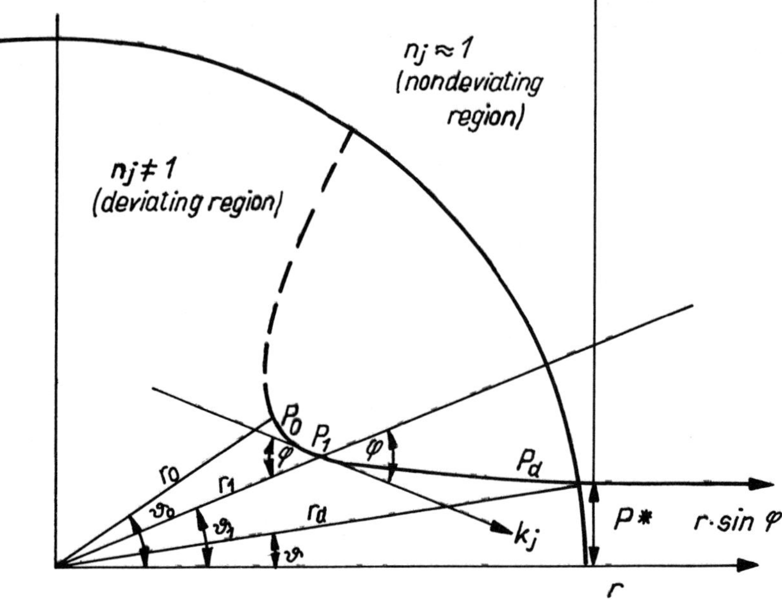

Fig. IV.8. Geometry of ray trajectories in a spherically symmetric region.

The equation of a ray path can be written as

$$\vartheta = p^* \int_0^\infty \frac{\mathrm{d}r}{r(n_j^2 r^2 - p^{*2})^{1/2}}$$

or

$$\vartheta = \int_0^u \frac{\mathrm{d}u}{(n_j^2/p^* - u^2)^{1/2}} \tag{IV.56}$$

where $u = 1/r$.

In practice it is sufficient to restrict the integration to a distance $r = r_d$ which is the radius of the 'deviating region', i.e. that region where the refractive index differs remarkably from unity.

At the turning point $P_0(r_0, \vartheta_0)$ of the ray trajectory one obtains

$$r_0 = p^*/n_{j,0} \qquad (\text{or } u_0 = n_{j,0}/p^*) \tag{IV.57}$$

and the path element is given by

$$\mathrm{d}s = \frac{r_0 \, \mathrm{d}r}{\left(1 - \dfrac{p^{*2}}{n_j^2 r^2}\right)^{1/2}} . \tag{IV.58}$$

In the simple case of spherical symmetry the ray trajectories are symmetrical to both sides of the radius vector through the turning point P_0. The first calculations concerning the quiet Sun applying this formalism were carried out by Burkhardt and Schlüter (1949), Jaeger and Westfold (1950), and Smerd (1950). More recently, a similar ray-tracing method adopted to microwave burst sources was also applied e.g. by Böhme *et al.* (1976).

Besides refraction, another process to be considered here is the scattering of radio waves. The scattering is particularly important for the study of solar phenomena in the meter and decameter range, e.g. of drift bursts and noise storms, but also of the scintillation of nonsolar radio sources caused by coronal inhomogeneities, etc. Scattering is a typical effect of inhomogeneous (fluctuating) media. Depending on the scale of the inhomogeneity in comparison with the wavelength, refraction and diffraction take part in the scattering.

Scattering processes are usually treated by statistical methods (cf. e.g. Scheffler, 1958, 1959; Uscinski, 1968). But since the aspects of scattering are manifold, we now pick up only those questions which are of interest from the view of ray paths.

Beginning with the work of Fokker (1965, 1967) models with isotropic inhomogeneities and effects of the directivity of point sources on the observed radiation were studied first. Later the regular refraction by a spherically symmetric corona was included (Steinberg *et al.*, 1971; Caroubalos *et al.*, 1972) and also extended to a consideration of radially elongated inhomogeneities (Riddle, 1972a). Deviations from the spherical symmetry were considered by Riddle (1972b).

An important result from this kind of investigation, which is characterized by numerical calculations in connection with a statistical treatment, is the following: If regular refraction in a spherically symmetric corona only is considered, the turning points of the ray trajectories are elevated above the local plasma level on the Sun except for the central ray. If scattering is now taken into account, the radiation can escape from the plasma level at positions almost anywhere on the Sun.

b. *Dispersion Relations*

In the foregoing section the knowledge of the refractive index

$$n_j = ck_j/\omega = c/v_{k,j} \tag{IV.59}$$

($v_{k,j}$ – phase velocity of the jth wave mode) was needed. This knowledge is obtained from the dispersion relation. In the cold plasma referring to radio waves we apply the *Appleton–Hartree formula*

$$n_j^2 = 1 - \frac{2X(1-X-iZ)}{2(1-iZ)(1-X-iZ) - Y_T^2 \mp [Y_T^4 + 4Y_L^2(1-X-iZ)^2]^{1/2}} \tag{IV.60}$$

where $Y_T = \omega_H \sin\theta/\omega$, $Y_L = \omega_H \cos\theta/\omega$, $j = 1$ (upper sign) and $j = 2$ (lower sign) refer to the extraordinary and ordinary (fast and slow) wave modes, respectively.

In a collisionless plasma ($Z = 0$) we have

$$n_j^2 = 1 - \frac{2X(1-X)}{2(1-X) - Y_T^2 \mp [Y_T^4 + 4(1-X)^2\, Y_L^2]^{1/2}}. \tag{IV.61}$$

Examples of such dispersion curves are shown in Figure IV.9.

Further simplifications are obtained in an anisotropic plasma by the approxima-

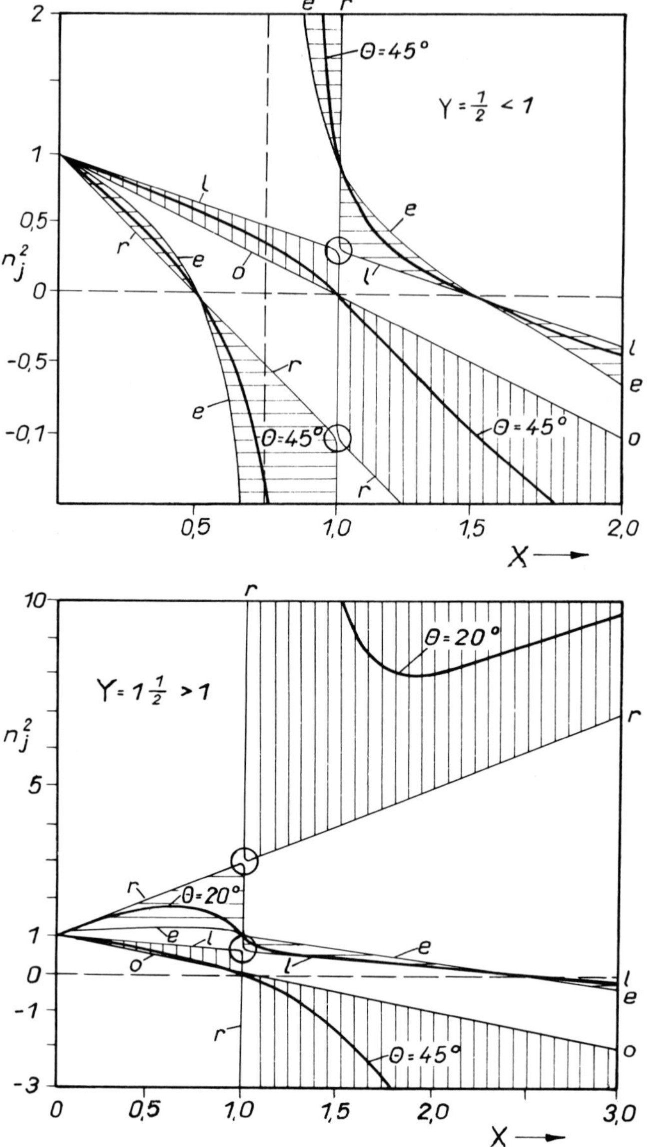

Fig. IV.9. Refractive index for electromagnetic waves in a cold and collisionless plasma for $Y < 1$ (top) and $Y > 1$ (bottom).

tions of 'quasi-longitudinal' (QL) (or 'quasi-circular') and 'quasi-transversal' (QT) (or 'quasi-linear') propagation (Ratcliffe, 1959; Rawer and Suchy, 1967). The conditions for the applicability of these approximations are

$$\text{QL:} \quad Y_T^4/4Y_L^2 \ll |(1 - X - iZ)^2|$$

$$\text{QT:} \quad Y_T^4/4Y_L^2 \gg |(1 - X - iZ)^2| \tag{IV.62}$$

or numerically, in the absence of collisions

$$\text{QL:} \quad |Y_L|/3 > Y_T^2/[2|(1 - X)|]$$
$$\text{QT:} \quad 3|Y_L| < Y_T^2/[2|(1 - X)|]. \tag{IV.63}$$

The simplest case of an isotropic plasma is represented by

$$n_j^2 = 1 - X. \tag{IV.64}$$

A compilation of several special cases of the refractive index is given in Table IV.5.

Equation (IV.60) contains certain singularities (poles) and zeros of the refractive index, which describe the resonance and cutoff characteristics of the cold plasma for radio waves. For arbitrary angles θ three zeros exist, which are reduced to two zeros if $\theta = 0$, and only one zero if $Y = 0$. The poles (resonances) of the refractive index are described by the relation

$$X = (1 - iZ)[(1 - iZ)^2 - Y^2]/[(1 - iZ)^2 - Y^2\cos^2\theta]. \tag{IV.65}$$

The zeros and poles are compiled in Table IV.6.

There are different possible forms of a representation of dispersion relations. For instance, contour maps of n_j and polar plots of $(n_j)^{-1} = v_{k,j}/c$ representing phase-velocity or wave-normal surfaces are instructive means. A detailed treatise on this subject can be found in Heald and Wharton (1965). The topology of the wave-normal

TABLE IV.5

Special cases of the dispersion relation for electromagnetic modes in a cold plasma

	$Y = 0$	$Y_T = 0$ ($\omega_H \sin\theta \ll \omega$)	$Y_L = 0$ ($\omega_H \cos\theta \ll \omega$)
$Z = 0$ $\left(\nu_{\text{coll}} \ll \dfrac{\omega}{2\pi}\right)$	$n^2 = 1 - X$	$n_j^2 = 1 - \dfrac{X}{1 \mp Y_L}$	$n_1^2 = 1 - \dfrac{X(1 - X)}{1 - X - Y_T^2}$ $n_2^2 = 1 - X$
$Z \neq 0$	$n^2 = 1 - \dfrac{X}{1 - iZ}$	$n_j^2 = 1 - \dfrac{X}{1 \mp Y_L - iZ}$	$n_1^2 = 1 - \dfrac{X(1 - X - iZ)}{(1 - iZ)(1 - X - iZ) - Y_T^2}$ $n_2^2 = 1 - \dfrac{X}{1 - iZ}$
	$X = 0$ ($\omega_P \ll \omega$)	$QL: \dfrac{Y_T^4}{4Y_L^2} \ll \lvert(1 - X - iZ)^2\rvert$	$QT: \lvert(1 - X - iZ)^2\rvert \ll \dfrac{Y_T^4}{4Y_L^2}$
	$n = 1$	$n_j^2 = 1 - \dfrac{X}{1 \mp Y_L - iZ}$	$n_1^2 = 1 - \dfrac{X(1 - X - iZ)}{(1 - iZ)(1 - X - iZ) - Y_T^2}$ $n_2^2 = 1 - \dfrac{X}{1 - iZ + (1 - X - iZ)\, Y_L^2/Y_T^2}$

TABLE IV.6

Zeros and poles of the refractive index for electromagnetic wave modes in a cold plasma

	Critical frequencies $n_j = 0$	Resonances ($n_j = \infty$) $\theta = 0$	$\theta = \frac{\pi}{2}$
$Z = 0$	$X = 1, \omega = \omega_p, \omega_H = 0$ $X = 1 \mp Y_L$ $\omega_j = \sqrt{\omega_p^2 + \frac{\omega_H^2}{4}} \pm \frac{\omega_H}{2}$	$[X = 1, \omega = \omega_p]$ $Y_L = 1, \omega = \omega_H$	$X = 1$ $X = 1 - Y_T^2$ $\omega_{uh} = \sqrt{\omega_p^2 + \omega_H^2}$
$Z \neq 0$	$X = 1 - iZ$ $\omega_j = \frac{i\nu_{coll}}{2} \pm \sqrt{\omega_p^2 - \frac{\nu_{coll}^2}{4}}$ $X = 1 - iZ \mp Y_L$ $\omega_j = \frac{i\nu_{coll} \mp \omega_H}{2} \pm$ $\pm \sqrt{\omega_p^2 + \frac{(i\nu_{coll} \mp \omega_H)^2}{4}}$	$X = 1 - iZ$ $\omega_j = \frac{i\nu_{coll}}{2} \pm \sqrt{\omega_p^2 - \frac{\nu_{coll}^2}{4}}$	$X = \frac{(1 - iZ)^2 - Y_T^2}{1 - iZ}$ $\omega_j = i\nu_{coll} \pm \sqrt{\omega_p^2 + \omega_H^2 - \nu_{coll}^2}$ $(Z \ll 1)$

surfaces with respect to the propagation angle θ of different wave modes is nicely demonstrated by the so-called *CMA-diagrams* (named after Clemmow, Mullaly, and Allis; cf. Stix, 1962; Allis *et al.*, 1963). In this way, in the form of a parameter diagram a synoptic picture of the possible wave modes, the shapes of the wave-normal surfaces, as well as of the resonances and cutoffs of the refractive index in different parameter regions is provided. Figure IV.10 gives an example of such a CMA-diagram in the cold plasma.

Turning to the warm plasma, the matter becomes more complicated. Here, in general, the basic dispersion equation describing the totality of the wave modes can no longer be factored into separate dispersion relations as in the case of the cold plasma. An exception is possible in the case of the isotropic plasma. But in general the anisotropy caused by the presence of an external magnetic field prevents the factorization of the dispersion relation into different parts describing pure transverse and longitudinal wave modes (which now come into play), except for very special propagation angles θ $(0, \pi/2)$ and wave modes. As an approximation only, the conditions

$$\mathbf{k} \cdot \mathbf{E} = 0, \quad \text{or} \quad \mathbf{k} \times \mathbf{E} = 0 \qquad \text{(IV.66–IV.67)}$$

properly applying to pure transverse and longitudinal waves, respectively, are used distinguishing both wave types with a more or less sufficient accuracy.

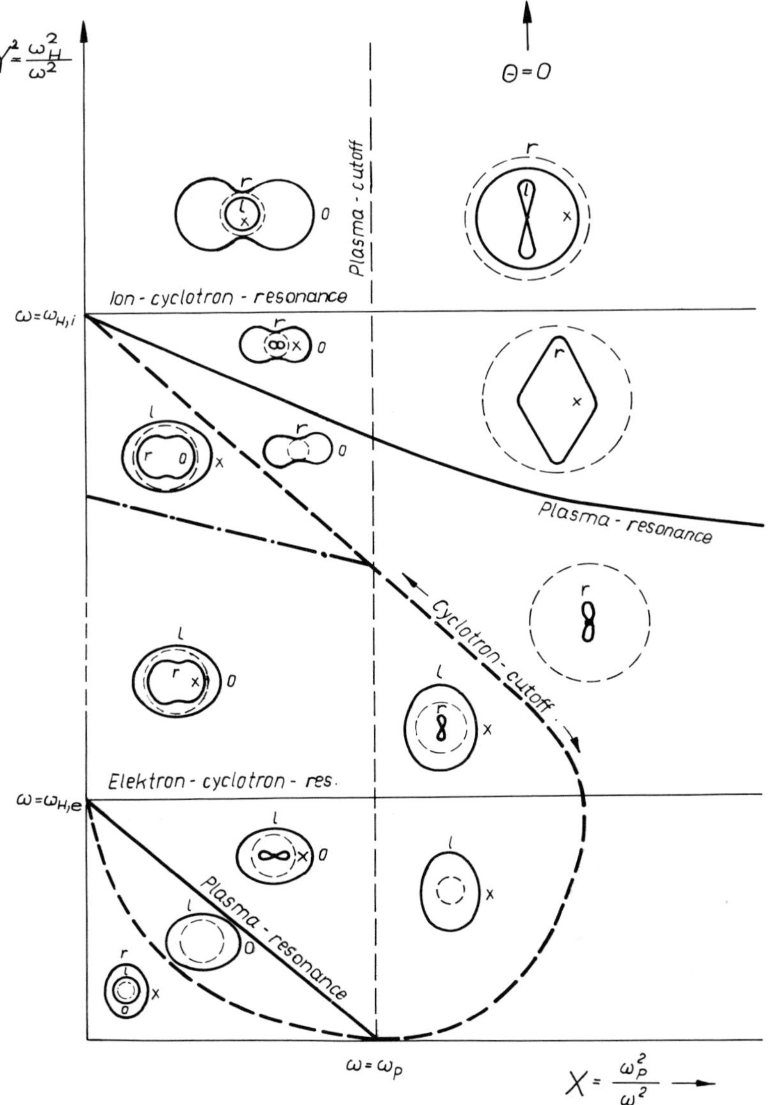

Fig. IV.10. CMA-diagram for a cold two-component magnetoplasma.

Another point to be noted is, that in contrast to the cold plasma, the warm plasma implies spatial dispersion. This means that the refractive index n_j now becomes a function of both ω and $\mathbf{k}$, so that dispersion both in time and space must be considered.

c. *Radiative Transfer*

The theory of radiative transfer, originally developed for the study of visible radiation from stellar atmospheres (cf. e.g. the textbook of Chandrasekhar, 1960), can be applied also for the spectral range of radio waves. Here the main complications, which make the work difficult in the optical range, are widely omitted: In the radio region, at least for the direct emission processes, the absorption and emission coefficients $\alpha_{\nu,j}$ and $j_{\nu,j}$ do not depend on the intensity $I_{\nu,j}$. On the other hand, the transfer of polarized radiation, which is of interest in radio physics, implies some complications demanding a tensor theory of radiation transfer.

Considering at first the intensity $I_{\nu,j}$, its exchange in a volume element along a path s is given by the equation of radiative transfer, which can be written for a refractive, nondispersive medium in the following form

$$n_j^2\, \mathrm{d}/\mathrm{d}s(I_{\nu,j}/n_j^2) = -\alpha_{\nu,j} I_{\nu,j} + j_{\nu,j}. \tag{IV.68}$$

This equation can be regarded as a one-dimensional approximation of a more general tensor relation, which will be considered below. If the medium is dispersive, the group velocity

$$\mathbf{v}_{g,j} = \mathrm{d}\omega/\mathrm{d}\mathbf{k}_j \tag{IV.69}$$

differs from the phase velocity $\mathbf{v}_{k,j}$, and the ray direction, i.e. the direction of the energy flux (Poynting vector), is different from the wave-normal direction. In this case the term n_j^2 is to be replaced by the expression $n_j^2 \cos\vartheta_j$, where

$$\vartheta_j = \sphericalangle(\mathbf{k}_j, \mathbf{v}_{g,j})$$

is the angle between the ray direction and wave-normal direction (cf. Figure IV.11).

Introducing the optical depth

$$\tau_{\nu,j} = \int_0^\infty \alpha_{\nu,j}\, \mathrm{d}s \tag{IV.70}$$

the equation of radiative transfer becomes

$$n_j^2 |\sec\vartheta_j| \frac{\mathrm{d}}{\mathrm{d}\tau_{\nu,j}} \left(\frac{I_{\nu,j}\cos\vartheta_j}{n_j^2} \right) = \frac{j_{\nu,j}}{\alpha_{\nu,j}} - I_{\nu,j}. \tag{IV.71}$$

The term $j_{\nu,j}/\alpha_{\nu,j}$ is called the *source function* or ergiebigkeit.

In the very simple case of thermal equilibrium the emission and absorption coefficients are connected by Kirchhoff's law

$$j_{\nu,j}/\alpha_{\nu,j} = I_\nu(\nu, T) \tag{IV.72}$$

and the source function is the Kirchhoff–Planck function

$$B_\nu(T) = 2\nu^2 KT/c^2 \tag{IV.73}$$

(K – Boltzmann's constant).

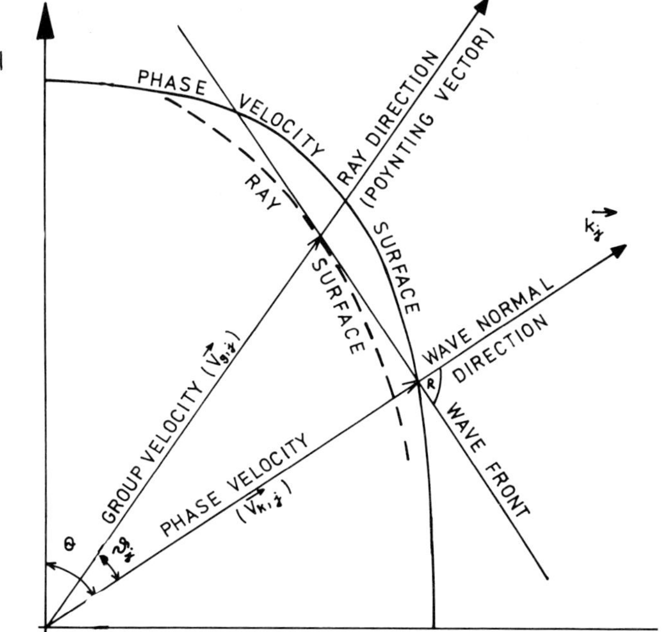

Fig. IV.11. On the geometry of the ray and wave-normal directions in a dispersive medium.

In the case $n_j = 1$ and $\cos\vartheta_j = 1$, one obtains the well known solution

$$I_\nu(\nu, T) = I_{\nu,0}e^{-\tau_\nu} + e^{-\tau_\nu}\int_0^{\tau_\nu}\frac{2\nu^2 KT}{c^2}e^{\tau'_\nu}\,\mathrm{d}\tau'_\nu \tag{IV.74}$$

where the index j has been omitted for simplicity.

Hence for an isothermal homogeneous layer with the optical depth τ_ν there is obtained

$$I_\nu(\nu, T) = I_{\nu,0}e^{-\tau_\nu} + (2\nu^2 KT/c)(1 - e^{-\tau_\nu}). \tag{IV.75}$$

Equations (IV.74) and (IV.75) can be also expressed in terms of the brightness temperature T_s:

$$T_s = T_{s,0}e^{-\tau_\nu} + e^{-\tau_\nu}\int_0^{\tau_\nu} T_s e^{\tau'_\nu}\,\mathrm{d}\tau'_\nu$$

$$T_s = T_{s,0}e^{-\tau_\nu} + T(1 - e^{-\tau_\nu}).$$

In the limiting cases of very large and very small optical depths, respectively, it follows:

(a) $\tau_\nu \gg 1$

$$I_\nu = B_\nu(T), \qquad T = T_s \tag{IV.76}$$

(b) $\tau_\nu \ll 1$

$$I_\nu = B_\nu(T)\cdot\tau_\nu + I_{\nu,0}(1-\tau_\nu)$$

$$T_s = T\cdot\tau_\nu + T_{s,0}(1-\tau_\nu). \tag{IV.77}$$

Considering now the transfer of polarized radiation, for each wave mode four equations are required instead of one for unpolarized radiation. In this case a description making use of the Stokes parameters is convenient. The general equation of radiative transfer of polarized radio waves in a magnetized plasma was described by Kawabata (1964). Later, Zheleznyakov (1968) gave some corrections to Kawabata's results. Details of the transport of polarized radiation through an anisotropic ('magnetoactive') plasma were investigated by Zheleznyakov (1966b), Sazonov and Tsytovich (1968), Sazonov (1969a, b), Zheleznyakov *et al.* (1974), and others. Reviews are due to Ginzburg *et al.* (1968), Zheleznyakov (1968), and Fung (1969b).

The Stokes parameters describing the state of polarization for a given wave mode in a particular frequency interval at any point of the space can be expressed by the components of a polarization tensor I_{iq} which is defined by

$$I_{iq} = \langle E_i E_q^* \rangle$$

or, more conveniently, by

$$I_{iq} = \langle D_i D_q^* \rangle \tag{IV.78}$$

where $i, q = x, y$ are Cartesian coordinates $(z \| \mathbf{k}_j)$, E_i, E_q – electric wave field components, D_i, D_q – the corresponding components of the electric induction tensor (which remains transverse in an anisotropic plasma, $D_z = 0$). The brackets $\langle\ \rangle$ denote averages over a time interval $\Delta t \gg 1/\Delta\omega$, so that I_{iq} becomes widely independent of time.

In this representation the Stokes parameters are given by

$$I_{iq} = \frac{1}{2}\begin{pmatrix} I+Q & U-iV \\ U+iV & I-Q \end{pmatrix} \tag{IV.79}$$

where

$$I = I_{xx} + I_{yy}$$

$$Q = I_{xx} - I_{yy}$$

$$U = I_{yx} + I_{xy}.$$

$$V = i(I_{yx} - I_{xy})$$

Accounting for emission, absorption, and Faraday rotation, but neglecting refraction, the equation of transfer of polarized radiation can be written in the following form

$$\mathrm{d}I_{iq}/\mathrm{d}z = S_{iq} - A_{iqlm}I_{lm} + R_{iqlm}I_{lm}, \tag{IV.80}$$

where the tensor S_{iq} represents the generalized source function and the fourth-rank tensors A_{iqlm} and R_{iqlm} describe the absorption and Faraday rotation, respectively.

Explicitly it can be written

$$A_{xxxx} = 2(b_1^2\alpha'_{v,1}(z) + b_2^2\alpha'_{v,2}(z)],$$
$$A_{yyyy} = 2[b_2^2\alpha'_{v,1}(z) + b_1^2\alpha'_{v,2}(z)],$$
$$A_{xxyy} = A_{yyxx} = A_{xyyx} = A_{yxxy} = 0,$$
$$A_{xyxy} = A_{yxyx} = \alpha'_{v,1}(z) + \alpha'_{v,2}(z),$$
$$A_{xyyy} = -A_{yxxx} = A_{xyxx} = -A_{yyxy} = A_{xxyx} = -A_{xxxy} = A_{yyyx} =$$
$$= -A_{yxyy} = ib_1b_2[\alpha'_{v,2} - \alpha'_{v,1}]. \qquad \text{(IV.81)}$$

$$R_{xxxx} = R_{yyyy} = R_{xxyy} = R_{xyyx} = R_{yxxy} = 0,$$
$$R_{xyxy} = -R_{yxyx} = i(b_1^2 - b_2^2)[k_1(z) - k_2(z)],$$
$$R_{xxxy} = R_{xxyx} = -R_{yyxy} = -R_{yyyx} = -R_{xyxx} = R_{xyyy} = -R_{yxxx} =$$
$$= R_{yxyy} = b_1b_2[k_1(z) - k_2(z)], \qquad \text{(IV.82)}$$

where

$$b_1 = (A + B)[2Y_L^2 + B^2 + AY_T^2/(1 - X)]^{-1/2},$$
$$b_2 = Y_L[2Y_L^2 + B^2 + AY_T^2/(1 - X)]^{-1/2},$$
$$A = (Y_L^2 + B^2)^{1/2},$$
$$B = Y_T^2/[2(1 - X)], \qquad \text{(IV.83)}$$

$\alpha'_{v,j}$ – amplitude absorption coefficients, k_j – wave numbers (Fung, 1969b).
Hence it follows

$$\mathrm{d}I/\mathrm{d}z = S_{xx} + S_{yy} - A_{xxxx}I_{xx} - A_{yyyy}I_{yy} + 2iA_{xyyy}V,$$
$$\mathrm{d}Q/\mathrm{d}z = S_{xx} - S_{yy} - A_{xxxx}I_{xx} + A_{yyyy}I_{yy} + 2R_{xxxy}U,$$
$$\mathrm{d}U/\mathrm{d}z = S_{yx} + S_{xy} - 2A_{xyyy}I_{xx} + 2R_{xxxy}I_{yy} - A_{xyxy}U,$$
$$\mathrm{d}V/\mathrm{d}z = i(S_{yx} - S_{xy}) + 2iA_{xyyy}(I_{xx} + I_{yy}) - A_{xyxy}V. \qquad \text{(IV.84)}$$

In the case of the *QL*-approximation one obtains

$$\mathrm{d}I/\mathrm{d}z = S_I - (\alpha'_1 + \alpha'_2)\,I + (\alpha'_1 - \alpha'_2)\,V,$$
$$\mathrm{d}Q/\mathrm{d}z = S_Q - (\alpha'_1 + \alpha'_2)\,Q + (k_1 - k_2)\,U,$$
$$\mathrm{d}U/\mathrm{d}z = S_U - (\alpha'_1 + \alpha'_2)\,U - (k_1 - k_2)\,Q,$$
$$\mathrm{d}V/\mathrm{d}z = S_V - (\alpha'_1 + \alpha'_2)\,V + (\alpha'_1 - \alpha'_2)\,I, \qquad \text{(IV.85)}$$

where the index v has been omitted for simplicity and the terms S_I, S_Q, S_U, S_V abbreviate the corresponding expressions for the source function.

The solutions for $S_{iq} = 0$ are

$$I = (1/2)(I_0 - V_0)\exp(-2\alpha_1' z) + (1/2)(I_0 + V_0)\exp(-2\alpha_2' z),$$
$$Q = \{Q_0\cos[(k_1 - k_2)z] + U_0\sin[(k_1 - k_2)z]\}\exp[-(\alpha_1' + \alpha_2')z],$$
$$U = \{U_0\cos[(k_1 - k_2)z] - Q_0\sin[(k_1 - k_2)z]\}\exp[-(\alpha_1' + \alpha_2')z],$$
$$V = (1/2)(V_0 - I_0)\exp(-2\alpha_1' z) + (1/2)(I_0 + V_0)\exp(-2\alpha_2' z) \quad \text{(IV.86)}$$

(Zheleznyakov, 1968).

4.3. Single-Particle Approximation: An Account of Direct Radio Emission Mechanisms

4.3.1. Coulomb Bremsstrahlung

a. *Thermal Free–Free Emission and Absorption*

The first basic process to be discussed in the framework of the single-particle approximation is the bremsstrahlung caused by an acceleration of charged particles (electrons) in the Coulomb field of other charged particles (protons, ions). Though there are different possibilities of encounters between various particles, the free–free transitions, particularly of electrons in the field of protons or positive ions, are of major interest for solar radio astronomy. This process can be understood as a kind of inelastic collision between the interacting particles.

In contrast to the gyro-synchrotron emission, the bremsstrahlung process is expected to dominate in regions where the magnetic fields are weak or missing. The mechanism is therefore predestined to be applied especially to the quiet-Sun emission and to the S-component outside the magnetic core regions. Modifications of the emission are to be anticipated near the gyrofrequency and its harmonics (so far as magnetic fields are considered), and near the plasma frequency, which, however, are beyond the scope of a mere single-particle description.

Calculations of the bremsstrahlung emission and absorption coefficients were carried out by several authors and are contained in many textbooks. Here we cite the work of Kramers (1923), Elwert (1948), Smerd and Westfold (1949), Denisse (1950), Oster (1959, 1961a, b, 1964), Scheuer (1960), as well as Landau and Lifshits (1962) and Shklovsky (1962). The methods and results of the various calculations are slightly different concerning the following aspects:

- Use of either a classical or quantum treatment (in particular adoption of the classical Lorentz theory or not),
- Primary derivation of the emission coefficient or of the absorption coefficient,
- Application of different ranges of the integration over the collision parameter and different frequency ranges, assumption of linear or hyperbolic electron orbits, etc.
- Assumption of thermal equilibrium or not.

The easiest approach is given by a calculation of the local rate of the spontaneous emission of a single electron as a function of its initial velocity and the closest distance (collision parameter) to the deviating ion (proton). The rate of emission is given in the classical theory by the acceleration of the electron in its hyperbolic path around the deflecting ion. The Fourier components of the acceleration yield the emitted spectrum as a function of the input parameters via Hertz's formula of the radiation of an electron. Averaging over a postulated velocity distribution delivers the radiation from all electrons in one volume element.

If thermal equilibrium is considered, the absorption coefficient can be obtained from Kirchhoff's law. This approach was widely adopted in the older literature.

Another possibility consists in the direct approximative calculation of the absorption coefficient e.g. by adoption of the Lorentz theory considering the perturbation of the wave motion of an electron under the influence of Coulomb collisions with positive ions. Then the applied equation of motion of an 'average' electron contains a 'friction' term, which is assumed to be proportional to the relative velocity of the electron turning out to be essentially a measure for the mean collision frequency (or free-path length). There results a complex dielectric constant, the imaginary part of which gives directly the wanted absorption coefficient.

Among the various developed formulae, which in principle differ more or less obviously in the expression for the Coulomb logarithm, we cite the equations of Oster (1961b) of the bremsstrahlung emission coefficient calculated by classical and quantum methods, respectively:

$$j_\omega^{(\text{class})} = N_i N_e \frac{8Z^2 e^6}{3\pi\sqrt{2\pi m_e^2 c^3}} \sqrt{\frac{m_e}{KT}} \ln\left\{\left(\frac{2KT}{\gamma m_e}\right)^{3/2} \frac{m_e}{\gamma Z e^2 \pi \nu}\right\} \qquad \text{(IV.87)}$$

$$j_\omega^{(\text{quant})} = N_i N_e \frac{8Z^2 e^6}{3\pi\sqrt{2\pi m_e^2 c^3}} \sqrt{\frac{m_e}{KT}} \ln\left\{\frac{4KT}{\gamma \hbar\omega}\right\} \qquad \text{(IV.88)}$$

where N_i, N_e – ion and electron number density, respectively, Z – charge number, e, m_e – charge and mass of an electron, respectively, c – speed of light, K – Boltzmann's constant, T – temperature, $\gamma = e^{\gamma^+} = 1.781$ ($\gamma^+ = 0.577$) Euler's constant, $\hbar = h/2\pi$ – Planck's constant.

Equations (IV.87) and (IV.88) were derived under the following assumptions:

(a) Validity of the approximation of binary encounters (multiple encounters are neglected; a justification for such a procedure was given by Scheuer (1960)).

(b) The energy loss of the electrons due to radiation should be negligible, i.e. the photon energy $h\nu$ and momentum $\hbar k$ are much less than the particle energy E and momentum p, respectively.

(c) The refractive index is unity.

It was first shown by Elwert (1948), that the quantum mechanical formula is well approximated by the classical formula. The classical treatment is permissible for radio astronomical applications, if the temperature is less about 5.5×10^5 K (Oster, 1961b).

From Equations (IV.87) and (IV.88) the absorption coefficient is easily derived by Kirchhoff's law, providing thermal equilibrium is assumed:

$$\alpha_\omega^{(\text{class})} = \frac{j_\omega^{(\text{class})}}{B_\omega} = \frac{N_e N_i}{\omega^2} \frac{32\pi^2 Z^2 e^6}{3\sqrt{2\pi m_e^3 c}} \left(\frac{m_e}{KT}\right)^{3/2} \ln\left\{\left(\frac{KT}{\gamma m_e}\right)^{3/2} \frac{m_e}{\gamma Z e^2 \omega}\right\} \tag{IV.89}$$

$$\alpha_\omega^{(\text{quant})} = \frac{j_\omega^{(\text{quant})}}{B_\omega} = \frac{N_e N_i}{\omega^2} \frac{32\pi^2 Z^2 e^6}{3\sqrt{2\pi m_e^3 c}} \left(\frac{m_e}{KT}\right)^{3/2} \ln\left\{\frac{4KT}{\gamma\hbar\omega}\right\}. \tag{IV.90}$$

(For deviations from the thermal equilibrium cf. Subsection c.) A more detailed survey of these matters considering asymptotic expressions for the Coulomb logarithm for different ranges of temperature and frequency was given by Bekefi (1966; p. 135).

b. *Influence of External Magnetic Fields, Polarization Effects*

In the presence of an external magnetic field the Coulomb bremsstrahlung is modified. For a growing ratio Y/Z the bremsstrahlung changes gradually into the gyro-synchrotron emission. A qualitative description of the influence of magnetic fields may be given by the cold-plasma refractive index n_j, which is given in the framework of the magneto-ionic theory by Equation (IV.60):

$$n_j = \mu_j - i\chi_j \tag{IV.91}$$

(μ_j – real part of the refractive index, χ_j – imaginary part, which is also called attenuation index or *absorption index*).

As indicated in the previous section, the absorption coefficient can be calculated from the absorption index by the relation

$$\alpha_{\omega,j} = \frac{\omega}{c}\chi_j \tag{IV.92}$$

which follows immediately from a consideration of the equation of the oscillation of the wave E-vector, e.g. in the form

$$E_x = E_{x,0} \exp\left[i\omega(t - n_j z/c)\right] \tag{IV.93}$$

where the term $\exp(-\omega\chi_j z/c)$ describes the spatial absorption.

It should be noted, that because $I_\omega \sim E^2$ sometimes the twofold value of Equation (IV.92) is also defined as the absorption coefficient by some authors.

Further, in dispersive media the right side of Equation (IV.92) is to be multiplied with the term $\cos\vartheta_j$.

In the case of wave propagation into the direction of an external magnetic field (QL-approximation) one obtains from

$$n_j^2 = 1 - X/(1 \mp |Y_L| - iZ) \tag{IV.94}$$

the expression

$$\alpha_{\omega,j} = \frac{\omega}{2c} \frac{XZ}{\left[(1 \mp |Y_L|)^2 + Z^2\right]\mu_j}. \tag{IV.95}$$

For $Z \ll 1$ this reduces to

$$\alpha_{\omega,j} = \frac{\omega}{2c\mu_j} \frac{XZ}{(1 \mp |Y_L|)^2}\,. \tag{IV.96}$$

Similarly for the QT-approximation (and assuming $Z \ll 1$, $X \ll 1$) one finds

$$\alpha_{\omega,1} = \frac{\omega}{2c\mu_j} \frac{XZ(3 - Y_T^2)}{(1 - X - Y_T^2)^2} \tag{IV.97a}$$

$$\alpha_{\omega,2} = \frac{\omega}{2c\mu_j} \frac{XZ}{(1 - Z^2)}\,. \tag{IV.97b}$$

Discussing now the polarization characteristics we take into consideration that

$$I_\omega = I_{\omega,1} + I_{\omega,2}\,. \tag{IV.98}$$

Then, assuming wave propagation in an isothermal, homogeneous plasma in the direction of the external magnetic field, the degree of circular polarization is given by

$$\begin{aligned} \rho_c &= \frac{I_{\omega,1} - I_{\omega,2}}{I_{\omega,1} + I_{\omega,2}} \\ &= \frac{e^{-\tau_{\omega,2}} - e^{-\tau_{\omega,1}}}{2 - e^{-\tau_{\omega,1}} - e^{-\tau_{\omega,2}}} \\ &= \frac{\sinh \Delta\tau_\omega}{e^{\bar{\tau}_\omega} - \cosh \Delta\tau_\omega} \end{aligned} \tag{IV.99}$$

where

$$\tau_{\omega,j} = \int_0^z \alpha_{\omega,j}\, \mathrm{d}z',$$

$$\bar{\tau}_\omega = (\tau_{\omega,1} + \tau_{\omega,2})/2$$

$$\Delta\tau_\omega = (\tau_{\omega,1} - \tau_{\omega,2})/2$$

(Gelfreikh, 1962).

If $Z \ll 1$, it follows

$$\tau_{\omega,1} = \frac{\tau_\omega^0}{(1 - |Y_L|)^2}, \qquad \tau_{\omega,2} = \frac{\tau_\omega^0}{(1 + |Y_L|)^2}$$

$$\bar{\tau}_\omega = \tau_\omega^0 \frac{1 + |Y_L|}{(1 - |Y_L|)^2}, \qquad \Delta\tau_\omega = \tau_\omega^0 \frac{2|Y_L|}{(1 - |Y_L|)^2} \tag{IV.100}$$

where τ_ω^0 denotes the optical depth in the case of absence of an external magnetic field.

For $\tau_\omega^0 \ll 1$ it follows that

$$\rho_c = 2Y_L/(1 + Y_L^2) \tag{IV.101}$$

which for $|Y_L| \ll 1$ reduces to

$$\rho_c = 2Y_L. \tag{IV.102}$$

If, on the other side, $\tau^0 \gg 1$, it follows that

$$\rho_c = 0, \tag{IV.103}$$

i.e. isothermal optically thick radiation is not polarized.

c. *Nonthermal Bremsstrahlung Effects*

We assumed that the elementary emission process of bremsstrahlung is accessible to a single-particle description. An integration over an assembly of particles which do not show collective interactions was found to be sufficient for a wide range of applications for deriving the emission coefficients. Calculating the absorption of Coulomb collisions, the equation of motion of an average particle, or, what is nearly the same, the fluid description of a cold plasma was applied. Turning now to effects of nonthermal particles, which are significant especially near the local plasma frequency, collective processes play a dominant role and a description in the frame of the kinetic theory of a warm plasma is adequate. Here we give only a few prospects of this topic.

Nonthermal effects are often treated by the assumption of an ensemble of superthermal ('energetic') particles (electrons), distributed isotropically in the velocity space or exhibiting a net streaming motion, coexisting with a thermal background plasma. In such a configuration, under special circumstances, the bremsstrahlung can be enhanced by several orders of magnitude at the frequencies ω_p and $2\omega_p$ as compared with the emission from a thermal plasma. Calculations of these emissions were made e.g. by Tidman and Dupree (1965) based on a formalism elaborated by Dupree (1963, 1964). As shown by these authors, it is convenient to divide the contributions to the bremsstrahlung into two parts, namely the range $|\mathbf{k}| > k_D$ and the range $|\mathbf{k}| < k_D$, where the collisional contributions and wave contributions become important, respectively. Here

$$k_D = [4\pi N_e e^2/KT]^{1/2} \tag{IV.104}$$

is the Debye wave number.

In the range of the wave contributions resonances occur in the calculated spectral density as a dominant feature, which correspond to longitudinal electron (and ion) plasma waves represented by an emission line at ω_p. The physical mechanism for this enhanced emission can be visualized as follows: The energetic electrons excite the wave-field part of the longitudinal fluctuation spectrum in the thermal plasma up to high amplitudes by the process of Čerenkov emission of electron plasma waves (cf. Section 4.4.3). Here the close connection between bremsstrahlung and Čerenkov emission becomes evident, which was previously outlined in Table IV.1. The generated electron plasma waves are partly transformed into electromagnetic radiation by the processes of 'collisions' between each other and with low-frequency ion density fluctuations as discussed in Sections 3.4.6 and 4.6. By the former process the second line emission at $2\omega_p$ is produced. The amplitude of the plasma waves is controlled by the balance with the absorption by Landau damping, which depends sensitively on the velocity distribution of the superthermal particles.

The presence of a weak external magnetic field gives rise to a splitting of the plasma emission lines by an amount of $\omega^2/2\omega_p$ for the fundamental and twice this value for the second harmonic, if an anisotropic distribution of the energetic electrons is assumed (Tidman *et al.*, 1966).

4.3.2. Cyclotron (Gyro-Synchrotron) Emission

a. *The Emission Equation*

After discussing the Coulomb bremsstrahlung, the next large and perhaps even more important group of primary emission mechanisms, the cyclotron or gyro-synchrotron emission (also called magnetic bremsstrahlung or gyro-magnetic emission), will be considered. This mechanism depends essentially on the presence of a magnetic field in the plasma. Then, if $\nu_H \gtrsim \nu_{\text{coll}}$, in the picture of classical physics instead of the Coulomb potential of positive ions or protons, the Lorentz force exerts the main influence on the orbits of the radiating particles. Depending on the pitch angle $\phi = \sphericalangle(\mathbf{v}, \mathbf{H})$ and on the magnetic field configuration an idealized trajectory of the gyrating motion of a single radiating electron will be either a *circle* around a field line ($\phi = 90°$) or a *helix* ('screw line') ($\phi \neq 90°$), or a helix with variable diameter, if the magnetic field is inhomogeneous. In any case the departure of the electron's orbit from a straight line causes an accelerated motion and hence an emission of electromagnetic waves.

For nonrelativistic velocities v the gyroradius r_H of a gyrating electron is determined by the magnetic field:

$$r_H = mcv \sin\phi/(eH) = \frac{v \sin\phi}{2\pi\nu_H} \tag{IV.105}$$

(e, m – charge and mass of the electron, respectively, c – speed of light in vacuum, $v \sin\phi$ – component of $\mathbf{v}$ perpendicular to the magnetic field $\mathbf{H}$, ν_H – gyrofrequency (cf. Equation (IV.2))).

The frequencies $\nu_s(\theta, \phi, \beta)$, which are emitted by an electron along a helical path as received from a stationary observer are given by the *emission equation*

$$\nu_s(\theta, \phi, \beta) = \frac{s\nu_H(1-\beta^2)^{1/2}}{1 - n_j\beta\cos\theta\cos\phi}. \tag{IV.106}$$

Here we denote by s – the harmonic number, ν_H – the nonrelativistic gyrofrequency, $\beta = v/c$, n_j – refractive index, θ – $\sphericalangle(\mathbf{k}, \mathbf{H})$.

The numerator of the emission Equation (IV.106) contains the relativistic mass correction term, whereas the denominator accounts for the Doppler effect of the moving electrons. A graphic representation of the emission equation was given by Krüger (1972).

Considering arbitrary integer values of the harmonic number s, three large domains can be distinguished:

(a) $n_j \beta \cos\theta \cos\phi < 1$, i.e. $s > 0$

referring to the *normal Doppler effect*, where $v \cos\phi < v_{k,j}$ is always satisfied. There occurs radiation corresponding to $\nu_s > s\nu_H(1 - \beta^2)^{1/2}$ and $\nu_s < < s\nu_H(1 - \beta^2)^{1/2}$ propagating into the hemisphere of the direction of the particle motion ($\beta_{||} \cos\theta > 0$) and opposite to this direction ($\beta_{||} \cos\theta < 0$), respectively.

(b) $n_j \beta \cos\theta \cos\phi > 1$, i.e. $s < 0$

(since ν_s must be positive).
This case refers to the so-called *anomalous Doppler effect*, taking place if $v \cos\phi > v_{k,j}$, which is positive only in a medium where $n_j > 1$ is fulfilled. Evidently, the condition for the anomalous Doppler effect holds only for forward emission ($\beta_{||} \cos\theta > 0$). This case has also some connections to the occurrence of negative absorption, which is to be discussed later.

(c) $n_j \beta \cos\theta \cos\phi = 1$, i.e. $s = 0$

This is the *Čerenkov effect*, which will be discussed in Section 4.3.3. Like the anomalous Doppler effect the radiation is emitted only into forward directions. The ranges of the angle ϕ referring to the anomalous and normal Doppler effects are located inside and outside the Čerenkov cone defined by the condition $n_j \beta \cos\theta \cos\phi = 1$. Hence the role of the Čerenkov radiation as a limiting case of the gyro-synchrotron emission (cf. Table IV.1) becomes evident.

b. *Basic Formulas*

A short historic review may facilitate the understanding of the matter. Two needs essentially influenced the development of the fundamentals of the gyro-synchrotron radiation. The first one came from experimental physics and was directly connected with the application to radiation of fast particles in accelerators (synchrotron, cyclotron, etc.) from which the names synchrotron or cyclotron radiation were originally derived. The second impulse came from astrophysics, where radiation originating from particles spiralling in magnetic fields in nonsolar and solar radio sources was anticipated to form an efficient emission mechanism (Kiepenheuer, 1946).

At present the literature on different aspects of the gyro-synchrotron radiation appears rather extensive. The majority of the mathematical derivations refer to classical or quantum-mechanical methods within the framework of the single-particle and cold-plasma approximations. The kinetic treatment, especially for nonequilibrium radiation, is rather complicated and yet incomplete.

With regard to solar physics the following main groups of a treatment can be noted:

(a) Based on the work of early forerunners (Schott, Larmor, cf. also Ivanenko and Pomeranchuk, 1944) the theory of gyro-synchrotron radiation from single electrons in vacuum was treated by Schwinger (1949) and subsequently generalized and adopted for the purpose of solar radio astronomy by Takakura (1960b). Takakura's results are contained in a number of reviews (Wild *et al.*, 1963; Kundu, 1965; Takakura, 1967).

Parallel to this development corresponding work was established by numerous authors (e.g. Vladimirsky, 1948; Ginzburg, 1951, 1953) partly under the aspect of an application to nonsolar relativistic radio sources. Reviews regarding this work were made e.g. by Ginzburg (1958), Ginzburg and Syrovatskij (1963), Ginzburg and Ozernoj (1966), and Ginzburg *et al.* (1968). Another series of excellent papers was initiated by Trubnikov (1958) and coworkers.

(b) A slightly different class of papers considered the problem of gyro-synchrotron radiation of single electrons moving in a cold-plasma medium. Here the pioneering work of Eidman (1958) and Twiss (1958) is to be noted, which was subsequently applied and improved by Liemohn (1965), Mansfield (1967), Zheleznyakov (1964), Fung and Yip (1966), Ramaty (1968b, 1969a), Ko (1973), and others. A comprehensive survey on the emission and absorption processes in a magnetized plasma was given by Melrose (1968).

(c) A third direction of an approach started with the calculation of the gyro-'resonance' absorption in a warm plasma. This question was successfully treated for the case of the thermal equilibrium (Sitenko and Stepanov, 1956; Gershman, 1960; Kawabata, 1964; Kai, 1965b; Yip, 1967; Ginzburg and Zheleznyakov, 1959; Kakinuma and Swarup, 1962). But a considerable mathematical awkwardness arises when the more interesting case of nonthermal energy distributions is taken into account (Jahn, 1970). This field is still wide open for future developments.

(d) Completing our brief survey on the widespread literature of gyro-synchrotron radiation we finally quote some general physical textbooks containing a consideration of the matter from different points of view: Ivanenko and Sokolov (1949), Heitler (1954), Landau and Lifshits (1962), Shklovsky (1962), Ginzburg and Syrovatskij (1963), Panofsky and Phillips (1955), Heald and Wharton (1965), Bekefi (1966), Sokolov and Ternov (1966).

It is an important fact that the frequency spectrum and the angular distribution of the gyro-synchrotron radiation depend sensitively on the energy of the radiating electrons. The limiting cases are given by very-low-energy particles at the one hand, and by ultra-relativistic particles at the other. Hence three characteristic energy ranges can be distinguished:

(1) Very-low-energy range referring to solely gyro emission,

(2) intermediate-energy range referring to gyro-synchrotron radiation at low and medium harmonic numbers (say, up to $s = 50–100$),

(3) (relativistic) high-energy range referring to synchrotron radiation (merging of high harmonic numbers).

The general treatment concerning a single-particle radiation (following group (a) above) may be sketched in the following way:

Concluding from the Maxwell–Lorentz theory the wave equation can be written

$$\Box A_\mu \equiv \left(\Delta - \frac{1}{c^2}\frac{\partial^2}{\partial t^2} \right) A_\mu = -\frac{4\pi e}{c} \int P(s)\,(x - \xi_\mu)\,\xi_\mu(s)\,\mathrm{d}s. \qquad \text{(IV.107)}$$

Hence D'Alembert's solution follows, from which for a point-like electron the retarded potentials

$$\mathbf{A} = A_\kappa = \frac{e}{c} \int v_\kappa(t') \frac{\delta\left(t' - t + \frac{r(t')}{c} \right)}{r(t')} \mathrm{d}t'$$

$$(\kappa = 1, 2, 3) \qquad \text{(IV.108)}$$

$$\Phi = A_4 = e \int \frac{\delta\left(t' - t + \frac{r(t')}{c} \right)}{r(t')} \mathrm{d}t'$$

follow and the integration over the time conveniently gives the Liénard–Wiechert form of the potentials. The different retardations observed at a fixed point r with regard to different points of the orbit explain the occurrence of harmonics for sufficiently high velocities v. This is described by the delta-function, which can be written as

$$\delta(t') = \frac{\omega}{2\pi} \sum_{s=-\infty}^{\infty} \exp(-is\omega t). \qquad \text{(IV.109)}$$

Bearing this in mind, the vector potential can be expressed by the sum $\sum A_s \exp(-is\varphi_s)$, where the coefficients A_s and the phases φ_s finally lead to the estimation of the emission probabilities or emissivities yielding the wanted emission coefficients.

Let $\eta_\omega(\mathbf{p}', \mathbf{r}, s)$ denote the emissivity, i.e. the differential rate at which the energy is emitted, spontaneously per unit frequency interval $\mathrm{d}\omega$, per unit time interval, and per unit solid angle, by one electron with a momentum between $\mathbf{p}'$ and $\mathbf{p}' + \Delta\mathbf{p}$, then the emission coefficient, i.e. the rate of emission per unit volume, is

$$j_{\omega,j} = \int \eta_{\omega,j}(p')\, f(p')\, \mathrm{d}^3 p'. \qquad \text{(IV.110)}$$

The absorption coefficient is given by

$$\alpha_{\omega,j} = \frac{8\pi^3 c^2}{n_j^2 \hbar\omega^3} \int \eta_\omega(p') \left[f(p) - f(p') \right] \mathrm{d}^3 p' \qquad \text{(IV.111)}$$

(p' and p refer to the states before and after emission, respectively).

If $\hbar\omega \ll KT$ and $f(p)$ is an isotropic distribution function (depending on p^2 only), one finds that

$$\Delta p = E\Delta E/(c^2 p) = \hbar\omega E/(c^2 p) \qquad \text{(IV.112)}$$

(E, p – energy and momentum state of the radiating particle, $E^2 = p^2c^2 + (m_0c^2)^2$), allowing a simplification of the expressions for j_ω and α_ω by expansion of $f(p')$ in a Taylor series

$$f(p') = f(p) + \hbar\omega\,\partial f/\partial E. \tag{IV.113}$$

Hence, and assuming $n_j = 1$, one obtains

$$j_\omega = \int \eta_\omega(p)\, f(p)\, \mathrm{d}^3 p \tag{IV.114}$$

$$\alpha_\omega = -\frac{8\pi^3 c^2}{\omega^2}\int \eta_\omega(p)\frac{\partial f(p)}{\partial E}\,\mathrm{d}^3 p. \tag{IV.115}$$

The cyclotron emissivity is given by

$$\eta_\omega = \frac{e^2\omega^2}{2\pi c}\sum_{s=1}^{\infty}\left\{\left(\frac{\cos\theta - \beta_\parallel}{\sin\theta}\right)^2 J_s^2(a) + \beta_\perp^2 J_s'^2(a)\right\}\delta(b) \tag{IV.116}$$

where

$$a = \frac{\omega}{\omega_H}\beta_\perp \sin\theta,$$

$$b = s\omega_H\sqrt{1-\beta^2} - \omega(1 - \beta_\parallel\cos\theta),$$

$$\beta = \frac{v}{c}, \qquad \beta_\parallel = \beta\cos\phi, \qquad \beta_\perp = \beta\sin\phi,$$

$J_s(a)$ – Bessel function of the order s and argument a.

In the presence of thermal equilibrium $f(p)$ is given by the Maxwell distribution

$$f_M(v) = \pi^{-3/2} v_{th}^{-3}\exp(-v^2/v_{th}^2) \tag{IV.117}$$

and the integration over the momentum p (or velocity v) can be carried out analytically yielding

$$j_\omega = \frac{\omega_p^2\omega\pi^{1/2}KT}{8\pi^3 v_{th}^2|\cos\theta|}\sum_{s=1}^{\infty}\left\{2\left(\frac{\sin^2\theta - s\omega_H/\omega}{\sin\theta\cos\theta}\right)^2 I_s(\mu) + \right.$$
$$\left. + \frac{v_{th}^2}{c^2}\left[\left(\frac{s^2}{\mu}\right)I_s(\mu) - 2\mu(I_s'(\mu) - I_s(\mu))\right]\right\}\exp(-\mu - z_s^2) \tag{IV.118}$$

where $I_s(\mu)$ are the modified Bessel functions and

$$\mu = \tfrac{1}{2}[\omega v_{th}\sin\theta/(\omega_H c)]^2,$$

$$z_s = c(1 - s\omega_H/\omega)/(v_{th}\cos\theta).$$

The absorption coefficient is given by

$$\alpha_\omega = \frac{8\pi^3 c^2}{\omega^2 KT}\, j_\omega \tag{IV.119}$$

which is identical with the Rayleigh–Jeans formula.

Considering now nonthermal energy distributions, two approaches were studied in the past. The first approach makes use of the assumption that a certain fraction of the nonthermal electrons attributed to a special energy distribution is coexisting with a pure thermal background plasma (the nonthermal energy distribution may be represented e.g. by a power law within a given range). Then the radiation from the nonthermal particles undergoes absorption by the nonthermal electrons themselves (called '*reabsorption*' or '*self-absorption*'), and by the thermal background plasma ('*gyro*'- or '*gyroresonance absorption*') (cf. e.g. Takakura, 1967).

The second approach makes use of one unique distribution function consisting of a Maxwell-distribution 'body' $f_M(v)$ and a nonthermal (e.g. power-law) 'tail' $f_p(v)$:

$$f(v) = A_e f_M(v)\,\theta(v_s - v) + (1 - A_e)\, f_P(v)\,\theta(v - v_s) \qquad \text{(IV.120)}$$

where $f_M(v) = A_M \exp(-v^2/v_{th}^2)$ and $f_P(v) = A_P(v_s/v)^\gamma$, θ denotes the step function; v_s is the velocity at the turning point between $f_M(v)$ and $f_P(v)$; A_e, A_M, and A_P contain the normalization constant since $f(v)$ is conveniently normalized to unity or the electron density (cf. Section 4.1.2.c).

Applying the distribution function (IV.120), the emission and absorption coefficients are given by

$$\begin{aligned} j_\omega = {} & A_e A_M \frac{\omega_p^2 \omega m}{4\pi N_e |\cos\theta|} \int_0^{v_s} dv\, v \sum_{s=1}^{\infty} \left\{ \left(\frac{\sin^2\theta - s\omega_H/\omega}{\sin\theta\cos\theta} \right)^2 J_s^2(a_0) + \right. \\ & \left. + \frac{v^2}{c^2}(1 - x_0^2)\, J_s'^2(a_0) \right\} \exp\left(-\frac{v^2}{v_{th}^2} \right) + \\ & + (1 - A_e)\, A_P \frac{\omega_p^2 \omega m v_s}{4\pi N_e |\cos\theta|} \int_{v_s}^{\infty} dv \sum_{s=1}^{\infty} \left\{ \left(\frac{\sin^2\theta - s\omega_H/\omega}{\sin\theta\cos\theta} \right)^2 J_s^2(a) + \right. \\ & \left. + \frac{v^2}{c^2}(1 - x_0^2)\, J_s'^2(a) \right\} \left(\frac{v_s}{v} \right)^{\gamma - 1} \end{aligned} \qquad \text{(IV.121)}$$

$$\begin{aligned} \alpha_\omega = {} & A_e A_M \frac{4\pi^2 \omega_p^2 c^2}{n^2 \omega v_{th}^2 N_e |\cos\theta|} \int_0^{v_s} dv\, v \sum_{s=1}^{\infty} \left\{ \left(\frac{\sin^2\theta - s\omega_H/\omega}{\sin\theta\cos\theta} \right)^2 J_s^2(a_0) + \right. \\ & \left. + \frac{v^2}{c^2}(1 - x_0^2)\, J_s'^2(a_0) \right\} \exp\left(-\frac{v^2}{v_{th}^2} \right) + \\ & + (1 - A_e)\, A_P \frac{4\pi^2 \omega_p^2 c^2}{n^2 \omega v_s N_e |\cos\theta|} \int_{v_s}^{\infty} dv \sum_{s=1}^{\infty} \left\{ \left(\frac{\sin^2\theta - s\omega_H/\omega}{\sin\theta\cos\theta} \right)^2 J_s^2(a) + \right. \\ & \left. + \frac{v^2}{c^2}(1 - x_0^2)\, J_s'^2(a) \right\} \left(\frac{v_s}{v} \right)^{\gamma + 1} \end{aligned} \qquad \text{(IV.122)}$$

where $a_0 = \omega v(1 - x_0^2)^{1/2} \sin\theta/(\omega_H c)$, $x_0 = c(1 - s\omega_H/\omega)/(v\cos\theta)$ (cf. Böhme *et al.*, 1977).

For the purpose of numerical computations the power-law distribution can be bounded by a cutoff at the high-energy side of the spectrum and the number of harmonics s is reduced to a practical number. Results of calculations are demonstrated by Figure IV.12, where the dependence of j_ω and α_ω on ω/ω_H is shown for selected angles θ.

More general expressions containing the refractive index, the distinction between the magneto-ionic wave modes, and applying the Lorentz factor

$$\gamma^* = (1 - \beta^2)^{-1/2} \tag{IV.123}$$

(referring to an energy distribution in the form $f(\gamma^*) = G(\gamma^* - 1)^{-\lambda}$) are as follows (Hildebrandt, 1976):

$$j_{\omega,j} = \frac{N_r e^2 \omega}{4\pi c |\cos\theta| (1 + T_j^2)} \sum_{s=1}^{\infty} \int_{\gamma^*} d\gamma^* (1 - \gamma^{*-2})^{1/2} (1 - y_s^2(\gamma^*)) \times$$
$$\times \left[- J_s'(z_j(\gamma^*)) + \left\{ a_j \frac{s}{z_j(\gamma^*)} + b_j \frac{y_s^2(\gamma^*)}{(1 - y_s^2(\gamma^*))^{1/2}} \right\} \times \right.$$
$$\left. \times J_s(z_j) \right]^2 f(\gamma^*) \tag{IV.124}$$

$$\alpha_{\omega,j} = \frac{2\pi^2 N_r e^2}{mc\omega n_j^2 |\cos\theta| (1 + T_j^2)} \sum_{s=1}^{\infty} \int_{\gamma^*} d\gamma^* (1 - \gamma^{*-2})^{1/2} (1 - y_s^2(\gamma^*)) \times$$
$$\times \left[- J_s'(z_j(\gamma^*)) + \left\{ a_j \frac{s}{z_j} + b_j \frac{y_s^2(\gamma^*)}{(1 - y_s^2(\gamma^*))^{1/2}} \right\} J_s(z_j) \right]^2 \times$$
$$\times \left[\frac{df(\gamma^*)}{d\gamma^*} - \frac{f(\gamma^*)}{\gamma^*} \left(2 + \frac{1}{\gamma^2 - 1} \right) \right] \tag{IV.125}$$

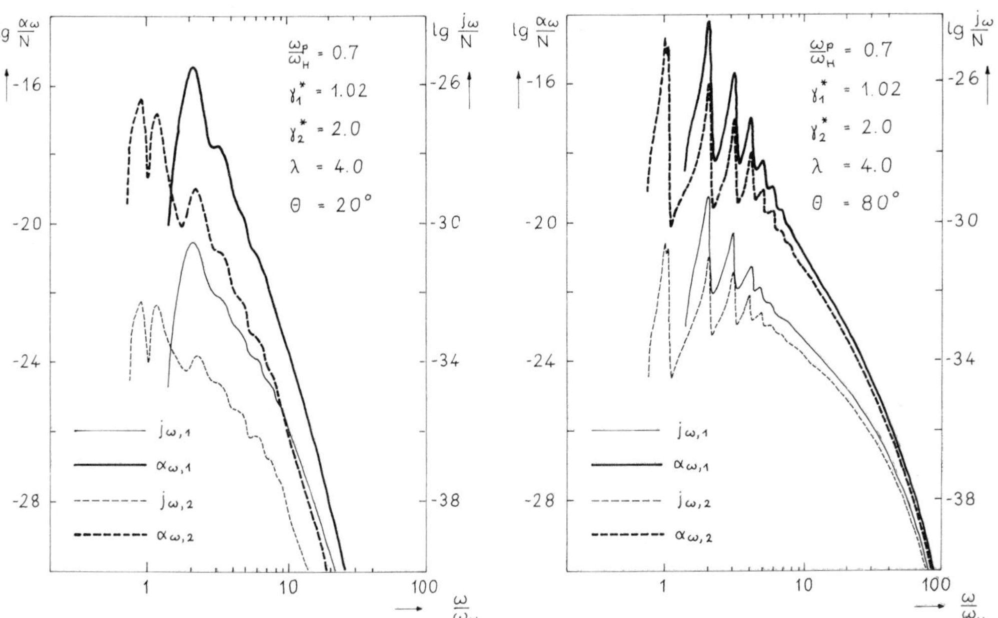

Fig. IV.12. Examples of nonthermal gyromagnetic emission coefficients (thin lines) and absorption coefficients (heavy lines) for the extraordinary mode (solid lines) and the ordinary mode (dashed lines) for different values of θ according to formulas (IV.124) and (IV.125). (Courtesy of J. Hildebrandt).

where partly the notations of Melrose (1968) have been used:

$$T_j = \frac{2(1-X)\,\omega\cos\theta}{\omega_H\sin^2\theta \mp [\omega_H^2\sin^4\theta + 4\omega^2(1-X)^2\cos^2\theta]^{1/2}}$$

$$K_j = \frac{2\omega(1-X-n_j^2)\sin\theta}{\omega_H\sin^2\theta \mp [\omega_H^2\sin^4\theta + 4\omega^2(1-X)^2\cos^2\theta]^{1/2}}$$

$$y_s(\gamma^*) = \frac{\gamma^* - s\omega_H/\omega}{n_j\sqrt{\gamma^{*2}-1}\cos\theta} \qquad \text{with} \quad y_s^2(\gamma^*) < 1$$

$$z_j(\gamma^*) = n_j Y^{-1}\sqrt{\gamma^{*2}-1}\,\sin\theta\sqrt{1-y_s^2(\gamma^*)}$$

$$a_j = T_j\cos\theta + K_j\sin\theta,$$

$$b_j = -T_j\sin\theta + K_j\cos\theta,$$

(N_r – density of radiating electrons).

The general expressions given above contain a set of corresponding formulas given by Takakura (1967) or Takakura and Scalise (1970) as a special case.

Finally we note the limiting cases for θ:

(a) $\theta = 0$

$$j_{\omega,2} = \alpha_{\omega,2} \equiv 0$$

$$j_{\omega,1}(\theta=0) = \frac{\omega\omega_p^2 m}{32\pi^2 c}\int_{\gamma_1^*}^{\gamma_2^*} \mathrm{d}\gamma^*(1-\gamma^{*-2})^{1/2}(1-y^2)\,f(\gamma^*)$$

$$\alpha_{\omega,1}(\theta=0) = \frac{-\pi\omega_p^2}{4\omega n_j^2}\int_{\gamma_1^*}^{\gamma_2^*} \mathrm{d}\gamma^*(1-\gamma^{*-2})^{1/2}(1-y^2)\times$$

$$\times\left[\frac{\mathrm{d}f(\gamma^*)}{\mathrm{d}\gamma^*} - \frac{f(\gamma^*)}{\gamma^*}\left(2+\frac{1}{\gamma^{*2}-1}\right)\right] \qquad \text{(IV.126)}$$

where

$$y = \frac{\gamma^* - \omega_H/\omega}{n_j\sqrt{\gamma^{*2}-1}}$$

(b) $\theta = 90°$

$$j_{\omega,1}(\theta=\pi/2) = \frac{\omega_p^2\omega n_1 m}{16\pi^2 c}\sum_s\int_{-1}^{1}\mathrm{d}y\left(-\frac{1}{\gamma_s}+\gamma_s\right)(1-y^2)\times$$

$$\times\left[-J_s'(z_1) + K_1\frac{s}{z_1}J_s(z_1)\right]^2 f(\gamma_s)$$

$$\alpha_{\omega,1}(\theta = \pi/2) = \frac{-\omega_p \pi}{2\omega n_1} \sum_s \int_{-1}^{1} \mathrm{d}y \left(\gamma_s - \frac{1}{\gamma_s} \right)(1 - y^2) \times$$

$$\times \left[-J_s'(z_1) + K_1 \frac{s}{z_1} J_s(z_1) \right]^2 \times$$

$$\times \left\{ \frac{\mathrm{d}f(\gamma^*)}{\mathrm{d}\gamma^*} \bigg|_{\gamma^* = \gamma_s} - \frac{f(\gamma_s)}{\gamma_s} \left(2 + \frac{1}{\gamma_s^2 - 1} \right) \right\}$$

$$j_{\omega,2}(\theta = \pi/2) = \frac{\omega_p^2 \omega n_2 m}{16\pi^2 c} \sum_s \int_{-1}^{1} \mathrm{d}y \left(\gamma_s - \frac{1}{\gamma_s} \right) y^2 J_s^2(z_2)\, f(\gamma_s)$$

$$\alpha_{\omega,2}(\theta - \pi/2) = \frac{-\pi\omega_p^2}{2\omega n_2} \sum_s \int_{-1}^{1} \mathrm{d}y \left(\gamma_s - \frac{1}{\gamma_s} \right) y^2 J_s^2(z_2) \times$$

$$\times \left\{ \frac{\mathrm{d}f(\gamma^*)}{\mathrm{d}\gamma^*} \bigg|_{\gamma^* = \gamma_s} - \frac{f(\gamma_s)}{\gamma_s} \left(2 + \frac{1}{\gamma_s^2 - 1} \right) \right\} \qquad \text{(IV.127)}$$

where

$$\gamma_s = s\omega_H/\omega, \qquad \gamma_s^{-2} < 1, \qquad \gamma_1 < \gamma_s < \gamma_2,$$

$$z_{1,2} = s n_{1,2} \left(1 - \frac{1}{\gamma_s^2} \right)^{1/2} (1 - y^2)^{1/2}.$$

c. *Correction Terms*

Though the subject of gyro-synchrotron radiation has been extensively treated in the literature for more than two decades, sometimes specific doubts arose on the validity or correctness of some derivations and a number of papers even of well established authors were subjected to reconsideration or refinement. In this connection there was a misunderstanding of the emission formulas: The emissivity is given by the mean energy loss P_{emitted} (radiated power) of an electron into unit frequency interval and unit solid angle, which, if the electron moves on a helical path, is different from the amount of radiation received by a distant observer P_{received} (Epstein and Feldman, 1967; Scheuer, 1968; Takakura and Uchida, 1968b). These authors give an approximate formula

$$P_{\text{emitted}} \approx P_{\text{received}} \times \sin^2\theta \qquad \text{(IV.128)}$$

which is valid for an isotropic, nondispersive medium with $n = 1$ (vacuum), $\beta \approx 1$, and $\theta \approx \phi$ (pitch angle). The difference between P_{emitted} and P_{received} is due to the nonstationary nature of the radiation field, whence the rate of the emitted energy is measured in terms of the electron's own time (retarded time) in contrast to the received power which is measured in the time of the observer.

However, as far as the radiation from a fixed volume in vacuum is concerned, by averaging over all directions of the pitch angle ϕ the above mentioned correction term (in Equation (IV.128)) is cancelled out in the equation for the emission coefficient as concluded by Scheuer (1968) and Ginzburg *et al.* (1968).

But if dispersion effects and an anisotropy of the medium are to be taken into account, the full correction term reads

$$P_{\text{emitted}} = P_{\text{received}} \left|1 - v \cos\phi \,|\partial(k \cos\theta/\partial\omega)|_{\theta\,\text{ray}}\right| \tag{IV.129}$$

which for an isotropic (nondispersive) medium is reduced to

$$P_{\text{emitted}} = P_{\text{received}} |1 - v \cos\phi \cos\theta \,\partial k/\partial\omega|$$

or

$$P_{\text{emitted}} = P_{\text{received}} |1 - (\beta/n) \cos\phi \cos\theta| \tag{IV.130}$$

(Ko and Chuang, 1973; cf. also McKenzie, 1964; Sokolov *et al.*, 1968; Sakurai, 1972c).

4.3.3. Time Dependencies of Radiation

In the foregoing discussion possible temporal variations of the radiation have been neglected. But temporal changes of the emitted radiation are of particular interest for gyro-synchrotron emission closely connected with fast changing phenomena of the active Sun, i.e. radio bursts. Unfortunately a complete, self-consistent theory of the temporal development of the particle energy spectra and the resulting radiation is still missing. Therefore only a few remarks on the subject may be given:

(a) The lifetimes of the gyro-synchrotron radiation were estimated by Takakura (1960b) under some simplifying assumptions. His result is reproduced in Figure IV.13, where the lifetime is plotted against the parameter $\beta = v/c$ (or the fundamental frequency) so far as collisions between nonthermal and thermal electrons and radiation losses are concerned. From this figure it becomes evident that radiation losses are most efficient at the higher levels in the corona. At the lower levels collisions become the dominating process, which is to be compensated by an instant energy supply in order to account for the observed burst durations in the microwave region.

(b) A discussion about the time variations of the energy spectrum after switching off of an acceleration process was made by Takakura and Kai (1966). In general, nonthermal initial energy distributions were considered mainly in the form of a power law, an exponential law, and a monoenergetic spectrum (delta function). The main energy loss processes to be involved there are gyro-synchrotron radiation, collision losses, and escape of trapped particles.

Other processes not extensively treated in this connection should also be included, e.g. heat condution, turbulent motions, and diverse nonelectromagnetic wave phenomena.

The time-dependent distributions resulting from the above mentioned initial conditions are not yet exactly comparable with the real observations, since in such a treatment only the effect of the nonthermal particles on the spectrum was considered, neglecting the Maxwell part of the thermalized background plasma.

The time-dependent changes of the gyro-synchrotron emissivity were calculated in a subsequent paper by Takakura *et al.* (1968) considering a variation of the spectral index of a power-law energy distribution. (It may be noted, that the results obtained by these authors originally contained an error by a factor $(s(\omega_H/\omega))(1 + p^2)^{-1/2}$

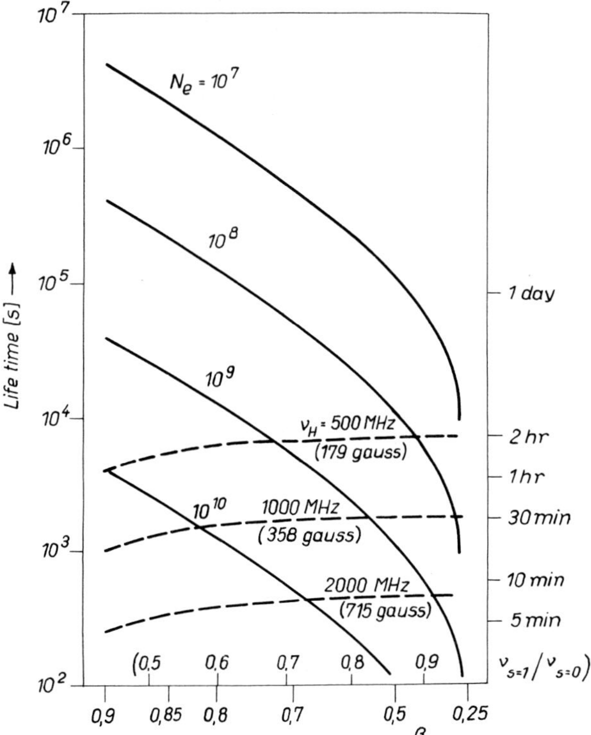

Fig. IV.13. Lifetimes of gyro-synchrotron radiation in dependence of $\beta = v/c$, N_e, and H if the process is controlled by collisions (solid curve) and radiation damping (dashed curve) (after Takakura, 1960).

(cf. Takakura and Uchida, 1968b) and later were, in principle, corrected by the calculations of Takakura and Scalise (1970).)

(c) A self-consistent description of the time variations of the solar radio emission would require the knowledge of all essential quantities contained in a general equation of energy balance, which can be written

$$\frac{\partial F(E, t)}{\partial t} + \frac{\partial}{\partial E}\left[F(E, t) \frac{\mathrm{d}E}{\mathrm{d}t} \right] = Q(E, t) \tag{IV.131}$$

where $F(E, t)$ is the energy distribution function and the term $Q(E, t)$ describes the external sources of energy. For illustration, a specialization of this equation may be given in the following form

$$\frac{\partial F(E, t)}{\partial t} + \left[\sum_{n=0}^{2} (A_n - B_n) E^{n-1} \right] F(E, t) = Q_1(E, t) - Q_2(E, t) \tag{IV.132}$$

where the coefficients A_n and B_n describe the energy gain and loss processes, respectively (e.g. A_0, B_0 refer to recombination and ionization processes, A_1 – Fermi acceleration, B_1 – collision losses, A_2 – Compton effect and synchrotron acceleration, B_2 – synchrotron radiation losses and inverse Compton effect losses, etc.). The terms $Q_1(E, t)$ and $Q_2(E, t)$ describe particle injection and ejection processes, respectively (cf. Krüger, 1972).

4.4. Cold-Plasma Approximation: Some Aspects of Synchrotron Radiation and Čerenkov Radiation

4.4.1. Magneto-ionic theory

As discussed in Section 4.2.2, the vacuum theory is to be considered only as a first approximation, which may well account for a principal understanding of the basic emission processes of bremsstrahlung and gyro-synchrotron radiation. But this approximation must fail if effects of a medium interacting with the radiation come into play. In such cases, as a next step of approximation, the influence of the medium is to be included, which in the simplest way can be done by assumption of the presence of a cold background plasma. In this picture the thermal motions of the heavier particles (ions, protons) are generally neglected (cf. Section 4.2.2). This can be treated applying a fluid theory which is involved in the magneto-ionic theory describing the basic properties of the propagation of the two modes of radio waves in a plasma (Ratcliffe, 1959; cf. also Section 4.2.3). The influence of the plasma is mainly expressed by the consideration of a refractive index $n_j \neq 1$, which for a cold plasma is given by the Appleton–Hartree formula (IV.60) or similar constructions (e.g. polarization coefficients) derived from the dispersion relation.

It became already evident in the foregoing sections that often it appears neither very easy nor necessary to separate strictly the vacuum approximation from a cold-plasma approximation, as it perhaps would be desirable from a sole theoretical point of view aiming to distinguish consistently between a 'one-particle' and a 'continuum' theory. The vacuum may be interpreted rather as the limiting case of a rarified medium and for each particular purpose it must be decided what degree of approximation appears just appropriate. In such a sense the following pages should not be understood as exclusively devoted to 'cold-plasma' phenomena only (in spite of the headings), but rather as a natural continuation of the foregoing sections, where the refractive index was also partly considered.

4.4.2. Synchrotron radiation

a. *Ultra-Relativistic Approximation*

We remind that in general the gyro-synchrotron radiation formulas can be applied for three different energy ranges (Section 4.3.2.b) which may be specified as follows

(a) The nonrelativistic range yielding 'gyro' emission characterized by $\beta \ll 1$ ($\beta^2 \to 0$) corresponding to electron energies $E \lesssim 1$ keV (i.e. $\beta < 0.1$–0.2).

(b) The mildly relativistic range yielding generally 'cyclotron' emission corresponding to electron energies between some keV and some hundreds of keV.

(c) The highly (or ultra-) relativistic range yielding 'synchrotron' radiation corresponding to $1 - \beta^2 \ll 1$ ($\beta^2 \to 1$, i.e. $\beta \gtrsim 0.8$).

While the ranges (a) and (c) in principle are accessible to convenient approximations of the radiation formulas, the range (b) – which is the most important range for solar radio applications – is the most cumbersome for practical computations (cf. Section 4.3).

In the following we are dealing briefly with the highly (ultra-) relativistic approximation. As a matter of fact, with increasing electron energies the number of gyroharmonics making contributions to the gyro-synchrotron emissivity rises continuously. Accordingly, also the efforts required for numerical computations increase rapidly (which is connected e.g. with the evaluation of the Bessel functions). Therefore it turns out to be highly useful that, in the ultra-relativistic range, the sums of Bessel functions can be replaced by simpler expressions containing the modified Hankel functions $K_{5/3}(z)$, which can be related to an Airy integral representation. Analytically, the synchrotron emissivity for highly relativistic energies is expressed by

$$\eta_\omega = \frac{\sqrt{3}\,e^2}{4\pi r_C}\gamma^{*4}\frac{\omega\omega_H \sin\phi}{\omega_c^2}\int_{\omega/\omega_c}^{\infty} K_{5/3}(z)\,dz$$

$$= \frac{\sqrt{3}\,e^3}{2\pi mc^2} H \sin\phi \frac{\omega}{\omega_c}\int_{\omega/\omega_c}^{\infty} K_{5/3}(z)\,dz \qquad \text{(IV.133)}$$

where

$$\omega_c = \tfrac{3}{2}\omega_H \sin\phi\gamma^{*2} = \tfrac{3}{2}\omega_{H,\mathrm{rel}} \sin\phi\gamma^{*3} \qquad \text{(IV.134)}$$

is the critical frequency of the synchrotron radiation, which is closely related to the position of the spectral maximum ($\omega_c \approx 0.3\omega_{\max}$, cf. below); $r_C = E/eH \sin\phi$.

$$\gamma^* = E/(mc^2) \qquad \text{(IV.135)}$$

is the 'Lorentz factor', which by Einstein's relation

$$E = mc^2/(1 - \beta^2)^{1/2} \qquad \text{(IV.136)}$$

is related to β:

$$\gamma^* = (1 - \beta^2)^{-1/2}. \qquad \text{(IV.137)}$$

Furthermore we have to note relativistic corrections for the gyroradius and the gyrofrequency:

$$r_{H,\mathrm{rel}} = \beta_\perp c/\omega_{H,\mathrm{rel}}$$

$$= mcv \sin\phi/[eH(1 - \beta^2)^{1/2}] \qquad \text{(IV.138)}$$

$$\omega_{H,\mathrm{rel}} = \frac{eH}{mc}(1 - \beta^2)^{1/2}. \qquad \text{(IV.139)}$$

The frequency dependence of the synchrotron radiation is expressed in Equation (IV.133) by the function

$$f(y) = y\int_y^{\infty} K_{5/3}(z)\,dz. \qquad \text{(IV.140)}$$

Numerical values for this function were calculated e.g. by Vladimirskij (1948) and Oort and Walraven (1956); a tabulation can be found in Pacholczyk (1970).

If the magneto-ionic wave modes are distinguished, instead of Equation (IV.133) it can be written

$$\eta_{\omega,j} \simeq \frac{\sqrt{3}\,e^3}{4\pi mc^2} H \sin\phi \frac{\omega}{\omega_c} \left\{ \left[\int_{\omega/\omega_c}^{\infty} K_{5/3}(z)\,\mathrm{d}z \right] \pm K_{2/3}\left(\frac{\omega}{\omega_c}\right) \right\}. \quad \text{(IV.141)}$$

In contrast to the low-energy gyromagnetic radiation, which has a comparatively low directivity according to the character of a dipole-field radiation, the synchrotron radiation is sharply beamed into the direction of the instantaneous particle velocity. For sufficiently high energies and not too small angles θ the radiation is emitted into a solid angle $4\pi\delta \sin\phi$ around the direction $\theta \approx \phi$. In this way, the emission towards a stationary observer, which may be located in the orbital plane of the gyrating particle, occurs in the form of individual pulses which are short in comparison with the fundamental revolution period. The Fourier representation of such pulses indicates the presence of a high content of harmonics of the fundamental frequency. For sufficiently high energies this leads to an essentially continuous spectrum, which is different from the more or less single-frequency radiation from single low-energy particles. Examples for the radiation patterns for the different energy ranges are shown in Figure IV.14 ($z \parallel \mathbf{v}$, $\mathbf{v} \perp \mathbf{H}$, ϕ, θ – polar coordinates).

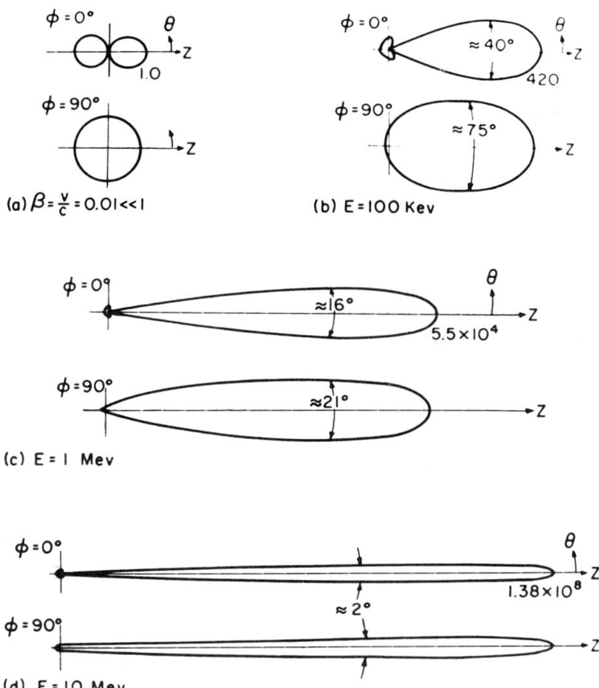

Fig. IV.14. Examples of total-power radiation patterns of gyro-synchrotron radiation (after Peterson and Hower, 1966).

Special reviews including numerical computations of different characteristics of the synchrotron radiation were made e.g. by Priester and Rosenberg (1965) and Peterson and Hower (1966). Some synchrotron emission spectra published by the latter authors are shown in Figure IV.15.

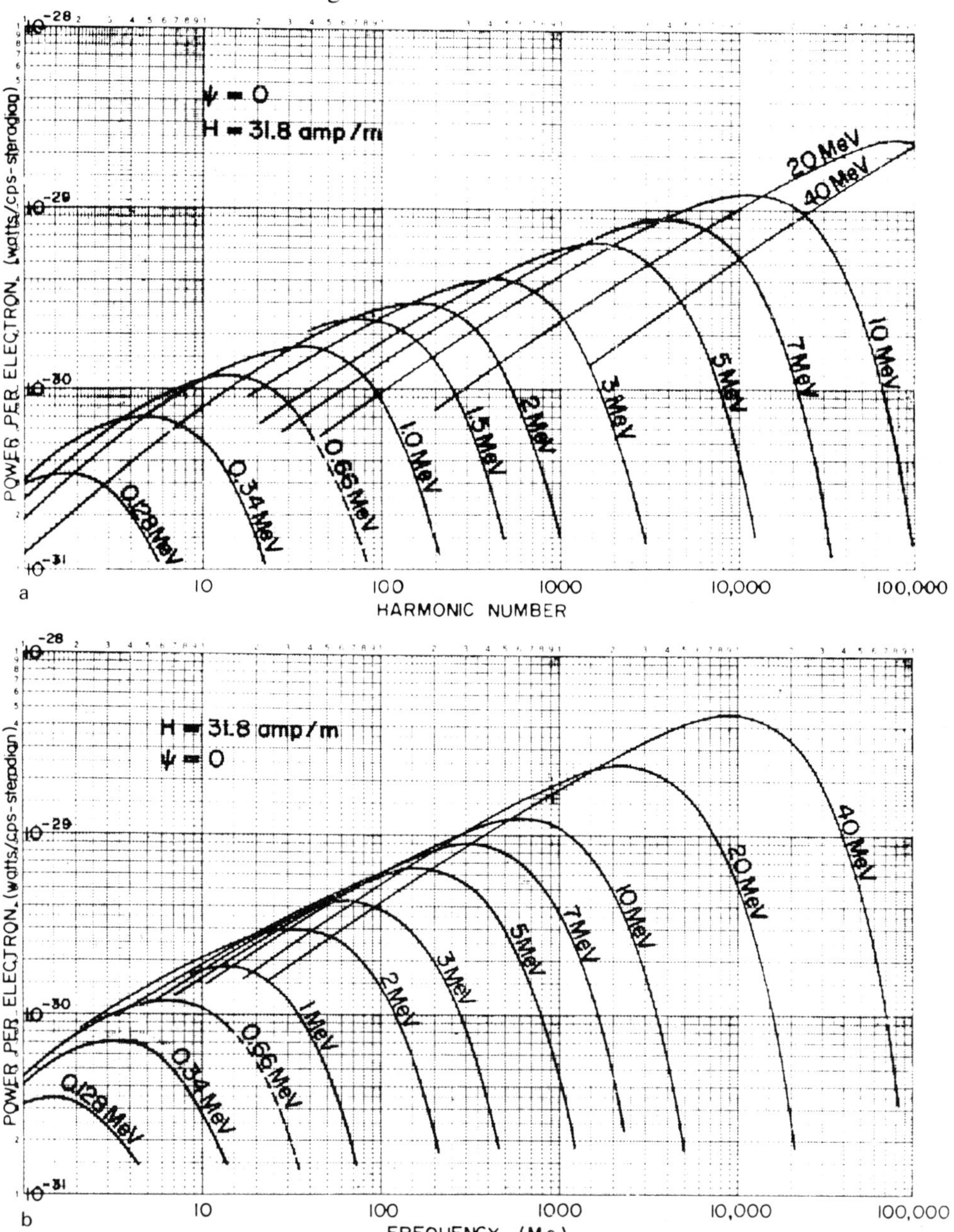

Fig. IV.15a,b. Examples of single-particle gyro-synchrotron radiation spectra in dependence of energy (after Peterson and Hower 1966).

We also mention the following practical approximations given by Jäger and Wallis (1961)

(a) $\omega_H \ll \omega \ll \omega_c \qquad (\omega/\omega_c \lesssim 10^{-2})$

$$\eta_\omega = \frac{2^{-1/2}3^{1/2}}{\pi}\,\Gamma\left(\frac{2}{3}\right)\frac{e^3}{mc^2}\,H\sin\phi\left(\frac{\omega}{\omega_c}\right)^{1/3}$$

$$= 8.21\times 10^{-23}H_{[\mathrm{gauss}]}\sin\phi\left(\frac{\omega}{\omega_c}\right)^{1/3}\qquad [\mathrm{cgs}]; \tag{IV.142}$$

(b) $\omega \gg \omega_c \qquad (\omega/\omega_c \gtrsim 10)$

$$\eta_\omega = \left(\frac{3}{8\pi}\right)^{1/3}\frac{e^3}{mc^2}\,H\sin\phi\left(\frac{\omega}{\omega_c}\right)^{1/2}\exp\left(-\frac{\omega}{\omega_c}\right)$$

$$= 4.66\times 10^{-23}H_{[\mathrm{gauss}]}\sin\phi\left(\frac{\omega}{\omega_c}\right)^{1/2}\exp\left(-\frac{\omega}{\omega_c}\right)\qquad [\mathrm{cgs}]. \tag{IV.143}$$

b. *Polarization*

The intensity of the synchrotron radiation is obtained from the emissivity by integration over the space volume and the energy distribution. For $n = 1$ and an isotropic energy distribution $F(E)$ this may be written in the following form

$$I_\nu = \frac{\sqrt{3}}{4\pi r^2}\frac{e^3}{mc^2}\left[\iint_{EV}\mathrm{d}E\,\mathrm{d}V\,F(E)\,H\sin\phi\,\frac{\nu}{\nu_c}\int_{\nu/\nu_c}^{\infty}K_{5/3}(z)\,\mathrm{d}z\right]. \tag{IV.144}$$

The volume element is given by $\mathrm{d}V = r^2\,\mathrm{d}r\,\mathrm{d}\Omega$ (r = distance from the observer, Ω = solid angle). Correspondingly the other Stokes parameters are given by

$$Q_\nu = \frac{\sqrt{3}}{4\pi r^2}\frac{e^3}{mc^2}\left[\iint_{EV}\mathrm{d}E\,\mathrm{d}V\,F(E)\,H\sin\phi\cos 2\chi\frac{\nu}{\nu_c}K_{2/3}\left(\frac{\nu}{\nu_c}\right)\right] \tag{IV.145}$$

$$U_\nu = \frac{\sqrt{3}}{4\pi r^2}\frac{e^3}{mc^2}\left[\iint_{EV}\mathrm{d}E\,\mathrm{d}V\,F(E)\,H\sin\phi\sin 2\chi\frac{\nu}{\nu_c}K_{2/3}\left(\frac{\nu}{\nu_c}\right)\right] \tag{IV.146}$$

$$V_\nu = 0 \tag{IV.147}$$

(Korchak and Syrovatskij, 1961). The latter equation denotes the absence of any part of circular polarization in the ultra-relativistic approximation of the synchrotron radiation, where only linearly polarized components are present.

Assuming an energy distribution in form of a power law

$$F(E)\,\mathrm{d}E = AE^{-\gamma}\,\mathrm{d}E \tag{IV.148}$$

in a certain range of energies $E_1 < E < E_2$, the integrals of Equations (IV.144–

IV.146) can be expressed by the Γ-function:

$$\int \mathrm{d}E\,E^{\gamma}\,\frac{\nu}{\nu_c}\int_{\nu/\nu_c}^{\infty} K_{5/3}(z)\,\mathrm{d}z = \frac{\gamma+\frac{7}{3}}{\gamma+1}\int_0^{\infty}\mathrm{d}E\,E^{-\gamma}\,\frac{\nu}{\nu_c}\,K_{2/3}\left(\frac{\nu}{\nu_c}\right)$$

$$= 2^{(\gamma-5)/2}\,\frac{\gamma+\frac{7}{3}}{\gamma+1}\,\Gamma\left(\frac{3\gamma-1}{12}\right)\Gamma\left(\frac{3\gamma+7}{12}\right)\times$$

$$\times\left(\frac{3eH\sin\phi}{4\pi m^3c^2\nu}\right)^{(\gamma-1)/2}, \qquad \text{(IV.149)}$$

so that

$$I_\nu = \frac{\gamma+\frac{7}{3}}{\gamma+1}\,B(\gamma)\,\nu^{(\gamma-1)/2}\int \mathrm{d}V(H\sin\phi)^{(\gamma+1)/2}, \qquad \text{(IV.150)}$$

$$Q_\nu = B(\gamma)\,\nu^{(\gamma-1)/2}\int \mathrm{d}V(H\sin\phi)^{(\gamma+1)/2}\cos 2\chi, \qquad \text{(IV.151)}$$

$$U_\nu = B(\gamma)\,\nu^{(\gamma-1)/2}\int \mathrm{d}V(H\sin\phi)^{(\gamma+1)/2}\sin 2\chi, \qquad \text{(IV.152)}$$

$$V_\nu = 0, \qquad \text{(IV.153)}$$

where

$$B(\gamma) = \frac{\sqrt{3}}{4\pi}\,2^{(\gamma-3)/2}\,\Gamma\left(\frac{3\gamma-1}{12}\right)\Gamma\left(\frac{3\gamma+7}{12}\right)\frac{A}{r^2}\,\frac{e^3}{mc^2}\left(\frac{3e}{4\pi m^3c^5}\right)^{(\gamma-1)/2}$$

$$\text{(IV.154)}$$

(cf. Korchak, 1963).

In a homogeneous magnetic field $\mathbf{H} = \mathbf{H}_0$ (ϕ = const, χ = const) the degree of polarization becomes

$$\rho_0 = \frac{\gamma+1}{\gamma+\frac{7}{3}}. \qquad \text{(IV.155)}$$

A more detailed treatise on this topic was given e.g. by Pacholczyk (1970).

c. *Influence of the Plasma Medium, Razin Effect*

In a medium with a refractive index deviating from unity, i.e.

$$0 < 1 - n_j \ll 1 \qquad \text{(IV.156)}$$

Equation (14.133) may be rewritten

$$\eta_{\omega,j} = \frac{\sqrt{3}e^3H\sin\phi}{2\pi mc^2}\left[1+(1-n_j^2)\gamma^{*2}\right]^{-1/2}\frac{\omega}{\omega_{n,j}}\int_{\omega/\omega_{n,j}}^{\infty} K_{5/3}(z)\,\mathrm{d}z$$

$$\text{(IV.157)}$$

whereas the critical frequency now reads

$$\omega_{n,j} = \omega_c[1 + (1 - n_j^2)\gamma^{*2}]^{-3/2} \tag{IV.158}$$

(Razin, 1960a,b; Ginzburg and Syrovatskij, 1963; Zheleznyakov, 1967a).

According to Equation (IV.157) some modifications of the spectrum of the synchrotron radiation due to the presence of a medium with $n_j \neq 1$ are to be noticed especially at the lower frequencies. The position of the spectral maximum is changed as indicated by the change of the critical frequency. Whereas for $n_j = 1$ the frequency of the spectral maximum is approximately given by

$$\omega_{\max} \approx (\omega_H/2)\gamma^{*2}\sin\phi, \tag{IV.159}$$

now, for $n_j \neq 1$, it becomes

$$\omega_{\max,n} \simeq \sqrt{2}\,\omega_p\gamma^* \tag{IV.160}$$

(Zheleznyakov and Trakhtengerts, 1965). Thus a shifting of the maximum of the synchrotron radiation spectrum towards higher frequencies and a suppression of the intensity at low frequencies is to be noted. This suppression of the spectrum of the synchrotron radiation at longer wavelengths is known as *Razin effect* or Razin–Tsytovich effect (Tsytovich, 1951; Razin, 1957, 1960a; Ramaty and Lingenfelter, 1967). It was argued by several authors (Zheleznyakov and Trakhtengerts, 1965; Ramaty and Lingenfelter, 1967; Boischot and Clavelier, 1967; Simon 1969a), that this effect may be of importance for the explanation of some spectral features of type IV bursts in the meter range.

According to condition (IV.156) the influence of the refractive index must be taken into account, if

$$1 - n_j^2 \gtrsim \gamma^{*-2} = 1 - \beta^2. \tag{IV.161}$$

In the spectral range of interest applying to synchrotron radiation the approximation

$$n^2 = 1 - X \tag{IV.162}$$

is sufficient. Hence it follows for the range where the influence of the medium is remarkable

$$\omega^2 \lesssim \omega_p^2\gamma^{*2} = \frac{2\omega_p^2\omega_c}{3\omega_H\sin\phi} \tag{IV.163}$$

or, using (IV.160),

$$\omega \lesssim \omega_{\max,n}/\sqrt{2}. \tag{IV.164}$$

Taking $\omega \lesssim \omega_c$ Ginzburg and Syrovatskij (1963, 1965) obtained the following criterion for an estimation of the frequency range, where the Razin effect should essentially influence (drop) the shape of the synchrotron radiation spectrum

$$\omega \lesssim \omega_R = \frac{\omega_p}{\alpha} = \frac{3}{2}\frac{\omega_H\sin\phi}{\alpha^2} = \frac{2}{3}\frac{\omega_p^2}{\omega_H\sin\phi} \tag{IV.165}$$

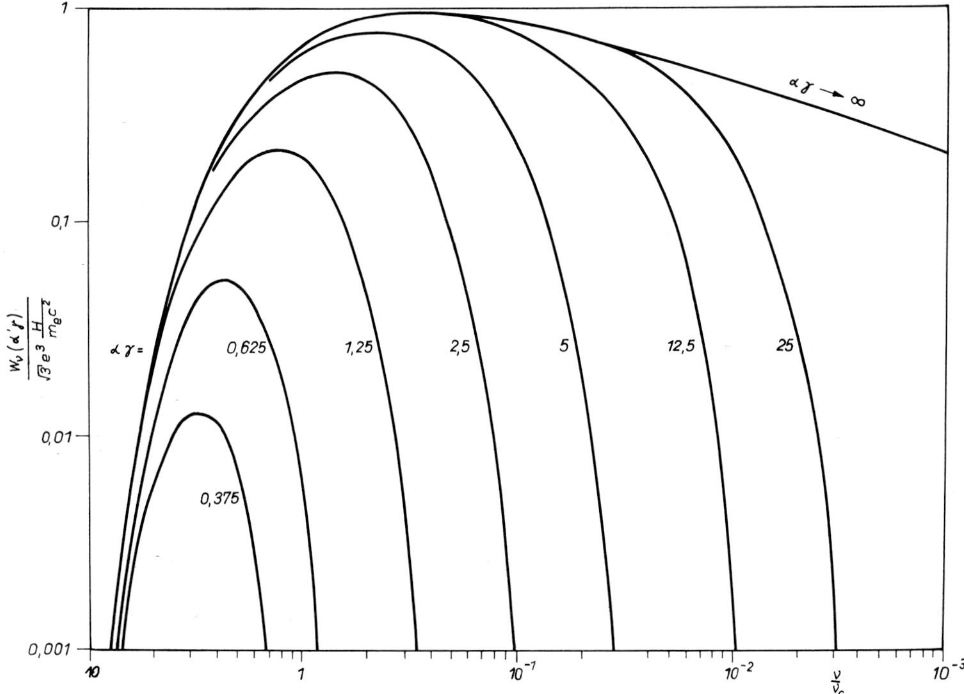

Fig. IV.16. Single-particle synchrotron radiation spectra for various values of the parameter $\alpha\gamma^*$ (after Ramaty and Lingenfelter, 1968).

where

$$\alpha = \frac{3\omega_H \sin\phi}{2\omega_p}$$

which yields numerically

$$\nu \leq \nu_R = \frac{4ceN_e}{3H\sin\phi} \approx 20\frac{N_e}{H_{[\text{gauss}]}\sin\phi} \qquad [\text{cgs}].$$

Emission spectra of monoenergetic electrons in a constant magnetic field were calculated e.g. by Ramaty and Lingenfelter (1967) and Ramaty (1968b, 1969b). Results are reproduced in Figure IV.16. It is obvious that for large values of the parameter $\alpha\gamma^*$ the influence of the medium becomes less important. In the solar corona α decreases with increasing height and values of $\alpha < 0.5$ and 0.66 can be expected at heights $h > 1.0$ and 0.7 $R_\odot$ above the photosphere, respectively.

d. *Synchrotron Reabsorption*

Absorption of synchrotron radiation can take place in vacuum by the emitting particles themselves which is called 'reabsorption' or 'self-absorption' and by the influence of a plasma medium (cf. Section 4.3.2.b, p. 202).

Usually the reabsorption coefficient is calculated by the method of the Einstein coefficients according to the principle of detailed balance. Considering again an isotropic energy distribution $F(E)$, one gets

$$\begin{aligned}\alpha_\nu &= \frac{1}{I_\nu}\frac{\mathrm{d}I_\nu}{\mathrm{d}s} = -\frac{c}{8\pi\nu^2}\int_E \mathrm{d}E\,\frac{\partial F(E)}{\partial E}\,\eta_\nu(E)\\ &= \frac{\sqrt{3}}{8\pi}\frac{e^3}{m\nu^2}H\sin\phi\int_E \mathrm{d}E\,E^2\frac{\mathrm{d}}{\mathrm{d}E}\left[\frac{F(E)}{E^2}\right]\frac{\nu}{\nu_c}\int_{\nu/\nu_c}^{\infty}K_{5/3}(z)\,\mathrm{d}z\\ &= \frac{\sqrt{3}}{8\pi}\frac{e^3}{m\nu^2}H\sin\phi\int_E \mathrm{d}E\,\frac{F(E)}{E^2}\frac{\mathrm{d}}{\mathrm{d}E}[E^2]\frac{\nu}{\nu_c}\int_{\nu/\nu_c}^{\infty}K_{5/3}(z)\,\mathrm{d}z. \end{aligned} \tag{IV.166}$$

It should be remarked, that the expression (IV.166) is always positive, so that an amplification of synchrotron emission ('negative reabsorption', cf. Section 4.7.4) cannot occur in vacuum.

Taking an energy spectrum of the form (IV.148), one obtains

$$\alpha_\nu = C(\gamma)\frac{e^3}{6\pi m}\left(\frac{3e}{2\pi m^3c^5}\right)^{\gamma/2}A(H\sin\phi)^{(\gamma+2)/2}\nu^{-(\gamma+4)/2} \tag{IV.167}$$

with

$$C(\gamma) = \frac{3\sqrt{3}}{4}\frac{\gamma}{\gamma+2}\Gamma\left(\frac{3\gamma+2}{12}\right)\Gamma\left(\frac{3\gamma+22}{12}\right)$$

(Ginzburg and Syrovatskij, 1963). Hence for synchrotron radiation with strong reabsorption (in the case of large optical depths) the dependence on frequency has the following form

$$I_\nu^{(\mathrm{syn})} \sim \nu^{(1-\gamma)/2}\nu^{(4+\gamma)/2} = \nu^{5/2}. \tag{IV.168}$$

In this particular case I_ν becomes independent of the exponent γ of the energy distribution.

e. *Relations Between the Particle-Energy Spectrum and the Synchrotron-Radiation Spectrum*

Under special conditions more or less simple relationships between the emitted spectrum of the synchrotron radiation and the originating particle energy spectrum can be established.

If, e.g., the power-law distribution of Equation (IV.148) is adopted, the intensity of the synchrotron radiation originating in an extended volume with the characteristic length L (where $n = 1$, $H = \text{const}$) is given by

$$I_\nu = \frac{\sqrt{3}}{4\pi}\frac{e^3}{mc^2}ALH\sin\phi\int_E F\left(\frac{\nu}{\nu_c(E)}\right)E^{-\gamma}\,\mathrm{d}E \tag{IV.169}$$

if reabsorption is neglected. For

$$F\left(\frac{\nu}{\nu_c(E)}\right) = \frac{\nu}{\nu_c}\int_{\nu/\nu_c}^{\infty}K_{5/3}(z)\,\mathrm{d}z \tag{IV.170}$$

the approximation of Wallis (1959) can be used

$$F\left(\frac{\nu}{\nu_c}\right) \approx q\left(\frac{\nu}{\nu_c}\right)^{\alpha^*} e^{-\nu/\nu_c} \tag{IV.171}$$

where numerically

$$q = 1.78, \qquad \alpha^* = 1/3.$$

Hence it follows

$$I_\nu = \frac{q}{4\pi} ACLH \sin\phi\, \nu^{1/3} \mu^{-1/3} \int_E E^{-\gamma-2/3} \exp\left(-\frac{\nu}{\mu E^2}\right) \tag{IV.172}$$

where

$$C = \frac{\sqrt{3}\,e^3}{mc^2}, \qquad \mu = \frac{3}{4\pi} \frac{eH \sin\phi}{m^3 c^5}.$$

Substituting

$$x = E\sqrt{\frac{\mu}{\nu}} \tag{IV.173}$$

it follows

$$I_\nu = \frac{q}{4\pi} ACLH \left(\frac{\nu}{\mu}\right)^{-(\gamma-1)/2} \int_x x^{-\gamma'-2/3} \exp\left(-\frac{1}{x^2}\right) \mathrm{d}x \tag{IV.174}$$

(Priester and Rosenberg, 1965).

This means that in the range of the validity of the above approximations the intensity of the synchrotron radiation is proportional to

$$I_\nu \sim \nu^{-(\gamma-1)/2}. \tag{IV.175}$$

If strong reabsorption is included, the intensity becomes independent of the particular exponent of the energy distribution as already remarked at the end of the foregoing subsection.

In more complicated cases the integral equation

$$I_\nu = \int_s \int_E \frac{\eta_\nu(E)}{4\pi} F(E, s)\, \mathrm{d}E\, \mathrm{d}s \tag{IV.176}$$

(s denotes the ray path) must be solved which under suitable conditions can be reduced e.g. to a Laplace transform.

4.4.3. The Čerenkov Effect

The role of the Čerenkov radiation as a limiting case of the synchrotron radiation as well as of the bremsstrahlung was already touched upon in a discussion of Table IV.1. The main importance of the Čerenkov effect for solar radio astronomy lies particularly in an efficient excitation of longitudinal wave components. Some aspects of this question will be discussed in the next section. Now we are merely dealing with the elementary process and its restricted relevance to the direct generation of electromagnetic waves.

The Čerenkov effect is given by the one-particle resonance condition

$$v_e \cos\psi = v_{k,j} \tag{IV.177}$$

where $\psi = \sphericalangle(\mathbf{v}_e, \mathbf{k}_j)$. Equation (IV.177) means, that radiation is emitted from a perturbing particle (electron) having a speed $\mathbf{v}_e$ greater than (or equal to) the phase velocity of a particular wave mode in a medium. In the literature the case $\mathbf{v}_e \| \mathbf{H}$ is mainly considered, so that the Čerenkov angle ψ, which is the angle between the electron's 'bow-shock' wave's normal direction and the electron's speed direction (cf. Figure IV.17) becomes equal to the angle $\theta = \sphericalangle(\mathbf{H}, \mathbf{k}_j)$. Then we have

$$\cos\theta = 1/(\beta n_j). \tag{IV.178}$$

At each point of the electron's path the radiation is emitted into a cone determined by the half-angle θ, called the *Čerenkov cone*, but the pulse radiated by a single electron is of finite duration. In a dispersive medium the emitted frequency spectrum as well as the angle θ depend on the wave frequency ω.

In an isotropic plasma the refractive index for electromagnetic waves (in contrast to that for plasma waves) is always ≤ 1, so that direct electromagnetic Čerenkov

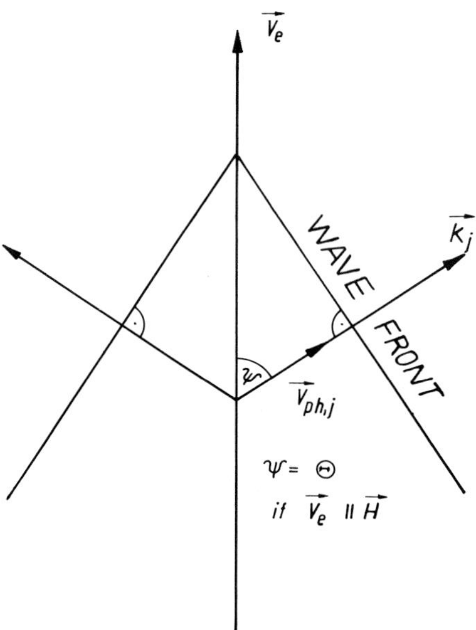

Fig. IV.17. Geometry of the Čerenkov cone.

radiation cannot exist here for any $\beta < 1$. Only in an anisotropic plasma under certain circumstances can Čerenkov radiation be emitted in the form of electromagnetic waves directly, but the propagation conditions are often unfavorable for leaving the Sun (cutoff of the extraordinary wave mode).

Inserting special expressions for n_j, e.g. using the Appleton–Hartree formula, into (IV.178), this equation can be resolved for $\cos^2\theta$ in dependence of β, ω, ω_H, and ω_p. A graphic representation of this relationship for a particular $\beta = 0.5$ according to Heald and Wharton (1965) is shown in Figure IV.18.

It may be noted that in a thermal plasma the Čerenkov radiation can never exceed the blackbody limit resulting from the inverse process (absorption), which is the Landau damping and making the effect indistinguishable from the influence of e.g. gyro-synchrotron emission. But the greater importance of the Čerenkov effect is in its relevance to superthermal particles. Here the Čerenkov emission corresponding to velocity components parallel to the direction of an external magnetic field is clearly distinguished from the synchrotron radiation, which preferably corresponds to particle speeds perpendicular to the magnetic field direction.

Detailed treatments of the Čerenkov radiation were given e.g. by Jelly (1958), Sollfrey and Yura (1965), McKenzie (1966), Seshadri and Tuan (1965), and others.

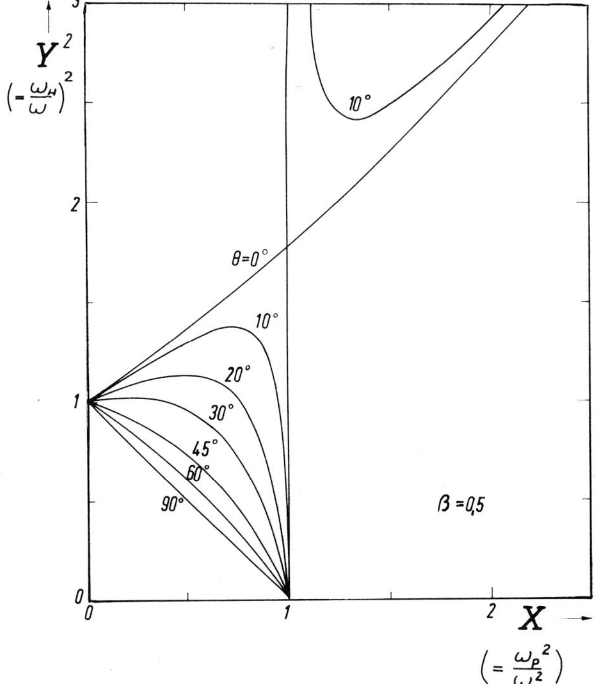

Fig. IV.18. Čerenkov angles θ for electrons moving in the direction of the magnetic field in dependence of ω_H^2/ω and ω_p^2/ω (after Heald and Wharton, 1965).

4.5 Warm-Plasma Effects: Gyroresonance Absorption and Plasma Waves

4.5.1. Kinetic Theory

a. *Introductory Remarks*

In contrast to the magnetohydrodynamics, which makes use of a macroscopic (fluid) description of the physical processes, the kinetic theory starts from a microscopic point of consideration. In this way equations for an assembly of electrons (and/or other particles) are considered which, in order to be tractable, must be treated with methods of statistical physics. The state (or 'phase') of a system of particles is commonly described by determination of their coordinates in a phase space. The following are to be distinguished:

(1) The *μ-space* is set up by six coordinates representing the instantaneous position and momentum vectors of a given particle. In a finite time interval each particle causes a single (in general moving) point (orbit) in this space. Instead of the determination of a multitude of such orbits a distribution function $f(\mathbf{r}, \mathbf{v}, t)$ is introduced describing the probability for a particle to have a certain position in a given volume element in the μ-space.

(2) As a refinement of the foregoing, the *Γ-space* is defined as being $6N$-dimensional where N is the total number of particles in the system. It consists of $3N$ coordinates determining the configuration space and $3N$ coordinates of the momentum space. Each state of the whole system is marked by one point in the Γ-space. Comparison of many systems leads to probability expressions (distribution functions).

In radio physics the kinetic theory offers useful applications especially for a treatment of longitudinal wave modes (plasma waves) and/or, generally, of radiation in the vicinity of critical frequencies.

b. *The Boltzmann–Vlasov Equation*

Let us remember that in the μ-space for each sort of particle α a distribution function $f_\alpha(\mathbf{r}, \mathbf{v}, t)$ is introduced, where

$$N_\alpha = f_\alpha(\mathbf{r}, \mathbf{v}, t)\, \mathrm{d}^3\mathbf{r}\, \mathrm{d}^3\mathbf{v} \tag{IV.179}$$

is the number of particles of the sort α being at the time t in a volume element $\mathrm{d}^3\mathbf{r}$ at $\mathbf{r}$ and in a velocity range $\mathrm{d}^3\mathbf{v}$ at $\mathbf{v}$. The motion of the particles in the plasma is then described by the Boltzmann equation

$$\frac{\partial f_\alpha}{\partial t} + \mathbf{v}\cdot\nabla f_\alpha + \frac{e_\alpha}{m_\alpha}\left(\mathbf{E} + \frac{1}{c}\mathbf{v}\times\mathbf{B}\right)\cdot\frac{\partial f_\alpha}{\partial \mathbf{v}} - \nabla\Phi_\alpha = \left[\frac{\partial f_\alpha}{\partial t}\right]_{\mathrm{coll}} \tag{IV.180}$$

(e_α, m_α = charge and mass of the particles of sort α, $\nabla\Phi_\alpha$ – external potential forces).

For the sake of self-consistency the electromagnetic fields must satisfy the Maxwell equations, e.g. in the form

$$\begin{aligned}
\operatorname{div}\mathbf{B} &= 0,\\
\operatorname{div}\mathbf{E} &= 4\pi(\rho + \rho_0),\\
\operatorname{curl}\mathbf{B} &= (4\pi/c)(\mathbf{j} + \mathbf{j}_0) + (1/c)\,\partial\mathbf{E}/\partial t,\\
\operatorname{curl}\mathbf{E} &= (1/c)\,\partial\mathbf{B}/\partial t.
\end{aligned} \tag{IV.181}$$

ρ and $\mathbf{j}$ denote the charge and current densities, respectively, determined by

$$\rho = \sum_\alpha e_\alpha \int f_\alpha \, \mathrm{d}^3\mathbf{v} \tag{IV.182}$$

$$\mathbf{j} = \sum_\alpha e_\alpha \int \mathbf{v} f_\alpha \, \mathrm{d}^3\mathbf{v}. \tag{IV.183}$$

(The subscript 0 denotes constant background fields.)

Accordingly the particle density and average velocity, respectively, are given by

$$N_\alpha = \int f_\alpha \, \mathrm{d}^3\mathbf{v} \tag{IV.184}$$

$$\mathbf{v}_\alpha = \int \mathbf{v} f_\alpha \, \mathrm{d}^3\mathbf{v} \Big/ \int f_\alpha \, \mathrm{d}^3\mathbf{v}. \tag{IV.185}$$

The conservation of particles and charge is described by the continuity equations

$$\partial N_\alpha/\partial t + \nabla\cdot(N_\alpha \mathbf{v}) = 0, \tag{IV.186}$$

$$\partial\rho/\partial t + \nabla\cdot\mathbf{j} = 0. \tag{IV.187}$$

The term $[\partial f_\alpha/\partial t]_{\mathrm{coll}}$ in Equation (IV.180) contains possible changes of f_α due to collisions and comprises all interactions that are not taken into account by the average fields of $\mathbf{E}$ and $\mathbf{H}$ on the left-hand side of Equation (IV.180). However, if a plasma is sufficiently dilute and warm or hot (the designation 'hot' plasma may be applied in the relativistic range of energies (cf. Heald and Wharton, 1965)) so that kinetic effects predominate, the collisions become relatively unimportant and the collision term may be neglected (this case is just opposite to the MHD-approximation, where the collisions play an important role).

The homogeneous, collisionless Boltzmann equation is called the Vlasov equation and is written for an electron gas in the following simplified form

$$\frac{\partial f}{\partial t} + \mathbf{v}\cdot\nabla f - \frac{e}{m}\left[\mathbf{E} + \frac{1}{c}\mathbf{v}\times\mathbf{B}\right]\cdot\frac{\partial f}{\partial\mathbf{v}} = 0 \tag{IV.188}$$

(Vlasov, 1945). A solution of the Vlasov equation was given by Landau (1946) using a perturbation method (cf. also Montgomery and Tidman, 1964). It is found, that due to the long-range Coulomb forces in a plasma, oscillations are always present giving rise to longitudinal plasma waves. The minimal wavelength of the plasma waves turns out to be the Debye length. The existence of plasma waves is a fundamental result in this theory. Thus it can be stated, that two basic kinds of particle motions are present in a plasma: *stochastic thermal motions* and *oscillatory plasma motions*. The range of the usefulness of the plasma approximation, i.e. the range where the plasma description is statistically meaningful, is given by the condition, that the number of particles in a Debye sphere must be large, i.e.

$$g \equiv 1/(Nl_D^3) \ll 1 \qquad \text{(IV.189)}$$

where the parameter g indicates to which degree collective effects dominate over single-particle characteristics.

The general principles of the kinetic theory were treated e.g. by Bogolyubov (1946). Specific applications to plasma physics were given by Klimontovich (1964), Wu (1966), and others. Introductions to the kinetic theory of ionized gases can be found in numerous textbooks, e.g. by Van Kampen and Felderhof (1967), Krall and Trivelpiece (1973), etc.

4.5.2. Electromagnetic Gyroresonance Absorption

The kinetic theory is suitable for application for a calculation of the gyroharmonic effects. It is obvious that near the gyroharmonics the cold-plasma theory is a very bad approximation, since the refraction index given in Equation (IV.61) shows only a pole at the fundamental gyrofrequency and no influence at its harmonics. Essential improvements are given by the kinetic theory of a warm plasma. Such calculations were carried out at first by Sitenko and Stepanov (1956), Gershman (1960), and some others. These authors considered small fluctuations of an electron plasma in a thermal equilibrium permeated by a constant, homogeneous magnetic field described by the linearized Vlasov equation

$$\frac{\partial f}{\partial t} + \mathbf{v}\cdot\frac{\partial f}{\partial \mathbf{r}} - \frac{e}{m}\mathbf{E}\frac{\partial f_0}{\partial v} - \frac{e}{m}(\mathbf{v} \times \mathbf{B})\cdot\frac{\partial f}{\partial \mathbf{v}} = 0 \qquad \text{(IV.190)}$$

where $f_0 = N_0(m/(2\pi KT))^{3/2}\exp(-mv^2/(KT))$ denotes the Maxwell distribution (normalized to N_0) and following from the Maxwell equations the self-consistent electric field is given by

$$\Delta\mathbf{E} - \operatorname{grad}\operatorname{div}\mathbf{E} - \frac{1}{c^2}\frac{\partial^2\mathbf{E}}{\partial t^2} - \frac{4\pi}{c^2}\frac{\partial\mathbf{j}}{\partial t} = 0. \qquad \text{(IV.191)}$$

Solutions of the system of differential Equations (IV.191) can be found in the form

$$f(\mathbf{r}, \mathbf{v}, t) = f(\mathbf{v}, \mathbf{k}, \omega)\, e^{i(\mathbf{k}\cdot\mathbf{r} - \omega t)} \qquad \text{(IV.192)}$$

$$\mathbf{E}(\mathbf{r}, t) = \mathbf{E}(\mathbf{k}, \omega)\, e^{i(\mathbf{k}\cdot\mathbf{r} - \omega t)} \qquad \text{(IV.193)}$$

leading to a dispersion equation

$$An^4 + Bn^2 + C = 0. \qquad \text{(IV.194)}$$

The constants A, B, and C can be expressed in terms of the components of the dielectric tensor ε_{il}:

$$A = \varepsilon_{11}\sin^2\theta + \varepsilon_{33}\cos^2\theta + 2\varepsilon_{13}\cos\theta\sin\theta,$$

$$B = 2(\varepsilon_{12}\varepsilon_{23} - \varepsilon_{22}\varepsilon_{13})\cos\theta\sin\theta - (\varepsilon_{22}\varepsilon_{33} + \varepsilon_{23}^2)\cos^2\theta + \varepsilon_{13}^2 - $$
$$ - \varepsilon_{11}\varepsilon_{22} - (\varepsilon_{11}\varepsilon_{22} + \varepsilon_{12}^2)\sin^2\theta, \qquad \text{(IV.195)}$$
$$C = |\varepsilon_{il}|;$$

$$\varepsilon_{11} = 1 - \frac{\omega_p^2 4z_0}{\omega^2\sqrt{\pi}}\sum_{s=-\infty}^{\infty}\frac{s^2}{\alpha^2}\int_0^\infty t^{e-t}\, J_s^2(\alpha t)\,\mathrm{d}t\int_y \frac{e^{-y^2}}{z_s - y}\,\mathrm{d}y$$

$$\varepsilon_{12} = -i\frac{\omega_p^2\, 4z_0}{\omega^2\sqrt{\pi}}\sum_{s=-\infty}^{\infty}\frac{s}{\alpha}\int_0^\infty t^2 e^{-t^2} J_s(\alpha t)\, J_s'(\alpha t)\,\mathrm{d}t\int_y\frac{e^{-y^2}}{z_s - y}\,\mathrm{d}y$$
$$= -\varepsilon_{21}$$

$$\varepsilon_{13} = -\frac{\omega_p^2\, 4z_0}{\omega^2\sqrt{\pi}}\sum_{s=-\infty}^{\infty}\frac{s}{\alpha}\int_0^\infty te^{-t^2}J_s^2(\alpha t)\,\mathrm{d}t\int_y\frac{ye^{-y^2}}{z_s - y}\,\mathrm{d}y$$
$$= \varepsilon_{31}$$

$$\varepsilon_{22} = 1 - \frac{\omega_p^2\, 4z_0}{\omega^2\sqrt{\pi}}\sum_{s=-\infty}^{\infty}\int_0^\infty t^3 e^{-t^2}J_s'^2(\alpha t)\,\mathrm{d}t\int_y\frac{e^{-y^2}}{z_s - y}\,\mathrm{d}y$$

$$\varepsilon_{23} = i\frac{\omega_p^2\, 4z_0}{\omega^2\sqrt{\pi}}\sum_{s=-\infty}^{\infty}\int_0^\infty t^2 e^{-t^2}J_s(\alpha t)\, J_s'(\alpha t)\,\mathrm{d}t\int_y\frac{ye^{-y^2}}{z_s - y}\,\mathrm{d}y$$
$$= -\varepsilon_{32}$$

$$\varepsilon_{33} = 1 - \frac{\omega_p^2\, 4z_0}{\omega^2\sqrt{\pi}}\sum_{s=-\infty}^{\infty}\int_0^\infty te^{-t^2}J_s^2(\alpha t)\,\mathrm{d}t\int_y\frac{y^2e^{-y^2}}{z_s - y}\,\mathrm{d}y \qquad \text{(IV.196)}$$

where

$$\alpha = \frac{k_x}{\omega_H}\sqrt{\frac{2KT}{m}}, \qquad z_s = \frac{(\omega - s\omega_H)}{k_z}\sqrt{\frac{m}{2KT}}.$$

Based on these expressions different approximations are possible leading to a practical calculation of the absorption of electromagnetic radiation. Such gyro-'resonance' absorption coefficients were explicitly given e.g. by Ginzburg (1960) and applied by Kakinuma and Swarup (1962). For the first few harmonics we quote the following formulas based on Ginzburg's monograph in a form applied by Yip (1967):

$$\alpha_\omega|_{s=1} = \sqrt{\frac{2}{\pi}}\frac{k_j\beta_{th}\cos\theta/(Xn_j)}{2X - 2 - \sin^2\theta + 2n_j^2\sin^2\theta}\{[1 - (1 - \tfrac{7}{4}\sin^2\theta)\,X]\,n_j^4 - $$
$$ - [2 + X(\tfrac{7}{4}\sin^2\theta - \tfrac{5}{2}) + \tfrac{1}{4}X^2(2\cos^2\theta - \tan^2\theta)]\,n_j^2 + $$
$$ + [1 - \tfrac{3}{2}\mathrm{m} + \tfrac{1}{2}X^2(1 - \tan^2\theta) + \tfrac{1}{4}X^3\tan^2\theta]\},$$

$$\alpha_\omega|_{s=2} = \sqrt{\frac{\pi}{8}}\frac{k_j X\beta_{th} n_j \sin^2\theta}{\cos\theta} B(Y)_{Y=1/2} \exp\left\{\frac{-(1-2Y)^2}{2\beta_{th}^2 n_j^2 \cos^2\theta}\right\}_{Y=1/2},$$

$$\alpha_\omega|_{s=3} = \frac{27}{8}\sqrt{\frac{\pi}{8}}\frac{k_j X\beta_{th}^3 n_j^2 \sin^4\theta}{\cos\theta} B(Y)_{Y=1/3} \exp\left\{\frac{-(1-3Y)^2}{2\beta_{th}^2 n_j^2 \cos^2\theta}\right\}_{Y=1/3} \tag{IV.197}$$

where

$$B(Y) = \frac{Y^2-1}{Y^2 n_j^2}\left\{\frac{1}{2} n_j^4 \sin^2\theta + \left[\left(\frac{1}{2} + \frac{1}{2}\cos^2\theta + \frac{\sin^2\theta}{1+Y}\right)X - \right.\right.$$
$$\left.\left. - \left(1 + \frac{1}{2}\sin^2\theta\right)\right] n_j^2 + (1-X)\left(1 - \frac{X}{1+Y}\right)\right\} \times$$
$$\times \left[2(1 - Y^2 - X - XY^2\cos^2\theta)\, n_j^2 - 2(1-X)^2 - \right.$$
$$\left. - (1+\cos^2\theta)\, XY^2 + 2Y^2\right]^{-1}.$$

The expressions (IV.197) are valid for

$$\sqrt{\frac{KT}{m}}\, k_j \cos\theta \gg \nu_{\text{coll}}, \qquad \omega \gg 2\pi\nu_{\text{coll}}$$

$$\frac{KT}{m}\frac{k_j^2}{\omega_H^2}\sin^2\theta \ll 1, \qquad \theta \neq \pi/2.$$

It must be remarked, however, that a consequent extension of these calculations to nonthermal energy distributions leads to expressions which are hardly tractable. With respect to these difficulties practically only little more is achieved in comparison to the results of the single-particle theory. But there are still other aspects of interest from the point of view of the kinetic theory. Considering departures from the thermal equilibrium collective processes give rise to instabilities and also to excitation of other than electromagnetic wave modes. Concerning these questions a wide field was opened, where the methods of the kinetic theory proved to be a powerful tool.

4.5.3. General Aspects of Radiation in a Turbulent Plasma

Turning now to a description of general wave phenomena in a plasma (not being restricted to electromagnetic wave modes) it is convenient to consider some definitions developed for a treatment of plasma waves. The energy density of a wave of any mode j is given by

$$W_j = \int W_{k,j}\, \mathrm{d}\mathbf{k}_j = \int W_{k,j}\, \mathrm{d}k_{j,x}\, \mathrm{d}k_{j,y}\, \mathrm{d}k_{j,z}. \tag{IV.198}$$

The quantity $W_{k,j}$ is called *spectral density* of the waves of type j.

Furthermore a *wave number density* (of 'plasmons', cf. below) $N'_{k,j}$ in the wave vector space or $N_{k,j}$ in the phase space is introduced by the relation

$$N'_{k,j} = N_{k,j}/\hbar, \qquad N_{kj} = (2\pi)^3\, W_{k,j}/\omega(\mathbf{k}_j). \tag{IV.199}$$

In the case of isotropy one finds

$$W_j = 4\pi \int W_{k,j} k_j^2 \, \mathrm{d}\mathbf{k}_j \tag{IV.200}$$

and

$$W_{k,j} = k_j^2 \omega(\mathbf{k}_j)\, N_{k,j}/(2\pi^2). \tag{IV.201}$$

Instead of a normalization to the wave number a normalization to the angular frequency yields

$$W_{\omega,j} = W_{k,j}\, \mathrm{d}\mathbf{k}_j/\mathrm{d}\omega \quad \text{or} \quad W_{k,j} = W_{\omega,j} v_{g,j}\,(\mathbf{k}_j). \tag{IV.202}$$

Aiming an application to electromagnetic waves the spectral density W_ω is connected with the intensity of the radiation by

$$W_\omega = \int I_\omega \, \mathrm{d}\Omega / v_g \tag{IV.203}$$

and for isotropy

$$W_\omega = 4\pi I_\omega / v_g \quad \text{or} \quad W_k = 4\pi I_\omega \tag{IV.204}$$

(the index j was omitted for simplicity).

It is customary to express the electromagnetic (RF) radiation by quantities related to the frequency (ω or ν) and the (non-RF) plasma-wave phenomena by quantities related to the wave number.

In order to have a unique scale for a comparison of different radiation it is convenient to use the concept of temperature even in those cases when a thermal equilibrium is widely absent. Thus for any wave mode an effective temperature T_{eff} is defined by means of its (monochromatic) intensity

$$I_\omega = \frac{n^2 \omega^2 K T_{\mathrm{eff}}}{4\pi^3 c^2} \tag{IV.205}$$

(index j omitted for simplicity), whereas it was assumed $\hbar\omega \ll K T_{\mathrm{eff}}$, $\vartheta = \sphericalangle(\mathbf{v}_k, \mathbf{v}_g) = 0$. The refractive index is given by

$$n = ck/\omega = c/v_k \tag{IV.206}$$

$$v_k = \omega/k \tag{IV.207}$$

$$v_g = \mathrm{d}\omega/\mathrm{d}k, \tag{IV.208}$$

so that in the case of isotropy

$$T_{\mathrm{eff}} = \frac{2\pi^2}{Kk^2} W_k = \frac{\omega}{K} N_k. \tag{IV.209}$$

Since, in a dilute plasma, the dissipation by collisions plays a minor role, conditions for the growth of small perturbations (instabilities) and drastic motions are achieved. Now a state of a plasma, where any wave modes (broad bands) are strongly developed (and propagate in all directions), is called *turbulent* (cf. Kadomtsev, 1976; Davidson, 1972; Tsytovitch, 1971).

Plasma waves have also been considered as quasi-particles or *plasmons* (Tsytovich, 1967). Originally the term 'plasmon' was introduced by Pines (1956) to describe small-amplitude plasma oscillations.

If the energy density

$$W_j \ll N_e K T_{\text{eff}} \tag{IV.210}$$

is small compared with the thermal energy density of the ambient plasma, the turbulence is called weak. In the other case

$$W_j \approx N_e K T_{\text{eff}} \tag{IV.211}$$

strong turbulence is achieved, where strong interactions of different wave modes are to be expected and a fast dissipation appears, called 'turbulent heating'. This process is characterized by a successive irreversible flow of energy from larger into smaller scales of fluctuations, finally resulting in a statistically randomized thermal plasma. For the purpose of radio astronomy it is important to consider the generation of RF waves from such turbulent plasmas. The radio waves resulting from a transformation of different wave modes in a plasma are concurrent with those waves directly generated by the processes discussed in Section 4.3. For a quantitative evaluation in analogy to Equation (IV.201) (isotropy) an *emission probability* u_k is introduced by

$$\eta_{k,j} = k_j^2 \omega(\mathbf{k}_j)\, u_{k,j}/2\pi^2 \tag{IV.212}$$

($\eta_{k,j}$ = emissivity of a mode j in the wave-vector space, i.e. the energy emitted by one particle (plasmon) in the units of wave-vector space and time).

4.5.4. Longitudinal Plasma Waves

a. *Plasma Waves in an Isotropic Medium* (*Langmuir Waves*)

Out of the large number of possible wave disturbances in a plasma – apart from pure transverse radio waves – we will discuss a few examples at first in the frame of *linear waves*. The term 'linear' refers to an approximation made for small-amplitude waves which result from a linearization of the basic kinetic equation (nonlinear wave phenomena excluded in this section would refer to shock and large-amplitude waves).

For an understanding of the nature of any wave mode, generally the knowledge of the dispersion relation

$$\omega = \omega(\mathbf{k}) \tag{IV.213}$$

is necessary, from which e.g. the phase velocity $v_{k,j} = c/n_j$ and the group velocity $v_{g,j}$ can be derived.

The hitherto best treated example of longitudinal waves in a plasma is given by the electrostatic plasma waves or space charge waves, which are also called Langmuir waves, as far as the electron component is considered. Their basic property is an emission at the local plasma frequency. Electron plasma waves can propagate only if the electrons exhibit a noticeable distribution of the velocities deviating from a temperature $T \approx 0$ or any other observable velocity component. Their existence follows from a general dispersion relation where effects of a finite temperature are taken into account (cf. e.g. Stix, 1962, Allis *et al.* 1963), otherwise plasma oscillations cannot propagate in the medium. Under certain simplifications (viz. existence of a finite temperature, $\omega_i \ll \omega_r$, $\lambda_D = \lambda_{D,e}$, $H = 0$) the complex dispersion relation is

$$\omega = \omega_r + i\omega_i \tag{IV.214}$$

with

$$\omega_r = \left(\omega_p^2 + \frac{3KT}{m} k^2\right)^{1/2} \tag{IV.215}$$

$$\omega_i = -\sqrt{\frac{\pi}{8}} \frac{\omega_p}{(kl_D)^3} \exp\left[-\frac{1}{2}(kl_D)^{-2} - \frac{3}{2}\right] \tag{IV.216}$$

(cf. e.g. Krall and Trivelpiece, 1973).

Hence both propagation and dispersion (Equation (IV.215)) as well as absorption (expressed by l_D in Equation (IV.216)) follow from the existence of a temperature $T \neq 0$. The absorption is due to the Landau damping, which is a collisionless effect and does not occur in the frame of a nonstatistical description. It has a direct relation to the Čerenkov effect: Electrons moving slower than the plasma waves can take energy from it (this is the case of Landau damping), while electrons moving faster than the waves can supply energy to the waves (which is the Čerenkov effect).

b. *Cyclotron Harmonic Waves* (*Bernstein Waves*)

Cyclotron harmonic waves or *Bernstein waves* are electrostatic waves at harmonics of the gyrofrequency. They can also be regarded as complementary longitudinal components of transverse electromagnetic cyclotron waves since both components are coupled in a warm plasma. Reviews about cyclotron harmonic waves were given e.g. By Crawford (1965a, b, 1968) and Baldwin *et al.* (1969).

If the propagation proceeds as $\exp[i(\omega t - k_x x - k_z z)]$, $\mathbf{H} \parallel z$, and the presence of a bi-Maxwell distribution is presumed ($T_\parallel$ corresponds to $v_\parallel = v\cos\phi$, $T_\perp$ to $v_\perp = v\sin\phi$, $\phi = \sphericalangle(\mathbf{v}, \mathbf{B})$), cyclotron waves are subject to Landau damping unless the condition

$$(\omega - s\omega_H)/k_z \gg (2KT_\parallel/m)^{1/2} \tag{IV.217}$$

is satisfied (cf. Tataronis and Crawford, 1970). The inequality (IV.217) implies that for a wave propagation perpendicular to an outer magnetic field the Landau damping does not occur under the above given conditions. Cutoffs ($k_\perp \to 0$) occur at the cyclotron harmonics $s\omega_H$ ($s = 2, 3, 4, \ldots$) and at the upper hybrid frequency $(\omega_p^2 + \omega_H^2)^{1/2}$. The dispersion curves show in a uniform, infinite, warm, anisotropic plasma the existence of an infinite series of propagating modes carrying either positive or negative

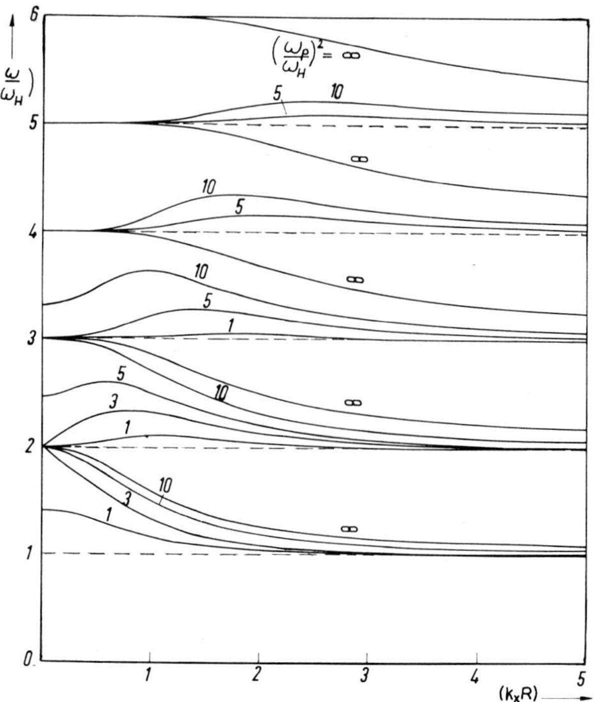

Fig. IV.19. Dispersion characteristics of perpendicularly propagating electron-cyclotron harmonic waves for a Maxwellian transverse electron-velocity distribution (after Tataronis and Crawford, 1970).

energy. The dispersion relation in the quasi-static approximation may be derived from the linearized Vlasov equation and Poisson's equation by a perturbation method:

$$\frac{\partial f_1}{\partial t} + \mathbf{v} \cdot \nabla f_1 - \frac{e}{m}(\mathbf{v} \times \mathbf{B}_0) \cdot \frac{\partial f_1}{\partial \mathbf{v}} = \frac{e}{m} \mathbf{E} \cdot \frac{\partial f_0}{\partial \mathbf{v}} \tag{IV.218}$$

$$\nabla \cdot \mathbf{E}(\mathbf{r}, t) = \frac{\rho_0(\mathbf{r}, t)}{\varepsilon_0} - \frac{N_0 e}{\varepsilon_0} \int_v \mathrm{d}^3\mathbf{v} f_1(\mathbf{v}, \mathbf{r}, t). \tag{IV.219}$$

Different types of a (time-independent) electron velocity distribution $f_0 = f_0(v_\perp, v_\parallel)$ can be used; f_1 denotes the lowest-order term of a perturbation expansion of the velocity distribution; ρ_0 = (external) charge density, $\mathbf{E}$ = perturbation field, ε_0 = = permittivity of the free space.

Resolving these equations by a Fourier transform in space and a Laplace transform in time, one obtains

$$\mathbf{E}(\mathbf{k}, \omega) = \frac{i\mathbf{k}\rho_0(\mathbf{k}, \omega)}{\varepsilon_0 k^2 \varepsilon(\omega, \mathbf{k})} \tag{IV.220}$$

where ε is the plasma equivalent permittivity (Tataronis and Crawford, 1970).

The dispersion relation is given by

$$\varepsilon(\omega, \mathbf{k}) = 0 \tag{IV.221}$$

describing the propagation of waves varying as $\exp[i(\omega t - \mathbf{k}\cdot\mathbf{r})]$.

For a propagation perpendicular to the external magnetic field it is obtained

$$\varepsilon(\omega, k_\perp) = 1 - \frac{\omega_p^2}{\omega_H^2}\sum_{s=-\infty}^{\infty} a_s(k_\perp)\frac{s\omega_H}{\omega - s\omega_H} = 0 \tag{IV.222}$$

where

$$a_s(k_\perp) = -\frac{\omega_H^2}{k_\perp^2}\int_v \mathrm{d}^3\mathbf{v}\,\frac{1}{v_\perp}\frac{\partial f_0}{\partial v_\perp}J_s^2\left(\frac{k_\perp v_\perp}{\omega_H}\right).$$

Considering a Maxwell distribution

$$f_0(v_\perp, v_\parallel) = \left(\frac{1}{2\pi v_{th}^2}\right)^{3/2}\exp\left(-\frac{v_\perp^2 + v_\parallel^2}{2v_{th}^2}\right) \tag{IV.223}$$

the dispersion characteristics shown in Figure IV.19 can be obtained.

4.5.5. Hydromagnetic Waves

Hydromagnetic waves also play a great part in the solar atmosphere both under normal and disturbed conditions. It was first suggested by Alfvén (1947), that a dissipation due to the passage of hydromagnetic waves may be important for the heating of the solar corona. Concerning solar radio astronomy, hydromagnetic waves are of interest, too, because of their modulating and scattering properties for radio waves.

Like the radio and X-ray astronomy, a new branch of 'solar hydromagnetic-wave astronomy' could arise providing new insights into solar physics. It was established in recent years, that hydromagnetic waves and discontinuities are likewise present in the interplanetary plasma and may significantly influence the thermal and dynamical properties of the solar wind. They also cause a scattering of solar and galactic cosmic rays (for references cf. e.g. Coleman, 1967; Belcher and Davies, 1971; Hollweg, 1972a, 1974). The basic equations describing small-amplitude hydromagnetic waves are the conservation laws of mass and momentum

$$\frac{\partial\rho}{\partial t} + \nabla\cdot(\rho\mathbf{v}) = 0 \tag{IV.224}$$

$$\rho\left(\frac{\partial}{\partial t} + \mathbf{v}\cdot\nabla\right)\mathbf{v} = -\nabla p + \frac{1}{c}\mathbf{j}\times\mathbf{B} \tag{IV.225}$$

(ρ = mass density, p = pressure) and Maxwell's equations (except Poisson's equation for the electric field, since the Debye length is small compared with the wavelength, so that the plasma is considered quasi-neutral)

$$\nabla\times\mathbf{B} = (4\pi/c)\,\mathbf{j} \tag{IV.226}$$

$$\partial/\partial t(\nabla\cdot\mathbf{B}) = 0 \tag{IV.227}$$

$$\nabla\times\mathbf{E} = -(1/c)\,\partial\mathbf{B}/\partial t. \tag{IV.228}$$

Ohm's law is given by

$$\mathbf{E} + (1/c)\,\mathbf{v} \times \mathbf{B} = 0. \tag{IV.229}$$

A linearization is made by the perturbation ansatz

$$\begin{aligned} \rho &= \rho_0 + \delta\rho, \\ \mathbf{v} &= \mathbf{v}_0 + \delta\mathbf{v}, \\ \mathbf{B} &= \mathbf{B}_0 + \delta\mathbf{B}, \end{aligned} \tag{IV.230}$$

distinguishing between the average values being characteristic for the background plasma (subscript 0) and fluctuating wave quantities (prefix δ).

From the basic equations one obtains

$$\frac{\delta\rho}{\rho_0} = \mathbf{k}\cdot\frac{\delta\mathbf{v}}{\omega} \tag{IV.231}$$

and

$$\delta\mathbf{v} = \frac{v_s^2}{v_A}\frac{\mathbf{k}}{\omega}\frac{\delta\rho}{\rho_0} + \mathbf{v}_A \times \left(\frac{\mathbf{k}}{\omega} \times \frac{\delta\mathbf{B}}{B_0}\right) \tag{IV.232}$$

where

$$v_A = B_0(4\pi\rho_0)^{-1/2} \tag{IV.233}$$

$$v_s = \left(\frac{\gamma p_0}{\rho_0}\right)^{1/2} \approx \left[\gamma K \frac{(T_e + T_p)}{m_p}\right]^{1/2} \tag{IV.234}$$

(the subscripts e and p refer to electrons and protons, respectively), if the fluid fluctuations are assumed to be either isothermal or adiabatic according to

$$p\rho^{-\gamma} = \text{const} \tag{IV.235}$$

with $\gamma = 1$ and $\gamma = 5/3$, respectively (γ = polytropic exponent).

The elimination of B in Equation (IV.232) yields

$$\begin{aligned} &(v_s^2 + v_A^2)(\mathbf{k}\cdot\delta\mathbf{v})\,\mathbf{k} - \omega^2\delta\mathbf{v} + (\mathbf{v}_A\cdot\mathbf{k}) \times \\ &\times [(\mathbf{v}_A\cdot\mathbf{k})\,\delta\mathbf{v} - (\mathbf{v}_A\cdot\delta\mathbf{v})\,\mathbf{k} - (\mathbf{k}\cdot\delta\mathbf{v})\,\mathbf{v}_A] = 0. \end{aligned} \tag{IV.236}$$

Hence it follows that the component $\delta\mathbf{v}_\perp$ of $\mathbf{v}$ perpendicular to the plane containing $\mathbf{k}$ and $\mathbf{v}_A$ is uncoupled to the components $\delta\mathbf{v}$ in that plane, according to the relation

$$[\omega^2 - (\mathbf{v}_A\cdot\mathbf{k})^2]\,\delta v_\perp = 0 \tag{IV.237}$$

or

$$\omega^2/k^2 = v_A^2\cos^2\alpha \tag{IV.238}$$

where α is the angle between $\mathbf{k}$ and $\mathbf{v}_A$. Equation (IV.238) is just the dispersion relation for the Alfvén waves ('intermediate mode'). For these waves one obtains

$$\delta p = 0, \qquad \delta B/B_0 = -\operatorname{sign}(\cos\alpha)\,\delta v/v_A, \tag{IV.239}$$

i.e. in this linearized approximation the waves are noncompressive and transverse, analogous to a transverse wave of a string, where the magnetic field lines provide the tension and the plasma the inertia.

In the dispersion relation (IV.236) two other modes are contained, characterized by motions $\delta\mathbf{v}$ not perpendicular to the $\mathbf{k}$-$\mathbf{v}_A$-plane. These modes are represented by

$$[k^2(v_s^2 + v_A^2) - \omega^2](\mathbf{k}\cdot\delta\mathbf{v}) - k^2(\mathbf{v}_A\cdot\mathbf{k})(\mathbf{v}_A\cdot\delta\mathbf{v}) = 0 \tag{IV.240}$$

and

$$v_s^2(\mathbf{v}_A\cdot\mathbf{k})(\mathbf{k}\cdot\delta\mathbf{v}) - \omega^2(\mathbf{v}_A\cdot\delta\mathbf{v}) = 0 \tag{IV.241}$$

reducing to

$$(\omega^2/k^2)_j = v_j^2 = (1/2)\{v_s^2 + v_A^2 \pm [(v_s^2 + v_A^2)^2 - 4v_s^2 v_A^2\cos^2\alpha]^{1/2}\}. \tag{IV.242}$$

The upper and lower signs define the fast and slow hydromagnetic wave modes, respectively (cf. e.g. Hollweg, 1974).

4.5.6. Summary of Warm-Plasma Wave Modes

Terminating this section we give a brief survey of the main wave types in a plasma. Table IV.7 characterizes roughly the domains of the three basic wave types, viz. radio waves, plasma waves, and hydromagnetic waves with respect to the density and electric (or temperature-) conductivity properties of a plasma.

Finally, in addition to Figure IV.6, Table IV.8 represents a gross compilation of the dispersion equations of some main wave types. Table IV.9 gives a summary of warm-plasma emission probabilities according to Kaplan and Tsytovich (1972).

TABLE IV.7

Occurence of basic wave types in different parameter regions and ranges of approximations

	Radio waves	Plasma waves	MHD waves
A. High density and conductivity	–	–	×
B. Small density and conductivity			
(a) cold plasma	×	–	×
(b) warm plasma	×	×	×

(× denotes the existence of propagating wave modes)

TABLE IV.8
Compilation of dispersion properties of some main wave types

Symbol	Wave type	Dispersion relation	Remarks
t	electromagnetic waves	$\omega(k) \simeq \sqrt{\dfrac{\omega_{pe}^2}{1 \pm \dfrac{\omega_{He}}{\omega}\cos\theta} + c^2k^2}$	$v_k \geq c$, $v_k \leq c$
p	transverse plasmons	$\omega(k) = \sqrt{\omega_{pe}^2 + c^2k^2}$	$v_k \approx \dfrac{\omega_{pe}}{k} \gg c$,
		$\approx \omega_{pe} + \dfrac{c^2k^2}{2\omega_{pe}}$	$v_g = \dfrac{c^2k}{\omega_{pe}} = \dfrac{c^2}{v_k} \ll c$
l	longitudinal plasmons	$\omega(k) = \sqrt{\omega_{pe}^2 + 3v_{Te}^2k^2}$	$v_k \approx \dfrac{\omega_{pe}}{k}$,
		$\approx \omega_{pe} + \dfrac{3v_{Te}k^2}{2\omega_{pe}}$	$v_g = \dfrac{3v_{Te}^2k}{\omega_{pe}} \approx \dfrac{3v_{Te}^2}{v_k}$
h	cyclotron plasmons	$\omega(k) \simeq \left[\omega_{He}^2\sin^2\theta + \omega_{pe}^2 + 4\left(\dfrac{\omega_{He}}{\omega_{pe}}\right)^2 v_{Te}^2k^2\cot^2\theta\right]^{1/2}$	for $\omega_{He} \gg \omega_{pe}$
w	whistlers	$\omega(k) = \dfrac{\omega_{He}\lvert\cos\theta\rvert}{\omega_{pe}^2 + c^2k^2}c^2k^2$	$v_k = \dfrac{\omega_{He}\lvert\cos\theta\rvert}{\omega_{pe}^2}c^2k^2$
		$\approx \dfrac{\omega_{He}\lvert\cos\theta\rvert}{\omega_{pe}^2}c^2k^2$	$\dfrac{c\omega_{Hi}}{\omega_{pi}} + v_k < \dfrac{c\omega_{He}}{\omega_{pe}}$
s	ion-sound waves	$\omega(k) = \dfrac{v_sk}{\sqrt{1 + (kl_{De})^2}} + \dfrac{3v_{Ti}^2k^2}{2\omega_{pi}}$	$v_s = \sqrt{KT/m_i}$, $T_e \gg T_i$
a	Alfvén waves	$\omega(k) = \dfrac{v_ak\lvert\cos\theta\rvert}{\sqrt{1 + \left(\dfrac{v_a}{c}\right)^2}}$	$v_a = \dfrac{c\omega_{Hi}}{\omega_{pi}} = \dfrac{H}{\sqrt{4\pi m_in_i}}$
		$\approx v_ak\lvert\cos\theta\rvert$	
mf	fast magnetic sound waves	$\omega(k) \simeq k\sqrt{v_a^2 + v_s^2\sin^2\theta k}$ $\approx v_ak$	for $\omega \ll \omega_{Hi}, v_s \ll v_a \ll c$
ms	slow magnetic sound waves	$\omega(k) \approx v_sk\lvert\cos\theta\rvert$	$v_A = c\omega_{Hi}/\omega_{pi} \gg v_s$

4.6. Wave-Mode Transformations: Wave–Particle and Wave–Wave Interactions

4.6.1. Scattering by Thermal and Superthermal Particles

a. *General Cases*

The idea that conversion processes of plasma waves may significantly contribute to solar radio emissions is not very new. It was discussed explicitly by Ginzburg and Zheleznyakov (1958a) who attempted to explain the radiation from type III bursts.

TABLE IV.9

Averaged (over Ω) emission probabilities of waves in a warm plasma (after Kaplan and Tsytovich, 1972)

Symbol	Wave type	Emission probability	Remarks
l	longitudinal plasma (Langmuir) waves		
	a) $\omega_{pe} \gg \omega_{He}$	$\bar{u}_l = \dfrac{\pi}{2} \dfrac{m_e}{N_e v} \dfrac{\omega_{pe}^3}{k^3}$	$k > \dfrac{\omega_{pe}}{v}$
	b) $\omega_{pe} \ll \omega_{He}$	$\bar{u}_{l,\max} = \dfrac{\pi}{4} \dfrac{m_e v^2}{N_e} \left(\dfrac{\omega_{pe}}{\omega_{He}}\right)^3$	$k \lesssim \dfrac{\omega_{He}}{v}$
h	Cyclotron harmonic (Bernstein) waves		
	a) $\omega_{pe} \gg \omega_{He}$	$\bar{u}_h = \dfrac{\pi}{4} \dfrac{m_e}{N_e v} \left(\dfrac{\omega_{He}}{k}\right)^3 \left(\dfrac{kv}{\omega_{He}}\right)^2$	$k < \dfrac{\omega_{He}}{v}$
	b) $\omega_{pe} \ll \omega_{He}$	$\bar{u}_h = \dfrac{2\pi}{3} \dfrac{m_e}{N_e v} \left(\dfrac{\omega_{pe}}{\omega_{He}}\right)^4 \left(\dfrac{\omega_{He}}{k}\right)^3$	$k > \dfrac{\omega_{He}}{v}$
w	whistler ($\omega_{pe} \gg \omega_{He}$)	$\bar{u}_w = \dfrac{\pi}{4} \dfrac{m_e}{N_e v} \left(\dfrac{\omega_{He}}{k}\right)^3 \left(\dfrac{kc}{\omega_{pe}}\right)^4 \left(\dfrac{\omega_{pe}^2 v}{\omega_{He} c^2 k}\right)^2$	$k > \dfrac{\omega_{pe}^2 v}{\omega_{He} c^2}$
s	ion-sound waves ($\omega_s > \omega_{Hi}$)	$\bar{u}_s = \dfrac{\pi}{2} \dfrac{m_i}{N_e v} \left(\dfrac{\omega_s}{k}\right)^3$	$v_s < v$
a	Alfvén waves ($v_s \ll v_a \ll c$)	$\bar{u}_a \simeq \dfrac{\pi}{16} \dfrac{m_i}{N_e v} \left(\dfrac{\omega_{pi}}{ck}\right)^2 v_a(v_a^2 + 2v^2)$	$v_a < v$
	$\dfrac{\omega_H}{v} < k < \dfrac{\omega_{pi}}{c}$	$\bar{u}_a \simeq \dfrac{\pi}{4} \dfrac{m_i}{N_e v} \left(\dfrac{\omega_{pi}}{ck}\right)^2 \dfrac{v^2}{v_a} (v_a^2 + 2v^2)$	$v_a > v$
mf	fast magnetic sound waves		
	$\dfrac{\omega_H}{v} < k < \dfrac{\omega_{pi}}{c}$	$\bar{u}_{mf} = \dfrac{\pi}{8} \dfrac{m_i}{N_e v} \left(\dfrac{\omega_{pi}}{ck}\right)^2 v_a(v^2 - v_a^2)$	$v_a < v$
ms	Slow magnetic sound waves		
	$\dfrac{\omega_H}{v} < k < \dfrac{\omega_{Hi}}{v_s}$	$\bar{u}_{ms} \simeq \dfrac{\pi}{4} \dfrac{m_i v_s^3}{N_e v}$	$v_s < v$
		$\bar{u}_{ms} \simeq \dfrac{\pi}{4} \dfrac{m_i v_s^3}{N_e v} \left(\dfrac{v}{v_s}\right)^2$	$v_s > v$

The early studies used the picture of a scattering of waves at local inhomogeneities. Various types of such inhomogeneities could be expected to occur in the solar atmosphere. Another suggestion of an action of wave-mode transformations was made by Denisse (1960) by a consideration of the coupling points in a dispersion diagram in order to interpret the radiation from noise storms. In principle, scattering processes may be due to interactions of waves with single particles, clouds of particles, and waves of either the same or another type. Accordingly different mechanisms were discussed in the past, which may result in a change of the ray direction and wave frequency, as well as the type (mode) of the scattered wave.

There were distinguished by:

(a) *Rayleigh scattering* of plasma waves due to density fluctuations at the frequency

$$\nu \simeq \left[\frac{\nu_p^2}{1 - \frac{3}{2}\left(\frac{v_{th}}{v_k}\right)^2} \right]^{1/2} \tag{IV.243}$$

(Bohm and Gross, 1949),

(b) *Combination scattering* of plasma waves due to space charge fluctuations (thermal plasma waves) yielding the second harmonic radiation of the plasma frequency,

(c) *Thomson scattering* at thermal or superthermal particles (electrons), whereas only the directions of the scattered wave, but not the frequency are altered. The scattering probability for *t*-, *p*-, and *l*-waves is given by

$$u_k \simeq \frac{\pi}{3} \frac{\omega_{pe}^4}{N_e^2 \omega^3} W_\omega, \tag{IV.244}$$

(d) *Compton scattering* at superthermal particles (electrons) changing the frequency distribution of the scattered waves drastically. For electromagnetic waves the the scattering probability is

$$u_k \simeq \frac{2(2\pi)^3}{3} \frac{e^4}{m_e^2} \frac{W'_{\omega'}}{\omega^3}, \tag{IV.245}$$

where $W'_{\omega'}$ is the spectral density of the wave before scattering.

In the case that for (low-frequency) waves in a volume element

$$\lambda^3 = (2\pi/k)^3 \tag{IV.246}$$

more than one electron is involved, the single electrons cannot be regarded as free and the scattering occurs at fluctuations of the electron density, i.e. at clouds of electrons caused by the action of positive ions. Apart from such a scattering at positive ions a scattering at single electrons is also possible. But the latter process only becomes remarkable, if the phase velocity satisfies the relation

$$v_k \lesssim (3m_i/m_e)^{1/5} v_{Te}, \qquad \text{i.e.} \quad v_k < 6v_{Te} \tag{IV.247}$$

(see Tsytovich, 1967).

Scattering of longitudinal plasma waves (Langmuir waves) as well as of cyclotron harmonic waves (Bernstein waves) and also of ion-sound waves by superthermal particles into high-frequency electromagnetic waves is of special interest for radio astronomy. In this respect considerable progress has been achieved in recent years, but the whole field is still developing. For rough estimates the transition probabilities averaged over all angles α and θ are reproduced in Table IV.10 according to Kaplan and Tsytovich (1972). From the transition probabilities u_σ for an isotropic velocity or momentum distribution, the emission coefficients $j_{\omega,t}$ can be calculated

TABLE IV.10

Average transition probabilities of nonlinear scattering of waves at superthermal particles (after Kaplan and Tsytovich, 1972)

Process	Transition probability	Remarks
$l + e \rightleftarrows t + e'$ (weak magnetic field $\omega_l \approx \omega_{pe}$)	$\bar{u}_{l\to t} = \frac{\pi}{10}\frac{\omega_{pe}}{N_e^2}\left(\frac{\omega_{pe}}{\omega}\right)^3 \frac{\omega_p}{k_l v_e c^2}\left(v_e - \frac{\omega^2}{k_l^2}\right)$	$\frac{\omega}{k_l} < v_e$
$l + i \rightleftarrows t + i'$ (weak magnetic field $\omega_l \approx \omega_{pe}$)	$\bar{u}_{l\to t} = \frac{\pi}{6}\frac{\omega_{pe}}{N_e^2}\left(\frac{\omega_{pe}}{\omega}\right)^3 \frac{\omega_{pe}}{k_l v_i}$	$\frac{\omega}{k_l} < v_i$
$h + e \rightleftarrows t + e'$ a) $\omega_{pe} \gg \omega_{He}$	$\bar{u}_{h\to t} = \frac{\pi}{15}\frac{\omega_{pe}}{N_e^2}\left(\frac{\omega_{He}}{\omega}\right)^3 \frac{\omega_{pe}}{k_l v_e c^2}\left(v_e^2 - \frac{\omega}{k_l^2}\right)$	$\frac{\omega}{k_l} < v_e$
b) $\omega_{pe} \ll \omega_{He}$	$\bar{u}_{h\to t} = \frac{\pi}{15}\frac{\omega_{pe}}{N_e^2}\left(\frac{\omega_{He}}{\omega}\right)^3 \frac{\omega_{pe}}{k_l v_e c^2}\left(v_e^2 - \frac{\omega}{k_l}\right)\left(\frac{\omega_{pe}}{\omega_{He}}\right)^4$	$\frac{\omega}{k_l} < v_e$
$h + i \rightleftarrows t + i'$		
a) $\omega_{pe} \gg \omega_{He}$	$\bar{u}_{h\to t} = \frac{\pi}{24}\frac{\omega_{pe}}{N_e^2}\left(\frac{\omega_{He}}{\omega}\right)^3 \frac{\omega_{pe}}{k_l v_i}$	$\frac{\omega}{k_l} < v_i$
b) $\omega_{pe} \ll \omega_{He}$	$\bar{u}_{h\to t} = \frac{\pi}{24}\frac{\omega_{pe}}{N_e^2}\left(\frac{\omega_{He}}{\omega}\right)^3 \frac{\omega_{pe}}{k_l v_i}\left(\frac{\omega_{pe}}{\omega_{He}}\right)^4$	$\frac{\omega}{k_l} < v_i$
$s + e \rightleftarrows t + e'$	$\bar{u}_{s\to t} = \frac{\pi}{10}\frac{\omega_{pe}}{N_e^2}\left(\frac{\omega_s}{\omega}\right)^3 \frac{m_i}{m_e}\frac{\omega_{pe}}{k_l v_e c^2}\left(v_e^2 - \frac{\omega^2}{k_l^2}\right)$	$\frac{\omega}{k_l} < v_e$
$s + i \rightleftarrows t + i'$	$\bar{u}_{s\to t} = \frac{\pi}{6}\frac{\omega_{pe}}{N_e^2}\left(\frac{\omega_s}{\omega}\right)^3 \frac{m_i}{m_e}\frac{\omega_{pe}}{k_l v_i}$	$\frac{\omega}{k_l} < v_i$

according to

$$j_{\omega,t} = \frac{\omega^3}{\pi^2 c^3}\int_0^\infty\int_0^\infty \bar{u}_{\sigma\to t}(\omega, k_\sigma, v)\, f_p W_{k,\sigma}\,\frac{dk_\sigma}{\omega_\sigma}\,dp. \tag{IV.248}$$

b. *Differential Scattering Within Individual Wave Modes*

Let k and k' refer to the scattered and unscattered waves, respectively; two cases can be distinguished: If

$$(k - k')/k \ll 1 \tag{IV.249}$$

the process is called 'differential scattering'. Otherwise, if

$$(k - k')/k \approx 1 \tag{IV.250}$$

the process is called 'integral scattering' (cf. Kaplan and Tsytovich, 1972).

The differential scattering is characterized by a small change of both ω and $|\mathbf{k}|$ (but not of the direction of $\mathbf{k}$). For an orientation, Table IV.11 gives a compilation of the most important realizations of differential scattering processes.

TABLE IV.11

Differential scattering of waves in a plasma (after Kaplan and Tsytovich, 1972)

Process	Change of spectral density	Conversion coefficient
$l + i \rightleftarrows l' + i'$, $l + s \rightleftarrows l'$	$\dfrac{\partial W_{k,l}}{\partial t} = \alpha_{l\to l'} W_{k,l} \dfrac{\partial W_{k,l}}{\partial k}$	$\alpha_{l\to l'} = \dfrac{\pi}{27} \dfrac{\omega_{pe}^3}{N_e m_i v_{Te}^4}$ for $T_e \gg T_i$
$l + i \rightleftarrows p + i'$, $l + s \rightleftarrows p$	$\dfrac{\partial W_{k,p}}{\partial t} = \alpha_{l\to p} W_{k,p} \dfrac{\partial (W_{k,p}\cdot k)}{\partial k} \cdot \dfrac{1}{k}$	$\alpha_{l\to p} = \alpha_{l\to l'}$
$p + i \rightleftarrows l + i'$, $p + s \rightleftarrows l$	$\dfrac{\partial W_{k,l}}{\partial t} = \alpha_{p\to l} W_{k,l} k \dfrac{\partial}{\partial k}\left(\dfrac{W_{k,p}}{k}\right)$	$\alpha_{p\to l} = \alpha_{l\to l'}$
$p + i \rightleftarrows p' + i'$, $p + s \rightleftarrows p'$	$\dfrac{\partial W_{k,p}}{\partial t} = \alpha_{p\to p'} W_{k,p} \dfrac{\partial W_{k,p}}{\partial k}$	$\alpha_{p-p'} = \dfrac{9v_{Te}^2}{c^2}\alpha_{l\to l'}$
$t + e \rightleftarrows t' + e'$, $t + l \rightleftarrows t'$	$\dfrac{\partial W_{\omega,t}}{\partial t} = \alpha_{t\to t'} W_{\omega,t} \dfrac{\partial}{\partial \omega}\left(\dfrac{W_{\omega,t}}{\omega}\right)$	$\alpha_{t\to t'} = \dfrac{\pi}{6} \dfrac{\omega_{pe}^3}{N_e m_e c^2}$
$w + i \rightleftarrows w' + i'$	$\dfrac{\partial W_{\omega,w}}{\partial t} = W_{\omega,w} \times \left(\alpha^{(1)}_{w\to w'}\omega^2 \dfrac{\partial W_{\omega,w}}{\partial \omega} + \alpha^{(2)}_{w\to w'}\omega W_{\omega,w}\right)$	$\alpha^{(1)}_{w\to w'} \approx \alpha^{(2)}_{w\to w'} \approx \approx \dfrac{\pi}{4} \dfrac{\omega_{pe}^2}{\left(1+\dfrac{T_e}{T_i}\right)^2 \omega_{He} N_e m_i c^2}$
$w + i \rightleftarrows s + i'$	$\dfrac{\partial W_{\omega,s}}{\partial t} = \alpha_{w\to s} W_{\omega,s}\omega^3 \dfrac{\partial W_{\omega,w}}{\partial \omega}$	$\alpha_{w\to s} \approx \dfrac{\pi}{2} \dfrac{1}{\left(1+\dfrac{T_e}{T_i}\right)^2 N_e m_i v_a^2}$
$s + i \rightleftarrows w + i'$	$\dfrac{\partial W_{\omega,w}}{\partial t} \approx \alpha_{s\to w} W_{\omega,w}\omega \dfrac{\partial}{\partial \omega}(\omega^2 W_{\omega,s})$	$\alpha_{s\to w} = \alpha_{w\to s}$
$s + i \rightleftarrows s' + i'$, $ms + i \rightleftarrows ms' + i'$	$\dfrac{\partial W_{k,s}}{\partial t} = \alpha_{s\to s'} W_{k,s} k^2 \dfrac{\partial}{\partial k}(k W_{k,s})$	$\alpha_{s\to s'} = \dfrac{4\pi}{15} \dfrac{T_i}{N_e m_i^2 v_s^3}$
$a + i \rightleftarrows a' + i'$, $m + i \rightleftarrows m' + i'$, $m + i \rightleftarrows a' + i'$	$\dfrac{\partial W_{\omega,a}}{\partial t} = \alpha_{a\to a'} W_{\omega,a}\omega^2\left(\omega \dfrac{\partial W_{\omega,a}}{\partial \omega} + W_{\omega,a}\right)$	$\alpha_{a\to a'} \approx \alpha_{m\to m'} \approx \alpha_{a\to m} \approx \alpha_{m\to a} \approx \dfrac{\pi}{4} \dfrac{1}{\left(1+\dfrac{v_a^2}{c^2}\right) N_e m_i v_a^2}$
$m + i \rightleftarrows ms + i'$	$\dfrac{\partial W_{\omega,ms}}{\partial t} = \alpha_{m\to ms}\omega^4 W_{\omega,ms}\left(\omega \dfrac{\partial W_{\omega,m}}{\partial \omega} + 2W_{\omega,m}\right)$	$\alpha_{m\to ms} \approx \alpha_{a\to ms} \approx \alpha_{ms\to m}$
$a + i \rightleftarrows ms + i'$	$\dfrac{\partial W_{\omega,ms}}{\partial t} = \alpha_{a\to ms}\omega^4 W_{\omega,ms}\left(\omega \dfrac{\partial W_{\omega,a}}{\partial \omega} + 4W_{\omega,a}\right)$	$\approx \alpha_{ms\to a} \approx \dfrac{\pi}{4} \dfrac{(T_i/T_e)^2}{\omega_{Hi}^2 N_e m_i v_a^2}$
$ms + i \rightleftarrows m + i'$	$\dfrac{\partial W_{\omega,m}}{\partial t} = \alpha_{ms\to m}\omega^4 W_{\omega,m}\left(\omega \dfrac{\partial W_{\omega,ms}}{\partial \omega} + 2W_{\omega,ms}\right)$	
$ms + i \rightleftarrows a + i'$	$\dfrac{\partial W_{\omega,a}}{\partial t} = \alpha_{ms\to a}\omega^4 W_{\omega,a}\left(\omega \dfrac{\partial W_{\omega,ms}}{\partial \omega} + 2W_{\omega,ms}\right)$	

4.6.2. Interactions of Different Wave Modes

The nonlinear interaction of waves, i.e. the merging or the decay of wave packets (as inverse processes to each other) was already touched in Section 4.2.1.c. A treatise on this matter was given by Tsytovich (1967) on the basis of a nonlinear description of plasma processes. In particular two cases were extensively treated in the literature. The first one is the interaction of two longitudinal plasma waves yielding a transverse electromagnetic wave at $\omega \approx 2\omega_{pe}$ in a weak external magnetic field:

$$l + l' \,_{(}\rightleftarrows_{)}\, t. \tag{IV.251}$$

The other case is the decay of a longitudinal plasma wave in a nonisothermal plasma (where $T_e \gg T_i$) into a longitudinal plasma wave and an ion-sound wave:

$$l \,_{(}\rightleftarrows_{)}\, l' + s. \tag{IV.252}$$

Some formulas describing these and also some other mode interactions including their transformation coefficients are listed in Table IV.12.

Beside the above mentioned cases interactions of higher orders may also occur, e.g., the interaction of four plasma waves:

$$l + l' \,_{(}\rightleftarrows_{)}\, l_1 + l_1', \tag{IV.253}$$

etc. In the case of weak turbulence, however, high-order interactions do not make essential contributions.

TABLE IV.12

Compilation of some important wave interaction processes (after Kaplan and Tsytovich, 1972)

Process	Change of spectral density	Transformation coefficients
$l + l' \to t$ $(\omega_t \approx 2\omega_{pe})$	$\dfrac{\partial W_t}{\partial t} = \beta_l \displaystyle\int \left(\dfrac{W_{k,l}}{k_l}\right)^2 \mathrm{d}k$	$\beta_l = \dfrac{4\sqrt{3}\pi}{5} \dfrac{\omega_{pe}^4}{N_e m_e c^5}$
$h + h' \to t$ $(\omega_t \lesssim 2\omega_{He})$	$\dfrac{\partial W_t}{\partial t} = \beta_h \displaystyle\int \left(\dfrac{W_{k,h}}{k_l}\right)^2 \mathrm{d}k$	$\beta_h = \dfrac{32\pi}{5} \dfrac{\omega_{pe}^3 \omega_{He}^2}{N_e m_e c^5}$
$l + s \to p$	$\dfrac{\partial W_{k,p}}{\partial t} = \beta_{ls} W_{k,l} W_{k,s}/k$	$\beta_{ls} = \dfrac{\pi}{2\sqrt{3}} \dfrac{\omega_{pe}^3}{N_e m_e v_{Te} c^3}$
$p \to l + s$	$\dfrac{\partial W_{k,p}}{\partial t} = -\beta_{ps} W_{k,p} W_{k,s}/k$	$\beta_{ps} = \dfrac{\pi}{18} \dfrac{\omega_{pe}^3}{N_e m_e v_{Te}^4}$
$l + w \to p$	$\dfrac{\partial W_{k,p}}{\partial t} = \beta_{lw} W_{k,l} W_{k,w}/k$	$\beta_{lw} = \dfrac{\sqrt{2}\,\omega_{pe}^{9/2}}{\omega_{He}^{3/2} N_e m_e c^4}$
$p \to l + w$	$\dfrac{\partial W_{k,p}}{\partial t} = -\beta_{pw} W_{k,p} W_{k,w}/k$	$\beta_{pw} = \dfrac{\pi}{8} \dfrac{\omega_{pe}^6}{\omega_{He}^3 N_e m_e c^4}$
$l + m \to p$	$\dfrac{\partial W_{k,p}}{\partial t} = \beta_{lm} W_{k,l} W_{k,m} k^{-3/2}$	$\beta_{lm} = \dfrac{1}{\sqrt{2}} \dfrac{\omega_{pe}^{3/2} \omega_{pi}^2}{N_e m_i v_a^{3/2} c^3}$
$p \to l + m$	$\dfrac{\partial W_{k,p}}{\partial t} = -\beta_{pm} W_{k,p} W_{k,m} k^{-3/2}$	$\beta_{pm} = \dfrac{\pi}{8} \dfrac{\omega_{pi}^2}{N_e m_i v_a^3}$

4.6.3. Some possible solar radio applications

Several applications of the processes treated in the foregoing sections appear meaningful in solar radio astronomy and the importance of plasma physics is recognized by a growing number of solar physicists. A standard example is given by the interpretation of fast-drift bursts, which was already described in Section 3.4.6. Recently a wide field was also opened by the attempts to explain the different features of type IV bursts. The detection of fine structure in the type IV burst emissions carrying a wealth of latent information was a particular challenge to develop an adequate theoretical interpretation. The whole field is still growing and new results can be anticipated. Some still existing proposals of an interpretation of different type IV burst features on the basis of various kinds of wave-mode conversion processes in a warm plasma are listed in Table IV.13.

4.7. Instabilities and Coherent Emission

4.7.1. General concepts

Plasmas are characterized by a multitude of instabilities which can occur if free energy, stored in any particular form, exceeds a critical level and becomes unidirectionally transformed. Relevant forms of energy may be thermal energy, kinetic energy

TABLE IV.13
Interpretation of some type IV burst phenomena by wave-mode transformations in a warm plasma

Phenomenon	Interpretation	References
Type IV dm bursts	Čerenkov plasma (Langmuir) waves, induced scattering on thermal ions	Kuijpers (1974), Benz and Tarnstrom (1976).
Type IV mA and IVμ bursts	Langmuir waves excited by Harris-type instabilities and nonlinear conversion into electromagnetic waves	Mollwo (1973), Mollwo and Sauer (1977).
Type IV mB and type I bursts	Coupling of electromagnetic wave modes for small angles θ at $X \approx 1$, $Y \gtrsim 1$ in a warm plasma	Mollwo (1970).
Zebra patterns	Coupling of electrostatic (upper-hybrid) waves and Bernstein waves,	Chiuderi *et al.* (1973),
	plasma waves at the upper hybrid frequency excited by loss-cone instability,	Kuijpers (1975b).
	(gyro-resonance absorption of plasma waves)	Zheleznyakov and Zlotnik (1971).
Tadpole bursts	Coherent Bernstein waves and nonlinear conversion into electromagnetic waves	Zheleznyakov and Zlotnik (1975).
Fiber bursts (intermediate-drift bursts)	Coupling of whistler and Langmuir waves excited by a loss-cone instability	Kuijpers (1975a).
Fast-drift absorption bands	Switching off of instabilities of plasma waves at the upper hybrid frequency and/or Bernstein waves due to the action of streams of fast particles	Zaitsev and Stepanov (1975), Benz and Kuijpers (1976).

of ordered motion, potential mechanical energy (pressure, gravitation), electrostatic and electromagnetic energy, etc.

The energy released by an instability may result in violent plasma motions and radiation finally leading via different scales of structures to a dissipation (randomization) of energy.

As is typical for instabilities, a small deviation x grows exponentially with time ($x \sim \exp(\gamma t)$), if the temporal change of x is taken proportional to the deviation itself, i.e. $\mathrm{d}x/\mathrm{d}t \sim \gamma x$, where γ is the growth rate of the instability.

One important aspect is that the energy exchange caused by an instability is much faster than the energy exchange e.g. due to random collisional processes which normally are present in a plasma, and the action of instabilities is immediately connected with the occurrence of turbulence. The latter is due to the fact that instabilities are effective in generating various kinds of waves, oscillations, and fluctuations.

A method for investigating plasma instabilities is the 'normal mode analysis'. This method makes use of a linearized wave equation (which can be obtained on the basis of a perturbation ansatz) yielding a dispersion relation

$$D(\omega, \mathbf{k}) = 0. \tag{IV.254}$$

Then a plasma is called *linearly unstable*, if the dispersion Equation (IV.254) has a solution for ω with a negative imaginary part for any real value of $\mathbf{k}$. On the other hand, evidence for stability is given, when ω is real for all values of $\mathbf{k}$. But there are naturally certain defects of a linear theory which cause a loss of information. In general the linear description may be sufficient, with some reservation, for a study of the regions of existence of instabilities. For a quantitative description of the amplitudes and growth rates, however, the consideration of the full nonlinear process is indispensable.

Though the whole field of plasma instabilities is still expanding, surveys on the topic are already contained in a number of reviews and monographs e.g. by Briggs (1964), Lehnert (1967, 1972), Wentzel and Tidman (1969), Mikhailovskij (1970, 1971), Hasegawa (1975), and Cap (1976). Furthermore many books on plasma physics contain more or less extensive representations of plasma instabilities, e.g. Cap (1970, 1972), Krall and Trivelpiece (1973), and Kadomtsev (1976). Here we can only try to give a concise extract of the matter mainly under the point of view of some applications to solar radio astronomy.

4.7.2. Classifications of Plasma Instabilities

According to the large number of degrees of freedom in a plasma many types of waves, oscillations, and fluctuations are to be considered which arise singly or by combination as coupled or uncoupled phenomena. Correspondingly also many types of instabilities exist which can be classified under quite different aspects listed below.

Concerning the character of the approximation made in the basic equations two large groups of instabilities are to be distinguished, viz. *MHD instabilities* and *kinetic instabilities*. The MHD treatment is justified for time scales greater than the order of ν_{coll}^{-1}.

A somewhat similar distinction can be made concerning *macroinstabilities* and *microinstabilities*.

The former are due to macroscopic nonequilibrium conditions in the configuration space while the latter are connected with deviations of the particle energy distribution from an isotropic Maxwell distribution, i.e. unequilibria in the velocity space.

We further mention the labels *explosive* instability and *nonexplosive* instability. An explosive instability is obtained when the growth rate γ is an increasing function of time.

Two other important types of instability can be distinguished according to their space-time characteristics: *absolute instabilities* and *convective instabilities*. If in an infinite system a pulse disturbance of initially finite spatial extent is growing in time at every point in space, the instability is called 'absolute'. Otherwise, if it propagates in the coordinate system with increasing and decreasing amplitudes at any fixed point in space, it is called 'convective'. In other words, if a spacelike wave packet associated with a certain mode of instability is not a timelike packet, then the instability is absolute (nonconvective). If, on the other hand, a spacelike packet associated with a mode of instability is also a timelike packet, the instability is called convective (Sturrock, 1961b).

Instabilities can be further divided into the following two classes: *electrostatic instabilities* and *electromagnetic instabilities*. Electrostatic instabilities are characterized by the conditions $\text{curl}\,\mathbf{E} = 0$, $\mathbf{E} = -\nabla\Phi$ (i.e. the electric field can be derived from a scalar potential Φ), whereas for electromagnetic instabilities one finds $\text{curl}\,\mathbf{E} \neq 0$.

Though there are still other possibilities of classifications and subclassifications of plasma instabilities (for details cf. e.g. Lehnert, 1972, Figures 2 and 3, and the *Thesaurus of the Plasma Physics Index*, monthly published by the Max-Planck-Institut fur Plasmaphysik (Garching/Munich)), we conclude our brief summary. A short list of some important groups of instabilities is given in Table IV.14. Some of them will be discussed in the next Section.

4.7.3. Some important instabilities

a. *Two-Stream Instabilities*

The two-stream instabilities probably have a great relevance to solar physics. They belong to the oldest known category of plasma instabilities and are often used as an example to demonstrate the action of a plasma instability in general. The subject is well investigated and covered by reviews and monographs (cf. e.g. Briggs, 1964).

According to various conditions two-stream instabilities can occur in different forms and are to be treated either kinetically or by means of MHD methods. The simplest case of a two-stream instability may be realized by a uniform stationary infinite plasma which is passed by a stream of electrons with a certain velocity v_D. If the stream becomes unstable, a number of different oscillations and waves can be driven (*beam–plasma instability*). The instability can be interpreted as the conse-

TABLE IV.14
Some major groups of instabilities

	Electrostatic	Electromagnetic
Macroinstabilities (mainly MHD)	Current-pinch instabilities, flute or interchange instabilities, thermal instabilities, Rayleigh-Taylor instabilities.	Collision-free shock instabilities, resistive tearing-mode instabilities, sausage instabilities, drift-wave instabilities, collisionless tearing-mode instabilities.
Microinstabilities (mainly kinetic)	Two-stream and beam-plasma instabilties, Harris-type (velocity anisotropy) instabilities, loss-cone instabilities, cyclotron instabilities, ion-wave instabilities.	Alfvén-wave instabilities, anisotropic-pressure instabilities, mirror instabilities, modulation instabilities.

quence of a coupling between a 'negative-energy wave' represented by the stream and, e.g., an electrostatic wave in the background plasma.

In a frequency range $\omega \gg \omega_{pi}$ the ion dynamics can be neglected. If $v_{th} \gtrsim v_D$ or $v_{k,j}$, a kinetic treatment is the adequate way of description. Otherwise, considering a cold plasma, the fluid description would be sufficient. In the following we sketch some points following the treatment of Simon (1969).

Aiming at electrostatic waves, a linearized Vlasov equation can be derived from the perturbation ansatz

$$f(v) = f_0(v) + f_1(v), \tag{IV.255}$$

where $|f_0(v)| \gg |f_1(v)|$, and combined with Poisson's equation. The zero-order solution refers to the state of equilibrium. The first-order solution may be given by an expression containing the permittivity ε from which the dispersion relation $\varepsilon(\omega, \mathbf{k}) = 0$ results, which in our special case is of the fourth order in ω. The region of occurrence of the instability follows from the dispersion relation for the fastest growing mode yielding

$$kv_D \lesssim \omega_{pe} \tag{IV.256}$$

with $v_D \gtrsim v_{Te}$.

In addition ion-acoustic waves, i.e. low-frequency electrostatic waves, also result.

A growth is obtained for

$$v_D > (KT_e/m_i)^{1/2} \tag{IV.257}$$

and $T_i \ll T_e$.

In a magnetized plasma additional effects occur (cf. e.g. D'Angelo, 1972). An important consequence of such an electron-ion two-stream instability is the generation of an electric-field component parallel to the magnetic field which is produced in connection with a field-aligned current. Further modifications arise by the action of inhomogeneous magnetic fields (trapping of particles, effects of magnetic mirrors). Beside the electrostatic waves also other waves, i.e. transverse modes can be driven by two-stream instabilities.

b. *Velocity-Anisotropy Instabilities*

For the evaluation of microinstabilities the special form of the energy distribution function plays a dominant role. A particular class of microinstabilities is due to anisotropies of the energy (velocity) distribution of the electron and ion components of a plasma. Clearly such anisotropies can arise under the influence of external magnetic fields, if $\nu_{\text{coll}} \ll \nu_H$, which is realized, e.g., in active regions of the Sun. The velocity-anisotropy instabilities are in some respect closely related to the two-stream instabilities discussed above.

A typical distribution function of interest may be given in the form

$$f_0(v_\perp, v_\parallel) = \frac{1}{(2\pi)^{3/2}} \frac{1}{v_{th,\parallel} v_{th,\perp}^2} \exp\left(- \frac{v_\perp^2}{2v_{th,\perp}^2} - \frac{v_\parallel^2}{2v_{th,\parallel}^2} \right)$$

$$\sim \exp\left\{ - \frac{m}{2K}\left[\frac{v_\perp^2}{T_\perp} + \frac{v_\parallel^2}{T_\parallel} \right]\right\}, \tag{IV.258}$$

which is a bi-Maxwell distribution. Here $v_{th,\parallel}$ and $v_{th,\perp}$ are the thermal velocities parallel and perpendicular to $\mathbf{H}$, respectively. Anisotropy arises for

$$v_{th,\parallel} \neq v_{th,\perp}$$

and the instabilities thus arising have attained much interest (cf. e.g. Crawford, 1968; Harris, 1969). Two general cases may be distinguished theoretically, namely $T_\perp < T_\parallel$ and $T_\perp > T_\parallel$. Both cases lead to instabilities. Electrostatic instabilities are generated for $T_\perp > T_\parallel$. Electromagnetic instabilities can occur in both cases $T_\perp \gtrless T_\parallel$.

The electrostatic instabilities produced by energy distributions of the type given in Equation (IV.258) ($v_{k,\perp} > v_{th,\parallel}$) were first considered in some detail by Harris (1959, 1961) and hence, are called 'Harris instabilities'. In that type of instability a coupling of different cyclotron harmonics is involved. A condition for instability is

$$T_\parallel / T_\perp < 1/(2s) \tag{IV.259}$$

where the unstable frequencies are distributed near $\omega \sim (s - 1/2)\,\omega_H$ with $s = 1, 2, 3, \ldots$ (positive integer), (Soper and Harris, 1965; Karpman *et al.*, 1973). This result is in nice contrast to the case of a ring distribution, where the instability occurs near the frequencies $\omega \sim s\omega_H$. The ring distribution is given by

$$f_0(v_\perp, v_\parallel) = (1/(2\pi v_{0,\perp}))\,\delta(v_\perp - v_{0,\perp})\,\delta(v_\parallel), \qquad \text{(IV.260)}$$

representing an extreme case of anisotropy. It describes electrons, which move only in the directions perpendicular to the external magnetic field being uniformly distributed on a circle with the radius $v_{0,\perp}$ in the velocity space. This case, similar to the other extremum represented by electrons moving only in directions parallel to **H**, embedded in a background plasma, would refer to two-stream instability.

If the wavelength of an excited perturbation is of the order of a characteristic scale length of local plasma inhomogeneities, additional effects come into play (cf. e.g. also Hasegawa, 1975):

Firstly, nonuniformities of the plasma (i.e. the presence of gradients in H, N, or T, etc.) are leading to particle drifts, drift waves, and drift-wave instabilities.

Secondly, bounce motions of trapped particles in a magnetic mirror configuration can lead to parametric effects, when the bounce frequency becomes comparable with the perturbation frequency. Furthermore, particles leaving the magnetic bottle (*pitch-angle diffusion*) produce a *loss-cone distribution*, which again is anisotropic and leads to instabilities.

c. *Tearing-Mode Instabilities*

The tearing-mode instability can be considered as an example of a macroscopic instability. It may be related perhaps to special stages of the development of solar flares as well as to certain processes in the magnetospheric tail (cf. Section 5.6). The instability is connected with a sheet current (e.g. along a neutral sheet in a magnetic field configuration) in a resistive plasma (or collisionless plasma as a second case) subject to instability (cf. Dungey, 1958; Furth *et al.*, 1963; Furth, 1969).

The driving force of the instability is given by a nonuniformity of the magnetic field developing e.g. in the initial stages of a θ-pinch configuration. The instability occurs for wavelengths greater than the thickness of the current sheet. It is typical for this kind of instability, that X-shaped neutral points arise, which tear the current sheet into smaller segments. In this way, the structure of the magnetic field is reduced into smaller pieces ('crumbled'), thus becoming 'digestible' and being transformed into kinetic energy of particle acceleration. A great number of papers is devoted to these questions which will be touched once more in Section 5.6.3 in connection with a discussion of possible flare mechanisms.

4.7.4. Amplification of Electromagnetic Waves

a. *Negative Absorption of Radio Waves*

With regard to radio waves the study of instabilities starts with a consideration of the absorption coefficient, which becomes negative for amplified radiation. Such a treatment was first proposed by Twiss (1958), who investigated the reabsorption of synchrotron radiation using the quantum method of the Einstein probabilities (an alternative, almost equivalent approach would consist in the application of the classical kinetic treatment, cf. e.g. Zheleznyakov, 1972). The results of Twiss were improved by Bekefi *et al.* (1961), Wild *et al.* (1963), and Smerd (1965) obtaining

$$\alpha_{\omega,j} = -\frac{2\pi^2 c^2}{\mu_j^2 \omega^2} \int_0^\infty \frac{d}{dE}\left[\frac{F(E)}{E^2}\right] E^2 Q_{\omega,j}(E)\,dE \tag{IV.261}$$

which yields, on integration by parts,

$$\alpha_{\omega,j} = \frac{2\pi^2 c^2}{\mu_j^2 \omega^2} \int_0^\infty \frac{F(E)}{E^2}\frac{d}{dE}[E^2 Q_{\omega,j}(E)]\,dE, \tag{IV.262}$$

where μ_j is the real part of the complex refractive index $n_j = \mu_j + i\chi_j$, $F(E)$ is the (isotropic) energy distribution function, and $Q_{\omega,j}(E)$ denotes the mean emissivity for synchrotron radiation of one single electron in a wave mode j at the energy E and unit frequency interval of ω. The resulting instability of the synchrotron radiation was discussed by numerous authors (Zheleznyakov, 1966a, 1967a, 1972; Fung, 1968a, b, 1969a, c; Zheleznyakov and Suvorov, 1968; Sazonov, 1969c).

From Equation (IV.261) it follows, that the synchrotron radiation cannot be amplified in vacuum. For relativistic electrons pervading a cold background plasma however, α_ω becomes negative, if

$$\left(\frac{\omega_p^2}{\omega}\right)^2 \gamma^{*2} \gg 1 \tag{IV.263}$$

and

$$\gamma^* > \frac{2\omega_H \sin\theta\, \omega_p^2}{\omega^3}. \tag{IV.264}$$

In this case, as well as for low gyro-harmonic radiation under special circumstances (beam parallel to **H**, velocity anisotropy) an amplification of the radiation also turns out to be possible (cf. e.g. Smerd, 1968).

More generally, without regard to one single emission mechanism, Equation (IV.261) can be written

$$\alpha_{\omega,j} = -\frac{2\pi^2 c^2 N_e}{\mu_j^2 \omega^2} \int_0^\infty \frac{dF(E)}{dE} Q_{\omega,j}(E)\, g(E)\,dE \tag{IV.265}$$

where $g(E)$ is the statistical weight of the continuous energy levels from which the volume emissivity is derived from the single-particle emissivity by

$$\eta_{\omega,j} = N_e \int_0^\infty F(E)\, Q_{\omega,j}(E)\, g(E)\,dE. \tag{IV.266}$$

Hence it is quite obvious that two necessary conditions must be simultaneously fulfilled for amplification of radiation in any region of E:

$$dF(E)/dE > 0 \tag{IV.267}$$

and

$$d/dE\{Q_{\omega,j}(E)\, g(E)\} < 0. \tag{IV.268}$$

The first condition characterizes an overpopulation of the higher energy levels, i.e. a state of disequilibrium. The second condition (IV.268) depends on the special emission mechanism requiring the existence of 'resonance' peaks or suitable gradients of the quantity $Q_{\omega,j}(E)\,g(E)$ on the energy axis.

It should be remarked that any linear theory as applied above cannot reproduce the development of instabilities and therefore fails to describe the stabilization processes and limiting amplitudes. A first nonlinear approximation is given by the quasi-linear theory leading to a plateau formation of a formerly double-humped energy distribution function. Higher approximations are mathematically formidable and only accessible by numerical methods. For this reason quantitative comparisons between theory and observations are as yet in the stage of more or less rough estimations only.

b. *Induced Scattering of Coherently Generated Plasma Waves*

In several cases, e.g. considering solar decimetric and metric type IV burst emissions, the direct generation of radio waves (coherently or incoherently) is felt to be less effective than the indirect production of radio waves by mode conversion (Kuijpers, 1974; Benz and Tarnstrom, 1976). Here especially the induced scattering of plasma waves turns out as one of the most promising mechanisms (Tsytovich, 1967).

Coherent plasma waves may arise from a loss-cone or bi-Maxwell distribution by Harris-type instabilities at multiples of the gyrofrequency in magnetically trapped regions. Two parameter ranges were tested in concrete models, $\omega_H/\omega_p \lesssim 1$ (Mollwo, 1971, 1973) and $\omega_H/\omega_p \ll 1$ (Kuijpers, 1974). Table IV.15 summarizes qualitatively some important instabilities for longitudinal and transverse wave modes with respect to the angles θ and ϕ.

The longitudinal waves may be subject to induced scattering, i.e. the conversion of the longitudinal into transverse (electromagnetic) wave modes is stimulated by already existing transverse waves (similar to a paramagnetic oscillator).

It is a remarkable fact that the amplitudes or the temperature

$$T_{\text{eff}}(\mathbf{k}) = (N(\mathbf{k})/K)\,\hbar\omega(\mathbf{k}) \tag{IV.269}$$

of the conversed waves can almost reach that of the original waves. The process is

TABLE IV.15
Summary of some important actions of instabilities

Velocity direction	Instability direction	
	$\parallel \mathbf{H}(\theta = 0)$	$\perp \mathbf{H}(\theta = 90°)$
$\parallel \mathbf{H}(\phi = 0)$ 'beam'	longitudinal waves	transverse waves
$\perp \mathbf{H}(\phi = 90°)$ 'gyrating motion'	transverse waves	longitudinal waves

described by the first term of the right-hand side of the kinetic equation

$$\frac{\partial N_t(\mathbf{k})}{\partial t} = N_t(\mathbf{k}) \sum_\alpha \int u_{\alpha,l\to t}(\mathbf{p}, \mathbf{k}, \mathbf{k}')\, \hbar(\mathbf{k} - \mathbf{k}') \cdot \frac{\partial f_\alpha(\mathbf{p})}{\partial \mathbf{p}} N_l(\mathbf{k}') \frac{\mathrm{d}^3\mathbf{p}\, \mathrm{d}^3\mathbf{k}'}{(2\pi)^6} +$$

$$+ \sum_\alpha \int u_{\alpha,l\to t}(\mathbf{p}, \mathbf{k}, \mathbf{k}')\, f_\alpha(\mathbf{p})\, N_l(\mathbf{k}') \frac{\mathrm{d}^3\mathbf{p}\, \mathrm{d}^3\mathbf{k}'}{(2\pi)^6} -$$

$$- N_t(\mathbf{k}) \sum_\alpha \int u_{\alpha,l\to t}(\mathbf{p}, \mathbf{k}, \mathbf{k}')\, f_\alpha(\mathbf{p}) \frac{\mathrm{d}^3\mathbf{p}\, \mathrm{d}^3\mathbf{k}'}{(2\pi)^6}, \qquad \text{(IV.270)}$$

where $u_{\alpha,l\to t}$ is the transition probability for particles of species α to scatter a plasma wave with wave vector $\mathbf{k}'$ into an electromagnetic wave with a wave vector $\mathbf{k}$ (Tsytovich, 1967).

CHAPTER V

INTEGRATION OF RADIO ASTRONOMY INTO SOLAR AND SOLAR-TERRESTRIAL PHYSICS

In this chapter we will consider some implications of solar radio physics (theory and observations) in the contemporary picture of solar physics as a whole. Solar radio waves carry much information about solar phenomena, processes, and parameters, which, to a certain extent, are difficult to obtain by other methods. On the other hand, it is sometimes difficult to quickly and easily use this information: Main difficulties consist e.g. of insufficient angular resolution of observations and of the complexity of theoretically treating instationary, nonthermal, and nonlinear models of radiative processes in a warm or hot plasma. Thus we are often confronted with the following restrictions:

(a) The number of physical parameters exceeds the number of independently observed quantities; and

(b) the evaluation of physical parameters strongly depends on the invoked working hypotheses or model assumptions.

To avoid these restrictions, work is progressing along the following general lines:

(1) Adoption of information from other (nonradio) methods;

(2) successive refinement of the radio observations; and

(3) successive improvement of the applied models and theoretical tools of radio and plasma physics.

Having these restricting remarks in mind, some possible conclusions on solar plasma parameters by different indicators (observations) are sketched in Table V.1. It is obvious that the conclusions are most valid where sufficient observations and reliable physical models exist. This may be the case for the quiet Sun, the *S*-component, and the microwave bursts. The deduction of physical quantities from different type IV burst and noise storm phenomena, which in principle should be very promising because of detailed information contained in the fine structures, has been omitted from Table V.1, since the physical models to be applied are not yet adequate.

5.1. Estimation of Solar Plasma Parameters

5.1.1. Particle Density and Magnetic Field

Two of the most important plasma parameters required to be accurately known are the electron density and the magnetic field strength, which can be obtained by several

TABLE V.1
On deductions of solar plasma parameters by radio methods

Phenomenon	Indicator	Process (model)	Parameters	Problems
Quiet Sun	Flux spectrum, center-to-limb variation, basic component and coronal holes	Thermal bremsstrahlung	$N_e V, T_e$	Accurate absolute calibration over long periods
S-component, μ-bursts	Spectrum and time variations of flux density, polarization, brightness distribution, oscillations	Gyro-synchrotron radiation, bremsstrahlung, magneto-ionic wave propagation, heat conduction, radiation and collision damping	$N_e V, f(E)$, $H \sin\theta$, (T), sign $(\sin\theta)$, L, σ	cm-mm heliography and spectrography, warm-plasma theory of non-thermal gyromagnetic absorption
Slow-drift bursts	Dynamic spectrum and source positions (herring-bone structure and band splitting)	MHD shock waves	$p, H^2/8\pi$, H, ρ	Nonlinear warm-plasma theory of type II burst emission
Fast-drift bursts	Dynamic spectrum and source positions, polarization, harmonic structure, frequency extent	Plasma-wave hypothesis	$N_e(h), T_e/T_u(?)$, $H(?)$	Improved theory of type III/V bursts, *in situ* measurement of the exciters

methods. A sketch of the density and the magnetic field versus height in the solar atmosphere expressed by the plasma frequency and gyrofrequency is shown in Figure V.1 as an example (distribution over the center of a strong active region). In the shaded area of Figure V.1 the ratio ω_p/ω_H is smaller than unity for high solar activity. For medium and small activity, ratios ω_p/ω_H greater or much greater than unity would be obtained. Correspondingly those regions, where $\omega_p = \omega_H$ which may play a key role in the interpretation of some plasma processes (e.g., of noise storms, cf. Mollwo, 1970), are likely to occur only in major active regions.

Using the distributions of ω_p and ω_H, for each electromagnetic wave frequency a trajectory can be drawn in a parameter diagram where ω_H/ω is plotted versus ω_p/ω (Figure V.2). In such a representation waves at any given frequency are propagating along these trajectories from the upper-right to the lower-left side, in principle corresponding to the Sun–Earth direction. In this connection two points are to be mentioned:

(a) The conditions really existing at the Sun cover only a restricted range of the full parameter diagram. Thus, not all known wave modes existing in the full region (e.g. CMA-diagram, cf Figure IV.10) may apply to solar physics.

(b) Only those waves which either 'leap' over the stop bands of the diagram (especially the barrier of the refractive index for the extraordinary electromagnetic wave mode) or which are simply generated in front of these barriers can reach

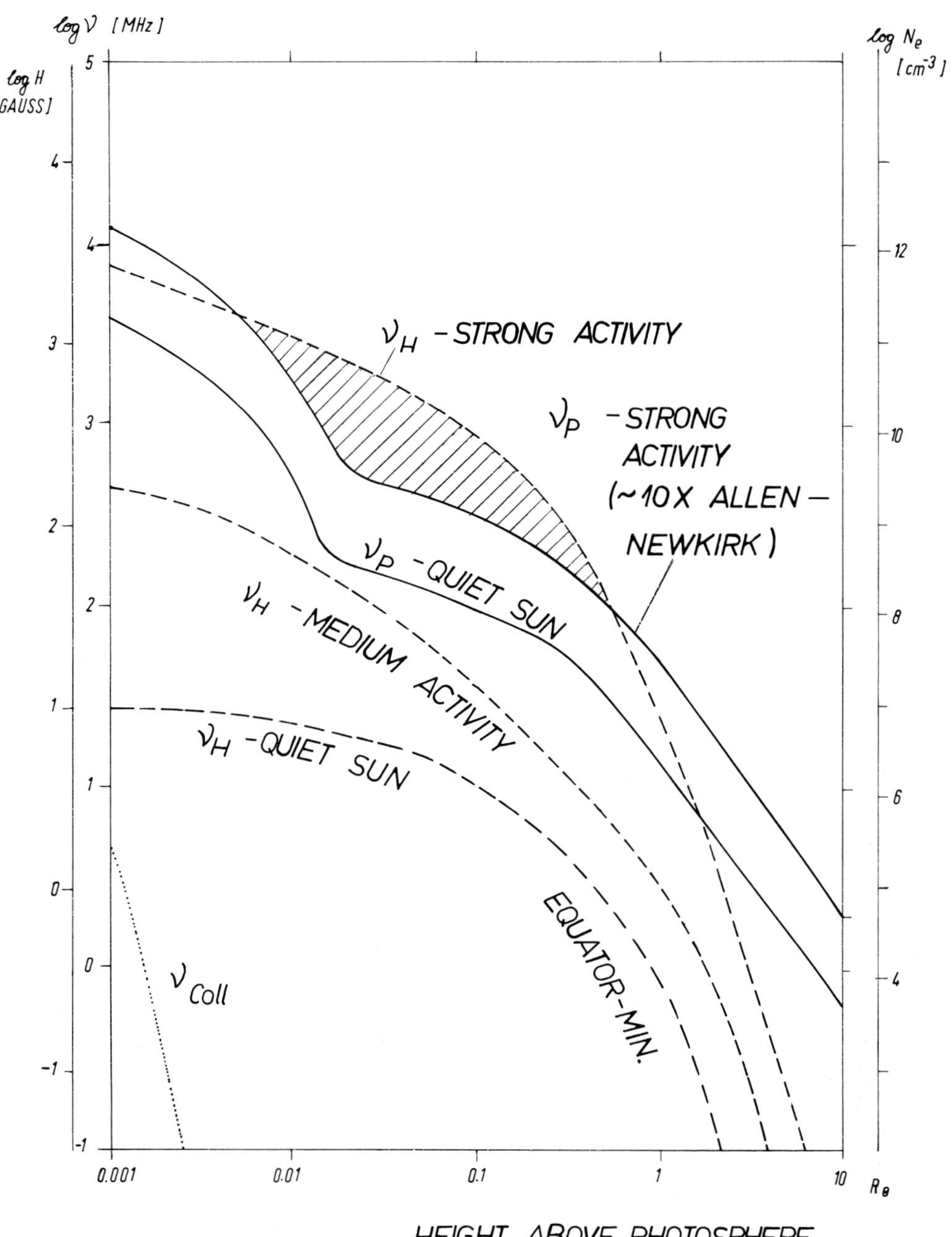

Fig. V.1. Mean distributions of the plasma frequency and the gyrofrequency in the solar atmosphere.

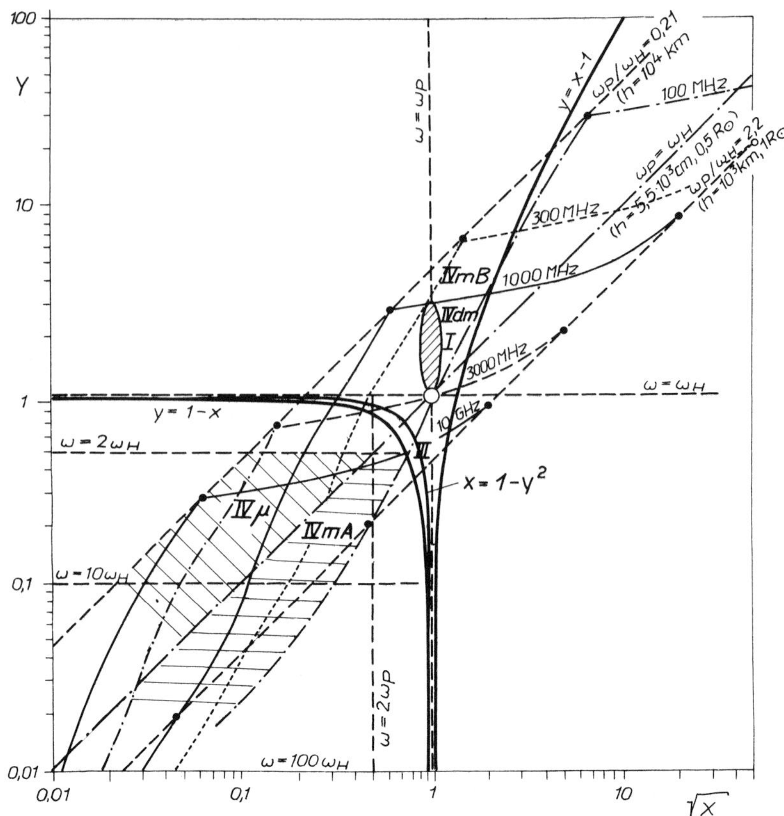

Fig. V.2. Parameter diagram for the solar atmosphere. There are indicated ray trajectories at different frequencies for strong activity (cf. Figure V.1.) and cutoff regions (barriers) of the cold plasma (ray direction from the Sun to the Earth from upper right to lower left). Emissions can be divided as to originate in front or behind the barriers corresponding to direct and indirect emission processes (mode transformation), respectively. Originating places for different radio-burst spectral types are tentatively marked in.

the observer. Examples for the first possibility (i.e. passing the stop bands by mode transformations) may be realized in emissions of type I, II, III, IVdm, and IVmB bursts as well as noise storms. Examples for the latter possibility are given by Coulomb bremsstrahlung and emissions at the harmonics of the gyrofrequency referring to the quiet Sun, the *S*-component, microwave bursts, and probably also to type IVμ and type IVmA bursts.

5.1.2. Energy Content

Though energetically neglible, the output of radio emission depends on the energy released and can be used as an important tool for a study of the energy release in different phenomena of solar activity.

At first it may be remarked that solar activity in general provides only a small modulation ($<1\%$) of the total permanent energy outflow from the Sun. The per-

TABLE V.2

Energetics of the solar wind (partly after Hundhausen, 1972)

	Low-speed	High-speed	Flare-induced
Flow speed	300–400 km s^{-1}	600–700 km s^{-1}	500–1000 km s^{-1}
Kinetic energy flux density	2.2×10^{-1} erg cm^{-2} s^{-1}	~ 1 erg cm^{-2} s^{-1}	(shock wave) $\gtrsim 1$ erg cm^{-2} s^{-1}
Magnetic energy flux density	3×10^{-3} erg cm^{-2} s^{-1}	$\sim 10^{-2}$ erg cm^{-2} s^{-1}	?
Proton temperature	4×10^4 K	3×10^5 K	—
Electron temperature	1.5×10^5 K	2×10^5 K	—
Total energy flux	10^{27} erg s^{-1}	2×10^{25} erg s^{-1}	10^{28} erg s^{-1}
Total energy flux (27 days)	10^{34} erg	5×10^{31} erg	10^{31}–5×10^{32} erg per event

manent energy output per day in the optical spectral range originating in the photosphere is of the order of 10^{38} erg per day nearly independently of solar activity. In comparison to this figure the total energy release even of the largest flares is smaller by five orders. The energies emitted in the radio region by the quiet Sun and the *S*-component are both of the order of about 10^{25} erg per day, which is comparable also with the magnitude of large radio bursts. Therefore, the effectivity of the radio emission does not consist in the amount of energy transported by it, but rather in its content of information about physical processes and conditions referring to the exciters of the radiation.

The energy output of the solar particle radiation is listed in Table V.2 for comparison. From a consideration of this table the predominance of the quiet, undisturbed energy flux over the slowly varying and sporadic fluxes of particles becomes apparent.

Tables V.3 and V.4 contain estimations of the energy ranges released by different types of flare and burst components. Roughly speaking, the energy is distributed into three main parts: Shock waves, optical electromagnetic radiation, and particle emission.

More detail will be discussed in the next section.

5.2. The Flare Phenomenon

5.2.1. The Low-Temperature Flare

Since the whole flare phenomenon covers a wide range of wavelengths (from about 2×10^{-3}Å up to ≥ 10 km) and emission heights in the solar atmosphere (photosphere – corona), there is a wide range of temperature and energy distributions belonging to different flare components. In principle three characteristic energy regions can be roughly distinguished, which are attributed to: the low-temperature flare, the high-temperature flare, and the (nonthermal) high-energy flare. (Additionally, for completeness beyond others a (nonthermal) low-energy flare could be introduced in order to describe some radio burst components (cf. Table V.6).)

TABLE V.3
Estimated mean energy content of strong flare radiations

	Hα	Optical, UV other lines (Lα)	continuum	X-rays hard	X-rays soft	Microwave bursts impulsive	Microwave bursts gradual	Type IV bursts IVμ	Type IV bursts IVmA,F
Particle energy E [erg]	10^{-12}–10^{-11}	10^{-12}–10^{-10}	10^{-10}–10^{-8}	10^{-8}–10^{-6}	10^{-9}–10^{-10}	10^{-8}–10^{-6}	10^{-9}–10^{-10}	10^{-7}–10^{-5}	10^{-7}–10^{-5}
Kinetic energy $(E \times E_e)$ [erg cm^{-}3]	10–10^2	10–10^2	10^2–10^3	10^{-2}–10^{-1}	10^{-1}–10	10^{-5}–10^{-1}	10^{-1}–10	10^{-4}–10^{-1}	10^{-5}–10^{-2}
Total energy $(E \times N_e \times \times V)$ [erg]	10^{31}	4×10^{31}	5×10^{31}	10^{27}–10^{30}	10^{29}–10^{31}	10^{24}–10^{29}	10^{29}–10^{30}	10^{26}–10^{27}	10^{26}–10^{29}
Radiated energy (total) [erg]	10^{31}	4×10^{31}	5×10^{31}	10^{25}–10^{29}	10^{26}–10^{30}	10^{22}–10^{24}	10^{23}–10^{24}	10^{24}–10^{25}	10^{23}–10^{24}
	$\lesssim 10^{32}$			$\lesssim 10^{30}$		$\lesssim 10^{25}$			

TABLE V.4
Tentive estimations of the total energy content of some indirectly generated radio burst types

Energy [erg]	Type I	Type II	Single type III	Type III storm	Type IV dm	Type IV mB	Type I storm
Exciter	10^{21}–10^{25}	10^{30}–10^{32}	10^{24}–10^{28}	10^{28}–10^{32}	?	10^{29}–10^{30}	10^{26}–10^{31}
Plasma-wave source	?	10^{30}–10^{32}	10^{23}–10^{27}	10^{27}–10^{31}	?	?	?
Radio waves	10^{18}	10^{22}–10^{23}	10^{21}	10^{25}	10^{22}–10^{24}	10^{23}	10^{23}–10^{24}

The low-temperature flare corresponds to that optical part of the flare radiation, which, consisting of the line emissions from H (Hα mainly), is naturally restricted to temperatures below the ionization temperature of H which is about 10^4 K. Thus the conditions where the low-temperature flare component develops are not directly comparable with those of the generation of the radio emission, which refers to high-temperature and high-energy flare plasmas. Nonetheless, as a consequence of local, small-scale inhomogeneities (by cooling mechanisms or ejecta from cooler and denser parts of the solar atmosphere and/or as an envelope of high-temperature regions) the low-temperature flare is intimately related to the high-temperature and high-energy parts of a flare. Moreover, the hitherto much better angular resolution of the optical flare pictures may supply useful information about source structure and positions relevant to a greater flare volume. From optical observations of the low-temperature flare a number of characteristics are to be noted which may be important for the study of the whole flare phenomenon:

(a) Flares occur most likely in strong, complex magnetic regions indicated e.g. by the presence of intense spot groups. However, simple bipolar configurations do not seem to be avoided by flares. (Fine structures of magnetic fields below 10″ and in greater than photospheric heights escaped hitherto a direct observation.)

(b) The projected flare areas tend to avoid a complete coverage of sunspots unless they belong to exceptionally strong events (Dodson and Hedeman, 1960).

(c) The brightness is not uniformly distributed over the whole flare area or a major part of it: There is evidence of the occurrence of (nonthermal) bright flare kernels and (thermal) flare knots (De Jager, 1967; Vorpahl and Zirin, 1970).

(d) Flares are typically situated at opposite sides of the zero line of the longitudinal magnetic field component.

(e) Especially large flares exhibit a characteristic double-ribbon structure which marks the foot-points of a system of tunnel-like magnetic field arches (Figure V.3). This field structure shows an expansion during the lifetime of a flare indicated by a separation of the flare ribbons with velocities ranging roughly between 1 and 50 km s^{-1}.

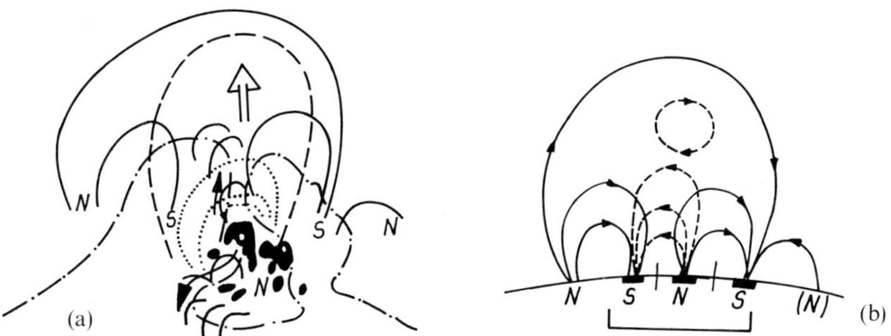

Fig. V.3a,b. Magnetic field configuration of the flare of August 7, 1972.

(f) Flares tend to occur preferably in active regions which show fast changes of its magnetic field structure with time. Magnetic field variations as a direct consequence of a flare event are difficult to detect.

(g) The electron density in low-temperature flare regions is about two orders of magnitude higher than in the corresponding undisturbed levels.

Though there is a good correlation on the average, a number of low-temperature flares occurs evidently without association of a strong high-temperature or high-energy flare component and vice versa (Hudson *et al.*, 1969; Drake, 1971; Thomas and Teske, 1971). The low-temperature flare can be regarded either as a more independent phenomenon or as a by-product of a high-temperature or a high-energy flare, depending on the height range of the actual flare volume and the resulting energy distribution.

From the same material, however, Švestka (1976) concludes that almost every Hα-flare would be found to be accompanied by a soft X-ray burst should the X-ray detector be sensitive enough. On the other hand, not every X-ray burst finds its counterpart in a well-reported flare or subflare, but it may quite well correspond to flare-like brightenings which are too small to be reported by the observing stations. Nevertheless, a large scattering of the intensity ratio between Hα and soft X-ray radiation as well as microwave radiation remains to be noted.

5.2.2. The high-temperature flare

High-temperature flare phenomena are displayed by soft X-ray, EUV, and microwave burst emissions. A clear distinction between the 'high-temperature' flare and the 'high-energy' flare (as well as the 'low-temperature' flare) components is sometimes difficult to achieve in particular cases.

We consider by definition all flare processes which in essence may correspond to a Maxwellian-like energy distribution referring to temperatures above the ionization temperature of hydrogen as the 'high-temperature' part of a flare. In comparison to this all those flare processes which are essentially determined by a nonthermal part of an energy distribution would be considered as belonging to the 'high-energy' flare.

Like the other flare components the high-temperature flare may also form a spatially bounded part of a flare and it may also represent a certain stage of the temporal development of a complex flare event. In some cases it may also exist as an isolated, independent phenomenon. An outline of the different main stages and components of complex flare events is shown in Table V.5.

The following features of the high-temperature flare are to be noticed:

(a) Photographs of soft X-ray flares reveal a loop-like structure with a maximum brightness at the top of this structure.

(b) The emission height of soft X-ray bursts scatter from about 10^4 to 10^5 km above the photosphere, with a maximum below 5×10^4 km. Typical source diameters are of the order of 10^4 km.

TABLE V.5
Characteristic stages of the development of solar flares

Flare component (energy range)	Indicator	Stage 'Gradual phase'	'Impulsive phase'	'Explosive Phase'
Low-temperature flare	Hydrogen lines (Hα)	slow rise of line width	flash phase	('Y'-phase (double-ribbon structure))
High-temperature flare	Soft X-rays, nonhydrogen lines (e.g. in UV), Radio waves	gradual burst (preheating)	impulsive burst	—
High-energy flare	Radio waves, hard X-rays, particles (protons)	—	impulsive burst, hard X-ray phase —	type IVμ burst — (?) explosive phase

(c) The flux of soft X-ray bursts is composed of a continuum ($> 90\%$) and line emissions of highly ionized atoms. The time profiles of soft X-ray burst fluxes exhibit gradual and impulsive forms.

(d) The peak temperatures of the soft X-rays range are typically about $(2–4) \times 10^7$ K, the emission measures reach values of $10^{48}–10^{49}$ cm^{-3} and more in exceptional cases.

(e) The duration of soft X-ray bursts exceeds that of the corresponding optical and radio (microwave) emission.

(f) The emission mechanism of the solar soft X-ray continuum is assumed to be thermal or quasi-thermal bremsstrahlung (free–free transitions) and/or radiative recombination (free–bound transitions).

Concerning the origin of the high-temperature flare, there are two groups of possible processes: The first is the production of mere thermal or quasi-thermal energy by heating. The second would be a source of nonthermal energy which produces heat by thermalization e.g. via collisional Coulomb interactions. There is strong evidence that both kinds may be realized on the Sun: A strong argument for the first possibility is the fact, that pure 'thermal' flare components (believed to be partly represented by gradual soft X-ray and microwave bursts) can occur without any – or at least with retarded – nonthermal phenomena.

On the other hand, also without doubt, nonthermal flare processes do exist in other cases, which must give rise to dissipative thermalization effects (cf. Section 5.2.3). In this way, both thermal and nonthermal processes are involved in solar flares and the share of the one or the other process may vary from one event to the other.

Finally, it should be mentioned, that the impulsive radio burst has a position somewhat in between thermal and nonthermal flare characteristics. It seems to be related to both types of processes (or both Maxwellian and nonthermal parts of the energy-distribution function) as it is indicated by its relation to both the soft and hard X-ray burst phenomena. This conclusion is not astonishing since, due to the generation mechanism (primary dependence on the magnetic field by the gyromagnetic emission), the frequency separation of thermal and nonthermal contributions in the radio emission is not as strongly expressed as is the case for the X-rays.

5.2.3. The high-energy flare

The high-energy flare comprises all electromagnetic flare radiations which refer to a nonthermal electron population above the level of the high-temperature flare. It is displayed by the occurrence of hard X-ray burst emissions ($\lambda \leq 1$ Å, $\varepsilon > 10$ keV) and different types of radio bursts.

Beside the above mentioned high-energy flare components (referring to the *radiation flare* – cf. also Section 5.3.2) still some other nonthermal parts of a flare can be recognized, e.g. the *dynamic flare*, consisting of shock waves (e.g. Moreton waves, blast waves) and plasma ejecta (sprays, surges, etc.) – cf. DeFeiter (1975), the *particle flare* consisting of fast protons and electrons, and the *nucleonic flare*, which occurs very rarely in cases of exceptionally strong events (Chupp *et al.*, 1973; Ramaty *et al.*, 1975).

Detailed reviews and results of recent studies are contained in Kane (ed.) (1975) and Švestka (1976). In the following we summarize some basic properties of the hard X-ray bursts and their relation to the radio bursts, in particular microwave bursts.

Records of solar hard X-ray and microwave burst fluxes often exhibit a remarkable correspondence even of certain details (Figure V.5b). These features are in favor of a unique interpretation of both hard X-ray and microwave bursts as originating in nearly identical source regions. First attempts of such an interpretation led to discrepancies of the required electron content for both components by several orders of magnitude due to an erroneous estimation of the radio emission (cf. Section 3.3.5), but a more detailed discussion removes the original discrepancy and makes the reconciliation of both emissions principally possible (Takakura, 1973; Böhme *et al.*, 1976, 1977). It must be taken into account that the X-ray emission comes from a more extended source volume (due to its low optical depth) than the microwave emission, which is assumed to originate at distinct low-harmonic gyro-resonance levels with a higher optical thickness.

Comparing hard and soft X-ray fluxes it can be stated that practically each hard X-ray burst is connected with a corresponding soft X-ray event, but there are many soft X-ray bursts without any association of a detectable hard X-ray burst. Furthermore, in the case of correlated events, the increase of the soft X-ray flux often precedes the well defined (more or less steep) onset of the hard X-ray burst (Figure V.5a). This feature gives a strong argument against the idea that the proper origin of a flare event may be situated in the hard X-ray source (or flare kernels). Obviously there are several stages of the development of flares (gradual, impulsive, explosive), whereas the gradual and the impulsive stages may occur either almost independently of each other or related, depending on the special case. But evidently the later stages comprising the high-energy flare occur as a consequence of the former ones. Analogously a nucleonic flare only develops on the basis of extraordinarily strong explosive events.

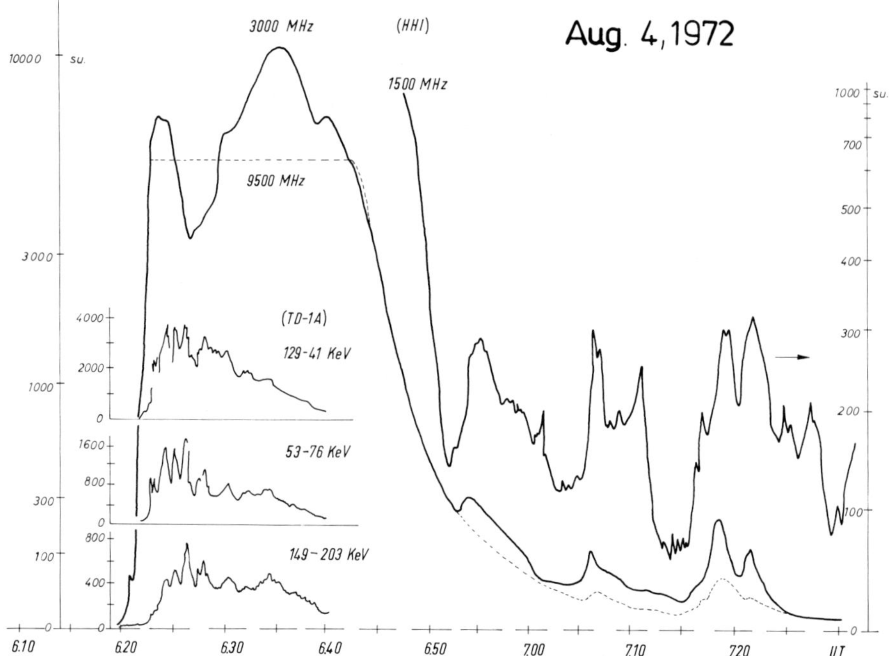

Fig. V.4. Example of time profiles of the type IV burst of August 4, 1972, in comparison with hard X-ray records (after Böhme et al., 1976).

It should be noted that the terms 'gradual' and 'impulsive' flare (burst) component do not necessarily seem restricted to a special energy or wavelength range and must not be interchanged simply with the terms 'high-temperature flare' and 'high-energy flare', respectively. In particular, impulsive flare (burst) components are observed in all energy ranges: the low-temperature, the high-temperature, and the high-energy domain.

Another, somewhat open point is the questionable association of the type IVμ burst component with hard X-rays. Though there is sometimes a small contribution of a hard X-ray burst emission during the period of the type IVμ burst (or 'explosive') phase (cf. Figure V.4 and also Frost, 1969), there remains a considerable lack of hard X-ray emission in comparison to the accompanying type IVμ burst radiation, if the latter is interpreted by the same electron population producing both hard X-ray and radio emission via gyro-synchrotron radiation. Thus it may be argued, that there is either a contribution of some other mechanism producing radio waves, (and) or there are very inhomogeneous source structures, which prohibit an estimation of the outcoming radiation fluxes by means of simple geometric models.

Now we summarize some further properties of hard X-ray bursts, which should be noticed in an interpretation of the high-energy flare:

Hard X-ray bursts have a relatively short duration of a few tens of a second up to a few minutes of time. The impulsive increase of the radiation flux is sometimes intersected by a complex structure (occurrence of short-lived spikes or 'elementary flare

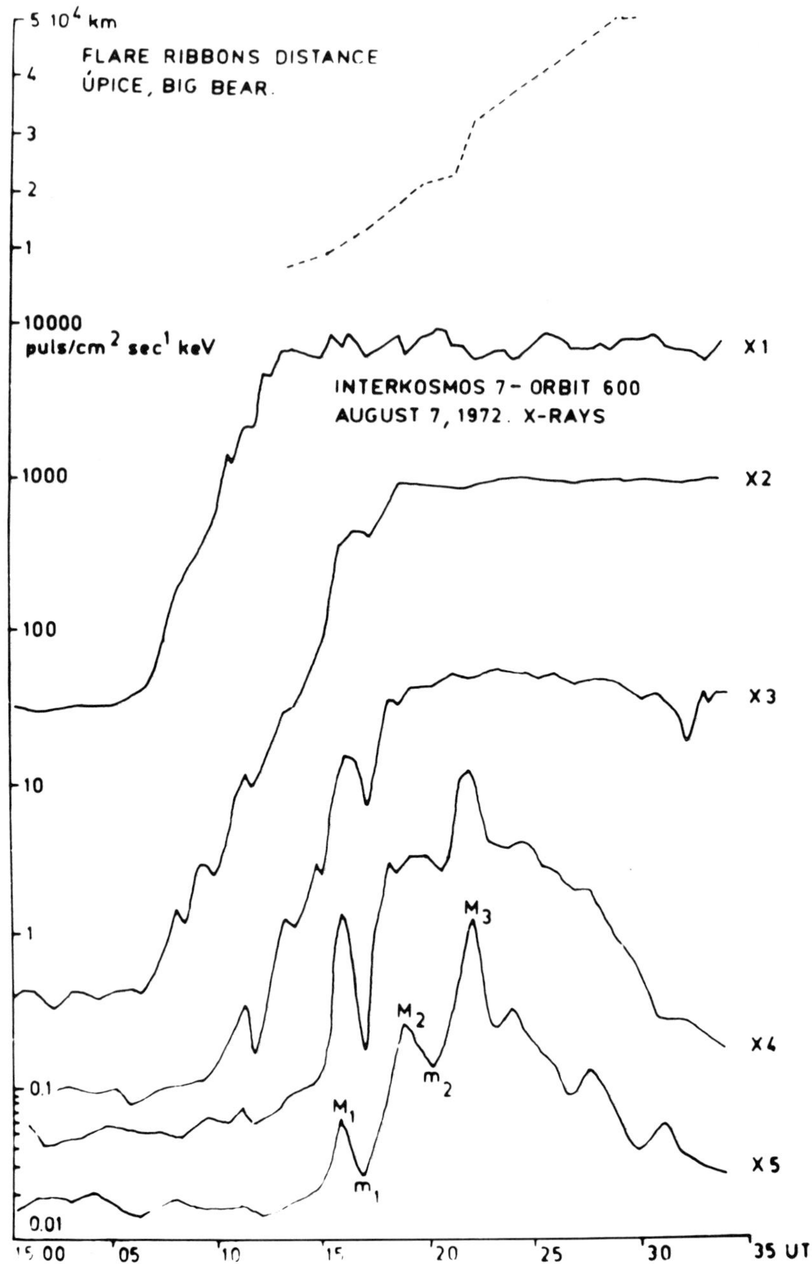

Fig. V.5a. Time profiles of the August 7, 1972, flare recorded at different X-ray channels demonstrating different onset and decline characteristics for soft and hard X-rays.

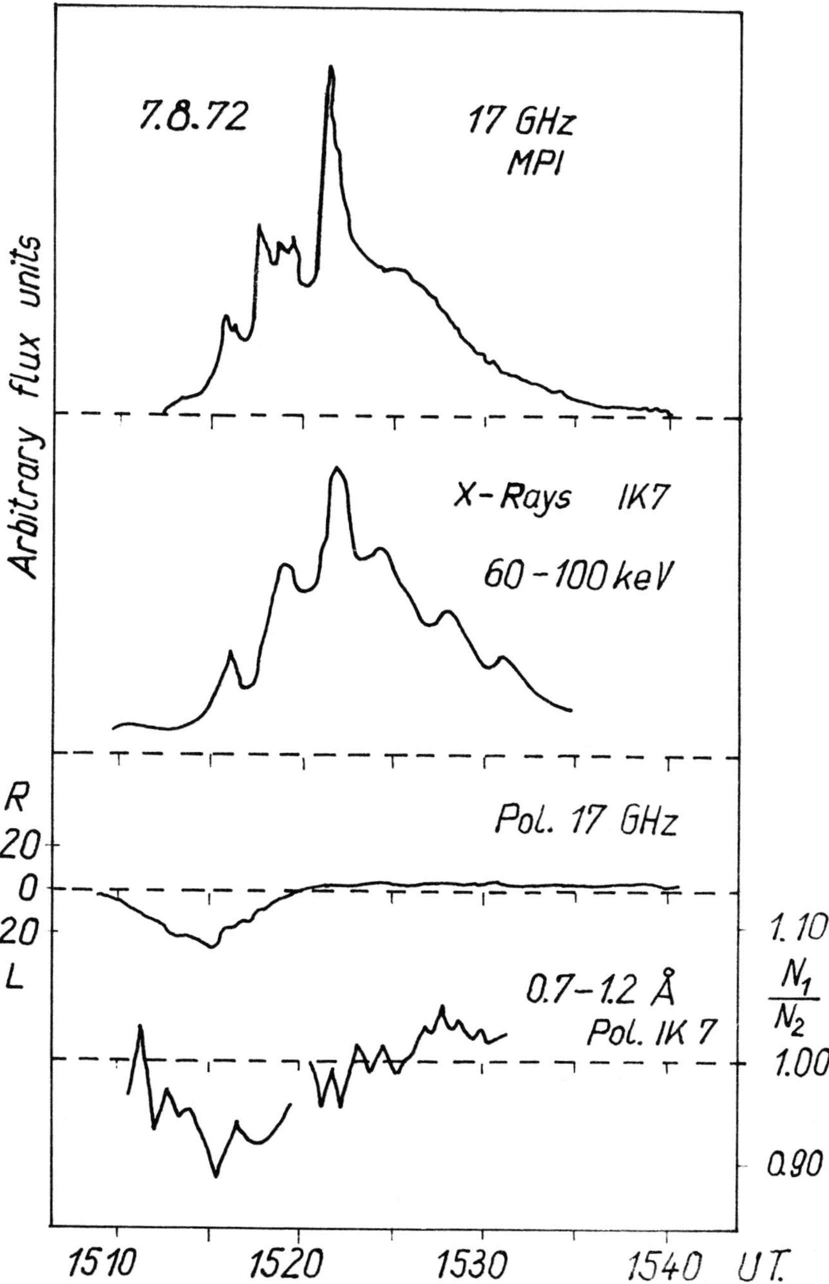

Fig. V.5b. Same event as Figure V.5a; comparison between X-ray and microwave fluxes and polarizations (after Křivský *et al.*, 1976).

bursts' in a time scale of one or a few seconds (Van Beek *et al.*, 1974b). Pulsations should not to be excluded, but yet it seems difficult to distinguish between quasi-periodic and pulse-like flux variations. In general the hard X-ray peaks precede the corresponding maxima of the Hα-radiation by an interval of up to a few minutes.

The (continuous) spectrum of the hard X-ray bursts extends up to energies of some hundreds of keV for strong events (extreme values were reported as 0.6 MeV in the events of July 7, 1966, and about 2 MeV in the event of August 4, 1972). Beside this, in rare cases γ-ray lines at 0.5 MeV and 2.2 MeV are superimposed on the hard X-ray continuum.

The observed hard X-ray spectrum fits rather well a power law (cf Sections 4.1.2.c and 5.3.3)

$$I(\varepsilon) = a\varepsilon^{-\lambda} \tag{V.1}$$

(ε–photon energy). Typical values of the exponent λ are 3–6. For higher photon energies a steepening of the spectrum can be observed. During the decay phase of a hard X-ray burst the spectrum is typically softening. Drastic changes of the photon spectrum with a time scale of a few seconds are quite possible especially during the rise phase of a burst event. The photon spectrum corresponds to a power-law velocity distribution

$$f(v) = A(v_s/v)^{\gamma} \tag{V.2}$$

where $A = (\gamma - 3)\,N/(4\pi v_s^3)$, $\gamma = 2\lambda$, v is the velocity above which the distribution can be well represented by a power law. A spectrum of the form of Equation (V.2) is a rather simple description and may be more likely than a thermal interpretation by a heterogeneous multi-temperature plasma exceeding 10^8 K (Milkey, 1971; Chubb, 1972) which theoretically would also be possible.

The sizes of the sources of the hard X-ray bursts are probably much smaller than those of the soft X-ray emission (and also smaller than the Hα-flare regions). A core-halo structure seems to be appropriate to describe the sizes of the hard and soft X-ray burst sources, respectively. From various lines of evidence values of less than 0.5 min of arc can be concluded for the core diameters of the hard X-ray radiation. An association with the small bright flare kernels visible in Hα (De Jager, 1967; Vorpahl and Zirin, 1970) seems very probable.

In some cases there is a good coincidence between the occurrences of hard X-ray bursts and white-light flares. Both phenomena are believed to be initiated by streams of particles accelerated at nearly the same place of origin.

Beyond other reasons the occurrence of polarized X-ray radiation can be regarded as further evidence for the action of a nonthermal, directive acceleration process. The polarization of solar X-ray bursts was detected by Tindo *et al.* (1970, 1972a, b), in the 0.6 to 1.2 Å band. These results were later questioned by Brown *et al.* (1974) but confirmed by Nakada *et al.* (1974). Interesting enough, the polarization is not sharply restricted to the impulsive rise of a burst, but also lasts for a short time after the burst maximum (cf. Figure V.5) in nice resemblance to the polarization of the accompanying microwave burst emission (cf. also Fürst *et al.*, 1973 and Tindo *et al.*, 1973).

Though a detailed discussion of the X-ray polarization features depends on a number of unknown parameters and therefore is somewhat complicated (Korchak, 1967, 1971; Elwert and Haug, 1971; Beigman, 1974), it can be assumed that the nonthermal acceleration process does in fact last a longer period than simply the impulsive rise time, thus supporting a certain post-maximum supply of fast particles.

5.3. Particle Acceleration and Energy Release

5.3.1. ACCELERATION MECHANISMS

a. *General Concepts*

Generally three groups of energy conversion processes are of interest with regard to the generation of excessive solar electromagnetic radiations, among them the radio emission. They are *plasma heating processes*, *particle acceleration processes*, and *turbulent wave generation.*

All these types of processes are interrelated and linked together. They lead *inter alia* either directly (heating and acceleration) or indirectly (plasma turbulence) to the emission of radio waves. The interrelations are schematically demonstrated in the flow diagram of Figure V.6. The heating and acceleration processes refer to the thermal and nonthermal parts of the particle energy distribution function, respectively. Excessive heating can lead to the onset of particle acceleration by instabilities. The turbulent wave modes contribute to the radio emission by different mode transformation processes, but may also cause a particle acceleration (and subsequent heating) in a plasma (Kaplan and Tsytovich, 1969).

Thus in accordance with Figure V.6 we state once more two basic classes of processes (cf. Section IV 4.2.1) leading to radio emission which are principally in competition with each other:

(1) particle acceleration and heating	→ ('direct') generation of RF waves → generation of plasma waves
(2) plasma wave turbulence	→ ('indirect') generation of RF waves → particle acceleration and heating.

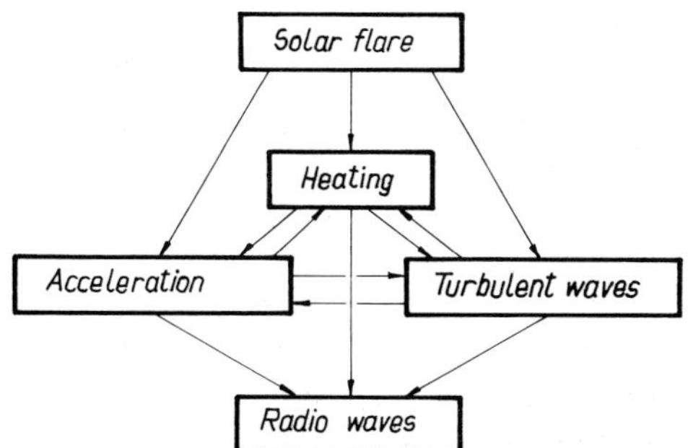

Fig. V.6. Basic types of flare-related energy conversion processes leading to radio emission.

The first class of processes is related to the primary emission mechanisms whereas the second class comprises the secondary generation of radio emission (cf. Section 4.2.1b).

Now let us summarize the major requirements to be posed on solar particle acceleration processes, which mainly refer to an interpretation of the impulsive and explosive flare phases, given by Lin (1974):

(a) The acceleration occurs typically during a period of the order of 10 s to a few minutes, most likely in groups of short pulses ('elementary acceleration processes').

(b) The time scale of a short pulse is typically < 1 s.

(c) The total amount of energy transformed by the particle acceleration reaches an appreciable fraction of 10^{32} erg in strong events.

(d) Electrons are typically accelerated in impulsive events to energies of 5–100 keV varying from event to event.

(e) Proton events are much rarer reaching energies of 10–60 MeV as typical for events exhibiting an explosive phase. Consequently, the acceleration occurs most likely as a two- (or more-?) step process with different energy outputs referring to the impulsive and explosive flare-burst stages.

The latter idea of a two-step acceleration was originally suggested by De Jager (1969) assuming a pre-acceleration to particle energies of about 100 keV during the flash phase providing the injection necessary for the action of the second-step acceleration, which was thought to be accomplished e.g. through a Fermi-process in shock waves.

In the following we consider different groups of acceleration processes.

b. *Acceleration by Electric Fields*

The classical acceleration mechanisms of potential interest for solar flares can be divided into electric and electromagnetic processes. The first attempts to explain solar particle acceleration by means of electric fields ('discharge theory') go back to several authors (Swann, 1933; Giovanelli, 1946, 1948; Dungey, 1958). The high conductivity of a fully ionized plasma, however, rules out a simple application of an acceleration by electrostatic forces in its original form (Sweet, 1958; Parker, 1960). But the observational evidence for the occurrence of strong electric currents in flare plasmas has stimulated a renewed discussion on the subject (Wentzel, 1963; Carlquist, 1969).

In the model of Alfvén and Carlquist (1967) the process of *current interruption* causes a space-charge concentration of high impedance. For a current of e.g. 3×10^{11} A in a plasma of $N = 10^8$ cm^{-3} and $T = 10^6$ K an energy of $E \lesssim 10^{32}$ erg would be released, theoretically, in a period as short as $\lesssim 7 \times 10^2$ s, in so far as no other faster developing instability (e.g. ion-sound instabilities) would prevent or suppress the effective acceleration by the current interruption (Smith and Priest, 1972). Unless the latter restrictions are sufficiently quantitatively explored (theoretically and experimentally) and the transport quantities are precisely known in solar flare plasmas, it is difficult to estimate the real importance of the current-interruption process (and herewith also of other mechanisms invoking electric-field actions) for the interpretation of acceleration processes in solar flares.

c. *The Fermi–Parker Mechanism*

The Fermi–Parker mechansim is one of the often cited processes of particle acceleration in connection with solar flares. It is based on the assumption of temporally changing magnetic fields reflecting charged particles running against them like a tennis ball at a tennis racket (Fermi, 1949, 1954).

The reflection is due to the adiabatic invariancy of the magnetic moment

$$\mu = mv^2/(2H), \tag{V.3}$$

which is proportional to the momentum p of the Larmor motion of a charged particle (Hertweck and Schlüter, 1957). Hence the reflection condition

$$\sin^2\phi/H \approx \text{const}, \tag{V.4}$$

(where ϕ is the pitch angle) is obtained when the value $\sin\phi = 1$ is attained for a special sufficiently high value of H. Encounters with such 'magnetic walls' moving with a velocity $\mathbf{w}$ in the configuration vector space cause an acceleration or deceleration of the reflected particles, if they have a component of their velocity $\mathbf{v}$ directed oppositely or in line with the direction of $\mathbf{w}$, corresponding to head-on or overtaking collisions, respectively. Repeated collisions, e.g. by approaching magnetic structures, decrease the pitch angle ϕ, so that the particles can escape in the direction of the magnetic field and are lost for a further, subsequent acceleration. For this reason two conditions must be realized to make the acceleration process efficient: Firstly, the head-on collisions must prevail over the overtaking collisions; and secondly, a permanent redistribution of the pitch angles ϕ is required.

Special realizations of these requirements were proposed by Parker (1958), Takakura (1961a), and Wentzel (1963). There are further numerous modifications of the mechanism, e.g. invoking a randomization of the pitch angle distribution by turbulence (Hall and Sturrock, 1967), resonant wave-particle interactions (Kulsrud and Ferrara, 1971; Cheng, 1972a; Melrose, 1974a), etc. Though the Fermi–Parker mechanism would be principally capable of providing an efficient acceleration, through its step-like action requiring a succession of repeated reflections, it has the basic disadvantage that it acts too slowly. Therefore, the mechanism can hardly explain the fast acceleration of the impulsive flare phase, so that it comes into consideration at best only for the second-stage acceleration during the explosive flare phase.

d. *Other Electromagnetic Acceleration Mechanisms*

Theoretically, the Fermi–Parker mechanism is only one among a great variety of possible acceleration processes. Other types of electromagnetic acceleration mechanisms are e.g., the magnetic pumping or the betatron process which are characterized by a cyclic action of the acceleration. The starting point is the principle that an increase of the magnetic field strength causes an enhancement of the kinetic energy of the Larmor motion according to

$$p^2\sin^2\phi \sim H,$$

which again is a consequence of the adiabatic invariancy of the magnetic moment of the Larmor motion. Application is made in the betatron acceleration by taking the magnetic field strength as proportional to the particle density in a first approximation, which provides an estimation of an upper limit of the efficiency of the mechanism (Wentzel, 1964). However this mechanism is not well favored in comparison to other mechanisms.

Another approach consists of a general estimation of the anticipated particle acceleration by Faraday's induction law

$$\left| \oint E_{ind} \, \mathrm{d}s \right| = (1/c) \left| \frac{\delta}{\delta t} \int H \, \mathrm{d}\Sigma \right| \tag{V.5}$$

where the path s encloses the region Σ (cf. De Jager, 1965a). A deeper consideration of the problem of particle acceleration on the Sun in connection with a possible conversion of magnetic field energy is involved in the different flare theories. This will be discussed in Section 5.6.

In essence, electromagnetic acceleration processes seem to be principally capable of explaining a second-stage acceleration, but their action depends on the presence of a suitable injection of particles, which could be produced by some kind of turbulence due to plasma instabilities.

5.3.2. Energy Release by Different Modes of Plasma Turbulence

After considering the classical electromagnetic acceleration processes the treatment of plasma turbulence brings new aspects to the question of the energy release and acceleration in astrophysical plasmas (Tsytovich, 1971; Kaplan and Tsytovich, 1972).

In principle, turbulent fields accelerating particles may be present either in the form of more slowly enhanced turbulence in solar active regions heating the corona (Pneuman, 1967; Schatzman, 1967), or in the form of rapid instabilities connected with impulsive flare events (Sturrock, 1966, 1968, 1974; Smith, 1974a).

Attempts to explain the acceleration of the impulsive flare phase by the action of microinstabilities (e.g. two-stream electron-ion instability) were made e.g. by Friedman and Hamburger (1969) and Coppi and Friedland (1971). These instabilities would be connected with the process of magnetic field reconnection producing intense currents and high-frequency electric fields. According to Sturrock (1974) such a process would be likely to occur during the first stage of particle acceleration (impulsive phase) in a flare. Such a mechanism would be consistent with the empirical fact, that impulsive events are mainly connected with an acceleration of electrons rather than of protons. For second- or third-stage acceleration sometimes a mechanism is expected which acts in association with the occurrence of a MHD shock front. Here the general possibility of a stochastic particle acceleration is to be considered, which also comprises the Fermi–Parker mechanism as a particular case (Hall and Sturrock, 1967; Newman, 1973). A stochastic acceleration according to MHD turbulence would be in favor of a resulting power-law particle-energy spectrum. Special estimations on this point were made by Sturrock (1974). However, it remains questionable whether processes after the formation of a shock wave are sufficient to explain the whole explosive phase acceleration. It seems more plausible that both

TABLE V.6

Survey of different stages of energy release and corresponding flare phases

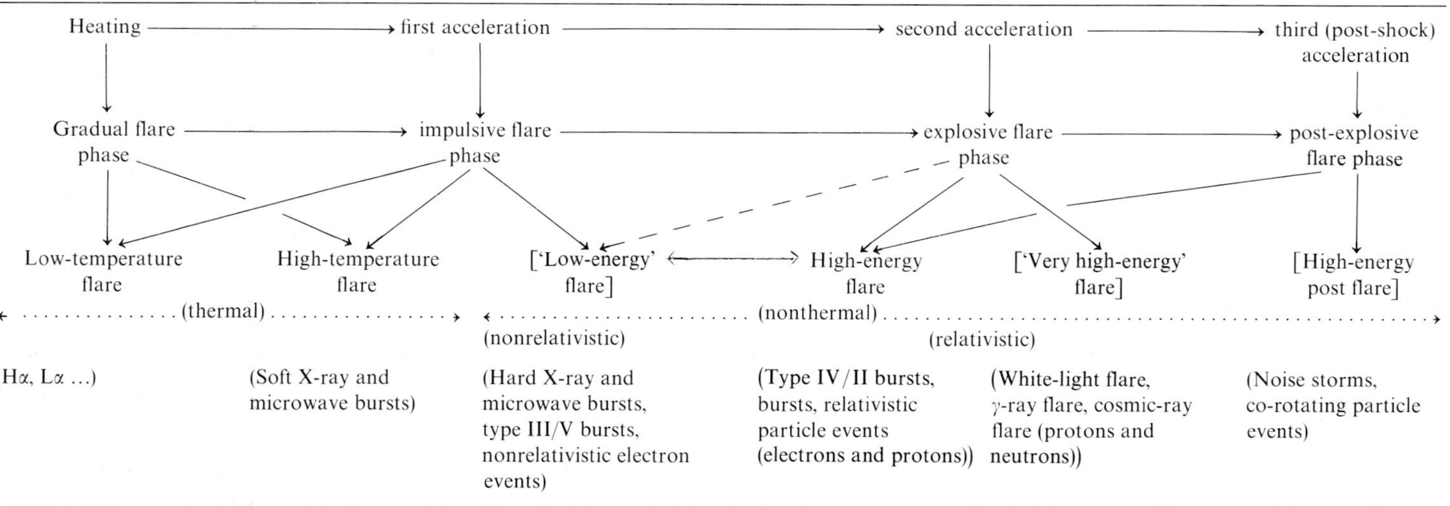

particle acceleration and the generation of a shock wave are products of an instability triggered in one and the same stage of the flare development, the explosive phase.

At the end of this section we try to summarize in the form of Table V.6 the different stages of flare heating and acceleration together with their connection to the resulting energy ranges of the exciters of different flare- and flare-like radiations and morphological forms. The aim is to point out in a simplified manner the complex interrelations between different actions of the radiation flare, dynamic flare, and particle flare.

The quantities in brackets, added in Table V.6 for completeness, express the personal view of the author and have not yet been generally introduced in the literature.

5.3.3. Relation between Electron Energy Spectrum and Emitted Photon Spectrum

Apart from more or less rough estimations direct quantitative evidence for the particle-energy spectra in solar source regions by radio methods alone can be obtained only in very particular cases. Sometimes an indirect way is followed by adopting the energy spectra derived from X-ray observations or also direct measurements in space and checking the compatibility of these data with suitable radio models.

In such a way one obtains the photon spectrum from observations of soft X-rays

$$\begin{aligned} I(\varepsilon) &= b\varepsilon^{-3/2} \exp(-\varepsilon/(KT)) \\ &= 1.05 \times 10^{-42}\, Y\varepsilon^{-3/2} \exp(-\varepsilon/(KT)) \\ &\quad [\text{photons s}^{-1}\,\text{cm}^{-2}\,\text{keV}^{-1}] \end{aligned} \tag{V.6}$$

(ε = photon energy, $Y = \int_V N_e^2 \,\mathrm{d}V$ – emission measure) yielding the temperature T and integral electron content of the emitting region.

In the case of an additional nonthermal (hard) X-ray burst component the observed photon spectrum can be fitted by a power law in a certain energy range $\varepsilon \gtrsim \varepsilon_0$

$$I(\varepsilon) = a\varepsilon^{-\lambda} \tag{V.7}$$

which corresponds to an electron energy spectrum

$$N(E)\,\mathrm{d}E = kE^{-\gamma}\,\mathrm{d}E \tag{V.8}$$

(cf. Section 4.1.2.c) or differential spectrum

$$\mathrm{d}N(E)/\mathrm{d}E = cE^{-\gamma'} \tag{V.9}$$

where $\gamma = \gamma' - 1$.

Corresponding power-law energy spectra were in fact obtained by direct observations of electrons in the interplanetary space (cf. the following section). Comparisons with the resulting microwave emission spectrum are generally not possible in an analytical form thus requiring numerical calculations (cf. Sections 3.3.5 and 4.3.2.b).

Nevertheless, a direct analytical connection between radio and particle-energy spectral indices could be established in the case of an ultra-relativistic isotropic synchrotron radiation (cf. Section 4.4.2.e), where it turned out that the intensity of synchrotron radiation without reabsorption is proportional to $\nu^{-(\gamma-1)/2}$. In the case of strong reabsorption no dependence from the exponent γ can be stated.

5.4. Particle Radiation and Radio Waves

5.4.1. The Solar Wind

Since there are close interrelations between the solar particle emission and the radio emission, a brief summary of some of the main facts concerning solar particles in space, so far as it may be of interest for radio astronomy, will be given.

Spacecraft observations have revealed the presence of a continuous flow of plasma from the Sun known as solar wind. Reviews on this topic were presented by Brandt (1970), Mackin and Neugebauer (1966), Hundhausen (1972), and Veselovskij (1974). The whole field of solar wind research is very extensive now and its thorough description would require its own series of monographs. It touches a multitude of interesting questions, e.g. propagation of different kinds of waves, fields, and particles in a plasma, instabilities and MHD discontinuities, etc. A wealth of *in situ* observations is available by space probes and satellites, so that the solar wind and the interplanetary medium have become at present the most extensively directly probed astrophysical plasma. Solar wind phenomena exhibit many variations of physical parameters with different time scales.

A detailed classification of solar wind phenomena has been proposed e.g. by Burlaga and Ness (1968), Burlaga (1969), and Hundhausen (1972). Here we mention only the following main phenomena:

(a) structureless ('quiet') low-speed solar wind;
(b) persistent solar wind from localized regions or sectors;
(c) flare-produced interplanetary solar wind disturbances.

The low-speed solar wind component is characterized by the following properties (Hundhausen, 1972):

Flow velocity, radial component:	300–325 km s^{-1}
nonradial component:	8 km s^{-1}
Proton density, electron density:	8.7 cm^{-3}
Average electron temperature:	1.5×10^5 K
Average proton temperature:	4×10^4 K
Magnetic field strength:	5γ
Proton flux density:	2.4×10^8 cm^{-2} s^{-1}
Kinetic energy flux density:	0.22 erg cm^{-2} s^{-1}.

Models to describe the feature of a steady, spherically symmetric coronal expansion were made on the basis of one-fluid or two-fluid theories. Effects of hydromagnetic waves, heat conduction, viscosity, and noncollisional energy exchange have been taken into account.

In practice the 'quiet' structureless coronal expansion is rarely seen in the solar wind observations. It is mixed with slowly varying high-speed plasma streams, which are related to unipolar regions known as magnetic sector structure. The properties of the high-speed solar wind component can be summarized as follows: (i) Enhancement of the flow speed up to $\approx$ 600–700 km s^{-1}; (ii) enhancement of the density (in particular at the leading stream edge) by several times the quiet level followed by a rare-faction; and (iii) enhancements of the proton temperature and of the magnetic field in connection with flow-speed variations.

Concerning the regions of origin of the high-speed solar wind component two extreme possibilities have been discussed:

As a first possibility the solar wind is assumed to originate in widespread areas of coronal expansion referring to sector structures ('mapping hypothesis'). The other possibility consists in the alternative assumption of an origin in a restricted number of highly inhomogeneously distributed discrete sources ('nozzle hypothesis').

The observed relationship between the high-speed interplanetary plasma streams and magnetic sector structures seems to be a strong argument for the mapping hypothesis. Similar arguments result from the attempts to identify the M-regions of recurrent geomagnetic activity with specific solar features, whereas no clear association with single features of solar activity could be established. On the contrary, a cone of avoidance of the particle emission could be suggested on the base of the steeper temperature gradients present in solar active regions (Billings and Roberts, 1964; Parker, 1965).

On the other hand, there are certain links between active regions and the sector structure, which are manifested by the existence of active longitudes, i.e. the locations of active regions with a strong source of the S-component, high flare rates, and strong noise-storm activity.

Regarding the latter point connections between noise storms and the solar particle emission are an interesting feature (cf. e.g. Martres *et al.*, 1970; Sakurai and Stone, 1971). The currently most probable geometric configuration describing the observations was proposed by Wilcox (1968) and Sakurai and Stone (1971). A combined version of the pictures given by these authors is sketched in Figure V.7.

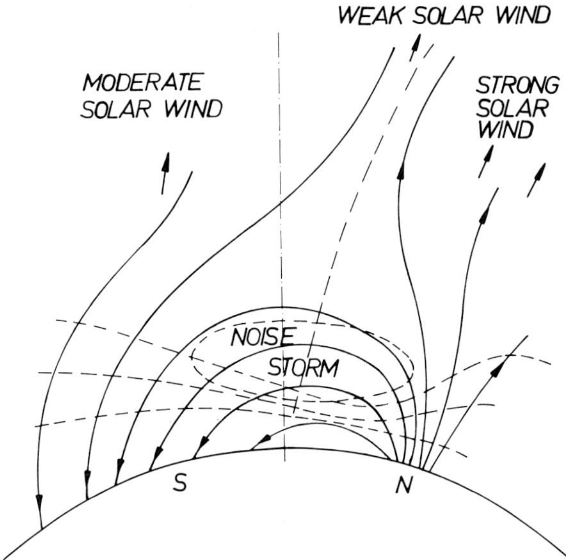

Fig. V.7. Schematic representation of different solar-wind outflow regions in relation to the sector boundary (vertical dashed line) and the position of a type I noise-storm region. Horizontally dashed lines refer to constant temperature.

The third large component of solar particle emission which is connected with the flare phenomenon will be considered in the next sections.

5.4.2. Electron Events

Radio and X-ray observations provide indirect evidence for the presence of accelerated electrons in the flare-burst plasma on the Sun. If these electrons travel into the interplanetary space, direct proof of the particle components, as well as conclusions on related physical processes, become possible by spacecraft observations.

Electron events observed in the interplanetary space reveal a wide range of electron energies. It is convenient to distinguish nonrelativistic energies from relativistic electron energies ranging from 1 to 100 keV and from 100 keV to 100 MeV, respectively. Reviews on these matters were given by Lin (1974) and Simnett (1974).

The first observation of relativistic electrons in space was not until 1960, when the galactic component of the cosmic rays was investigated in this respect, initiated by predictions due to early radio observations (Earl, 1961; Meyer and Vogt, 1961; cf. also Alfvén and Herlofson, 1950; Kiepenheuer, 1950). Later the relativistic electrons from the Sun were detected at balloon altitudes, the existence of which was concluded earlier by radio investigations (Boischot and Denisse, 1957).

Nonrelativistic solar electrons were first observed by the space probe Mariner 4 (Van Allen and Krimigis, 1965). Since that time many flare-produced electron events have been observed, in particular by the IMP (Interplanetary Monitoring Platform) spacecraft series.

In the following we summarize some important aspects of nonrelativistic and relativistic electron events. The nonrelativistic electron component is mainly connected with small and moderate flare events. The relativistic electron component is only present in cases of large flare events and exhibits a violent increase of energetic electrons in space.

a. *Nonrelativistic Electrons*

The occurrence of nonrelativistic electrons in space corresponds well with impulsive microwave and hard X-ray bursts of moderate strength. Obviously a certain amount of the nonrelativistic electrons accelerated on the Sun during flares is escaping into the interplanetary space. In contrast to the relativistic particles which undergo a diffusive propagation, the nonrelativistic electrons are characterized by large (scatter-free) mean free paths (of the order of 1 AU).

Flare-produced nonrelativistic electron events show the following characteristics:

(a) The time profile is in general consistent with an impulsive flare-related particle injection on the Sun.

(b) Depending on the (energy-dependent) amount of scattering the sharp impulsive onset of an electron event can be flattened.

(c) There is a dispersion in energy space.

(d) The transit times to the Earth orbit are less than 30 min.

(e) Electron events in space are typically associated with impulsive microwave and X-ray bursts and type III bursts.

Most nonrelativistic electron events are related to minor flares (often subflares) without associated relativistic electron events. Only 2.5% of all flares have been found to be associated with nonrelativistic electron events in space, whereas 0.4% of all flares were accompanied by relativistic electron events.

The exponents of the nonrelativistic electron energy spectra range from 2 to 5. In the case of an absence of related relativistic electrons a rapid dropping of the spectrum near about 100 keV is to be noticed. Also at energies below 5 keV the spectrum departs from a power law.

In the cases of mixed events consisting of nonrelativistic and relativistic electrons and protons the nonrelativistic electrons are observed simultaneously or before the onset of the relativistic electrons and protons. It may be concluded that the nonrelativistic electrons are released from the Sun some time ($\lesssim$10 min) earlier than the relativistic particles accelerated in the explosive flare phase.

Furthermore the presence of type I bursts seems to indicate situations where electrons are emitted from the Sun in a relatively broad cone of longitudes. These so-called *co-rotating events* form a separate class of electron events apart from the impulsive type. For the interpretation either a continuous, locally extended acceleration zone at the Sun, and/or a particle storage with a long-period leakage are required.

b. *Relativistic Electrons*

Beside the 'pure' electron events in the nonrelativistic energy range the solar events in the relativistic range are to be noted as a physically different class, which is typically associated with proton events. The most important features of these relativistic electron events are as follows:

(a) The time profile of an *impulsive-diffusive event* displays rise times of typically 1–4 h followed by a slow decay lasting one or two days.
(b) Departures from a classical impulsive-diffusive time profile occur due to propagation effects outside a direct magnetic connection between the observer and the flare site.
(c) In some cases co-rotating (sometimes recurrent) events are observed covering periods longer than one week with almost symmetric time profiles not identifiable with one particular optical flare on the visible hemisphere of the Sun. These events are thus in favor of a storage of electrons in the solar corona which may be related to noise-storm activity.
(d) The spectral index of relativistic electron events is typically about 3 with a more or less small scatter between 2 and 5 (or even steeper). Co-rotating events exhibit a larger scatter of the spectral index than flare-associated events. Sometimes the spectrum becomes steeper towards higher energies above 10 MeV.
(e) The time delay between the optical flare onset and the first arrival of the relativistic electrons at about 1 AU from the Sun is about 30 min.
(f) Beside the impulsive-diffusive and co-rotating events there also exist so-called *quiet-time electron increases* of 1 to 2 weeks' duration, which may be possibly due to a varying input of galactic electrons modulated by the Sun.

5.4.3. Proton events

In conjunction with the occurrence of solar relativistic electrons energetic protons are observed in interplanetary space. In contrast to the electron-energy spectrum which is almost constant for flare-produced events the proton-energy spectrum varies much more. Consequently, the electron-to-proton ratio also varies by more than two orders of magnitude for different events.

Differing from solar electrons, which can be detected only in space, the solar proton events can be divided into the following classes:

(a) *Ground-level effects* (GLE), which are very rare (20 events between 1942 and 1972) requiring energies above 500 MeV according to 'cosmic-ray flares.'
(b) *Polar-cap absorptions* (PCA) corresponding to energies in excess of 10 MeV recorded via cosmic radio noise absorption at the Earth's polar caps by riometers. This effect is due to an anomalous ionization of the polar ionospheric D-layer by the impinging protons.
(c) *Satellite-sensed events* (SSE) corresponding to energies in excess of 0.3 MeV.

Commonly the term 'proton flare' is used for proton events with energies in excess of 10 MeV, but sometimes it is extended also to SSEs. A collection of proton flares is contained e.g. in the Catalogue of Solar Particle Events edited by Švestka and Simon (1975).

In summary, the following properties of proton flares may be noted:

The time delay between the flare onset and the arrival of particles at 1 AU as well as other characteristics (e.g. the time difference between the onset and maximum of proton flux) are strongly dependent on the flare position at the Sun. The time differences have a minimum for flare longitudes of about 40° W.

There are several lines of evidence that protons must be stored near the Sun for several hours or days, but it is still difficult to understand the actual mechanism.

The spectral index varies between 1.3 and about 5 in the 20 to 80 MeV region. The hardness of the spectrum is not necessarily proportional to the total number of particles leaving the Sun.

Proton flares are typically accompanied by type II, III, and IV bursts. For the related type IV bursts the combination of IVμ and IVmF(A) components is very characteristic, evidently indicating the acceleration and release of particles, whereas the escape of electrons of lower energy is indicated by repeated groups of type III bursts superimposed upon the type IVmF (or IVmA) burst. The storage of particles seems to be associated with the subsequent formation of a type IVmB burst component.

It is very likely, but not generally accepted, that the associated centimetric and metric type IV burst components are the product of two different acceleration regions corresponding to a two-step acceleration as originally suggested by De Jager (1969). As already mentioned before, several authors ascribe the secondary acceleration taking place during the explosive phase to post-shock processes. But principally there is also the alternative possibility of interpreting the occurrence of shock waves as a *result* of the explosive phase. In the latter case the shock wave would occur after the acceleration (and release) of the bulk of the energetic particles and produce only a tertiary acceleration (cf. Section 5.3.2).

5.5. Shock Waves and Magnetospheric Disturbances

5.5.1. Flare-produced Interplanetary Shock Waves

The connection between the occurrence of geomagnetic storms and major flares on the Sun has been recognized for a long time. In particular the sudden commencement (sc) of geomagnetic storms is attributed to the passage of an interplanetary shock wave expelled from the Sun. From the average delay time between the flare onset and the sc of about 55 h a mean shock speed of roughly 730 km s^{-1} can be inferred.

The high correlation with type IVμ and moving type IV bursts indicates that the interplanetary shock wave can be regarded as a direct continuation of the exciters of the radio bursts in space. This has been proved by spaceborne measurements tracking the type II burst through the interplanetary medium (cf. Figure III.22). It should be mentioned, however, that not all geomagnetic storms and sc's are caused by flare-produced hydromagnetic shock waves: Tangential discontinuities of the solar wind without distinct shock fronts may also cause a number of preferably weaker sc's and recurrent geomagnetic storms (cf. Gosling *et al.*, 1967).

Together with the passage of an interplanetary shock front a number of variations in the particle flux is to be noted:

(a) Behind the shock a plasma cloud is moving outwards observable as a solar-wind enhancement lasting for several hours after the shock passage. It consists of a shell of compressed ambient solar wind surrounding a blob of flare ejecta.

(b) Low-energy particles (<20 MeV) are swept ahead of the shock front so that an increase of the particle flux up to 1 h in advance can announce the subsequent arrival of an interplanetary shock front.

(c) A propagating shock shifts and deforms the interplanetary magnetic field lines thus displacing the path of the high-energy particles in space.

(d) An interplanetary shock causes deflections of the galactic cosmic radiation displayed by a decrease of the ground-based neutron-monitor counts known as the Forbush effect.

Morphologically the shock waves can be classified into two main types labeled 'R' and 'F' events, which are characterized by a more continuous **r**ise and an abrupt jump of the flow speed followed by a steady **f**all, respectively. These two types may correspond to theoretical models of 'driven' waves and 'blast' waves (Hundhausen, 1972).

5.5.2. Analogies between Solar and Magnetospheric Processes

Obviously there is a certain correspondence between the flare phenomenon at the Sun and some magnetospheric processes. Special attention has been paid to the consideration of analogies between solar flares and magnetospheric substorms (called also 'auroral flares'), which are basically expressed by similar magnetic field configurations (helmet structure – magnetospheric tail). Though the dimensions and scales of the physical parameters may differ by some orders of magnitude, the similarities between the solar and magnetospheric phenomena are rather striking (Alfvén, 1950; Piddington, 1969). These topics are subject to efforts of interdisciplinary studies putting together both solar and magnetospheric physicists (Flare Build-up Study, cf. De Feiter, 1975; Obayashi, 1975; Nagata, 1975; Dungey, 1975).

Looking into the details, similarities may be noted with respect to the following points:

(1) Storage of the main part of energy by magnetic fields;

(2) Subsequent conversion into kinetic particle energy by acceleration and heating;

(3) Analogies in the morphological development of both solar and auroral flares:
  (a) preheating (pre-flash phase – growing phase),
  (b) flash or impulsive phase – expansion phase, and
  (c) recovery;

(4) Possibility of reconnection processes of magnetic field lines;

(5) Generation of plasma and electromagnetic waves;

(6) Occurrence of nonlinear wave–wave and wave–particle interactions;

(7) Occurrence of field-aligned currents, dissipation or electromagnetic energy, and precipitation of particles; and

(8) Occurrence of hydromagnetic shock waves triggering instabilities.

Since the auroral flare is observable by *in situ* measurements, one can hope to gain useful hints also for solar flare physics from magnetospheric measurements in spite of the fact that there may also be great differences between both phenomena (cf. below). In an auroral flare intense currents are observed to flow along the geomagnetic field lines causing a substorm in the polar regions of the Earth. Westward and eastward currents belonging to bands of the electron and proton aurora, respectively, lead to the geomagnetic bay effects in opposite directions. Simultaneously with the rapid onset of the expansion phase ('auroral break up') a number of associated effects can be observed, viz.: ionospheric substorms displayed by cosmic-noise absorption (CNA) effects; sudden increase of the VLF hiss emission followed by VLF chorus; enhancement of the ULF activity corresponding to the occurrence of magnetic pulsations of the PiB type followed by PiC-type pulsations; and X-ray substorm due to impacts of auroral electrons with neutral particles.

The VLF hiss emission can be interpreted probably as Čerenkov radiation and/or gyro-resonance emission directly produced by auroral electrons with energies as low as 1 to 10 keV precipitating through the magnetospheric plasmasphere (and magnetotail). It is followed by the auroral VLF chorus emission which is related to plasmaspheric electrons of higher energies (several tens of keV) trapped in closed magnetic fields and may be explained, in principle, by some kind of pitch-angle instability. The PiB and PiC pulsations, characterized by periods of about 5 and 10 s, respectively, are closely related to the auroral X-ray pulsations which may be due to precipitating electron beams.

The VLF emissions have been ascribed to ion-cyclotron resonance phenomena caused by injected protons. It can be hoped that the detailed study of magnetospheric processes, in particular of those connected with the substorms, will provide an important opportunity for a contribution to a deeper understanding of the basic physical processes related to the flare problem and *vice versa.*

But there are several specific differences between the physical conditions on the Sun and the Earth, which cannot be overlooked, e.g.:

(1) In the solar atmosphere H is the major constituent while in the magnetosphere N_2 and O prevail;

(2) Different ranges of the released energy ($< 10^{32}$ erg in comparison to $< 10^{22}$ erg);

(3) Differences in the magnitude and details of the configuration of the magnetic fields in complex solar active regions and in the magnetosphere;

(4) Different causes of magnetic field changes (penetration of new emergent magnetic flux from subphotospheric regions at the Sun and the drag of the solar wind from outward on the geomagnetic tail);

(5) Differences of certain other plasma parameters such as particle density, emitted frequencies, and the character of shock waves;

(6) Different time scales of the development of solar and auroral flares;

(7) Differences in some details of certain flare-burst components and the components of a substorm; and

(8) To a certain extent there are also entirely different methods of observation of solar and magnetospheric phenomena yielding different observed quantities.

For all these reasons an oversimplified translation from solar to terrestrial conditions and *vice versa* appears to be dangerous.

5.6. Burst Origin and Flare Theories

5.6.1. Basic problems

The problem of the origin of the different radio burst emissions is closely related to, and a part of, the general problem of the origin and 'mechanism' of the whole flare phenomenon comprising all its different components. Actually the 'flare problem' exists since the first solar flare observations of Carrington in 1859, but even now it is not yet completely solved. The problem can be divided into a number of fundamental questions, e.g.:

(a) What is the source of the flare energy?

(b) How and where is the energy stored and which are the basic flare build-up processes in an active region?

(c) What is happening during the pre-flare phase?

(d) How is the energy released in the impulsive flare phase and where is the location of the primary instability?

(e) What are the basic processes during the explosive and post-explosive flare phases?

(f) How is the energy transported in the flare volume and which means of communication are effective combining different components of a flare plasma?

a. *The Energy/Time Problem*

How can an energy as large as $<10^{33}$ erg be released in a time as short as about 10^3 s? From the classical hydromagnetic diffusion equation for a stationary conducting fluid

$$4\pi\,\partial\mathbf{B}/\partial t = (1/\sigma)\,\nabla^2\mathbf{B} \tag{V.10}$$

(σ – electric conductivity) one obtains the dissipation time of magnetic fields

$$t_d = 4\pi L^2 \sigma \tag{V.11}$$

(L – characteristic length) yielding much too long dissipation times (10^3 yr) for a highly conducting plasma and reasonable extensions of the flare volume (e.g. $L \approx 10^{10}$ cm). Hence the necessity of reducing either the dimensions of the dissipation region (and) or of the electric conductivity follows. Special realizations of these requirements have been proposed by various authors in various ways, e.g. invoking the action of: plasma waves (Scarf *et al.*, 1965); instabilities e.g. tearing mode (Furth *et al.*, 1963) and finite conductivity (Jaggi, 1963); MHD waves (Petschek, 1964); filamentary structures (Wentzel, 1964); neutral atoms (Gold and Hoyle, 1960); tensor character of the electric conductivity and boundary layers (Piddington, 1967); and stochastic plasma turbulence (Rädler, 1968; Romanchuk and Krivodubskij, 1974).

b. *The Mass Problem*

The mass problem was formulated by Bruzek (1967). The number of particles involved in the flare process can be estimated as being of the order of $N = 10^{41}$. From this number the part 10^{40} may be ejected. This corresponds to an output of 10^{23} particles per cm^2, but normally only 10^{19} cm^{-2} are available in the solar corona, which seems to indicate a particle supply from deeper levels.

c. *The Spatial Communication Problem*

In the solar atmosphere the regions of maximum energy storage (by magnetic fields) do not seem to be identical with the regions of maximum energy release. This poses the question of a sufficient fast energy transport within the flare volume and also of the transport of certain trigger actions for different flare components. Potential means of such a communication are shock waves, electric currents, different kinds of plasma waves, particle and mass transport, heat conduction, and electromagnetic waves.

An increasing number of flare models dealing with these questions stress special aspects of the points listed above. Reviews have been given e.g. by Sweet (1969), De Jager (1969), Švestka (1972), and Vasyliunas (1975).

5.6.2. Possible Sources of the Flare Energy

a. *Energy Storage by Particles and Nonelectromagnetic Force Fields*

In principle there is little doubt that the energy converted by solar flares essentially goes back to an origin in the fusion reactions in the interior of the Sun. The subsequent stages of the flare energy, however, up to the storage in active regions and the sudden release are not so clear. Though today it is more and more generally accepted that the flare energy is most likely stored in magnetic fields of active regions, a number of interesting proposals were made differing from that general assumption. Such a proposal invokes a storage of kinetic energy of high-energy (up to 0.5 MeV) protons in the corona, which would hardly be detected by γ-rays, neutrons, or other emissions prior to the energy release by particle ejection (Warwick, 1962; Elliot, 1964, 1969). The accumulation of the energy should be continuously acting by a slow-acting acceleration process which was specified by Schatzman (1967), where again transformations of magnetic energy come into play.

Another, in some respect similar model, was proposed by Carmichael (1964) assuming a temporary store of mechanical (solar wind) energy in trapping magnetic fields. In spite of the acceptability of the storage of mechanical energy from the physical point of view, there is a serious lack of evidence which could support the idea.

A further also interesting but even more questionable suggestion is based on the action of gravitational forces expressed in the inversion model of Sturrock and Coppi (1964).

b. *Energy Transport by Hydromagnetic Waves*

Energy fluxes from subphotospheric layers may also be present in the form of hydromagnetic waves. Then the presence of magnetic fields in stronger active regions has essentially two effects: making the regions 'transparent' for the passage of MHD waves which otherwise are strongly attenuated at the photospheric level (Osterbrock, 1961) and confining of MHD waves (in particular Alfvén waves and slow magneto-acoustic waves) in closed magnetic configurations and dissipation therein.

These waves generated in the convection zone below the photosphere may possibly contribute to solar flares in two alternative ways:

i. The flare energy is supplied immediately during the progress of a flare event. Thus a direct subphotospheric origin of the flares should be assumed in this case, which can hardly be checked by observations (Piddington, 1973).
ii. The flare energy emerges in the form of MHD waves more or less continuously during the pre-flare stage providing an energy storage above the photosphere (Pneuman, 1967; cf. also Wentzel, 1964; Parker, 1964a).

In principle, an influence of the above described processes (especially of the latter one) cannot be excluded. However, it remains doubtful, whether hydromagnetic waves will be sufficient to explain the essence of the flare phenomenon. Thus they may contribute to some heating processes in active regions but their proper role in flare events remains open.

c. *Electromagnetic Flare Build-Up Processes*

Considering all possible energy sources it turns out that the bulk of the flare energy is most probably explained by an intermediate storage, in the form of magnetic fields. There are many theories as to how this storage and release of the magnetic energy can be realized by processes connected with the penetration of a magnetic field into a plasma.

The energy supplied by magnetic fields of active regions deliver an almost force-free pattern characterized by

$$\nabla \times \mathbf{H} = \alpha \mathbf{H}. \tag{V.12}$$

But evidently not all of the magnetic field energy is available for a conversion into kinetic and wave energy by the flare process. This corresponds to the fact that the gross magnetic field structure of an active region caused by subphotospheric processes is not essentially altered by dissipative field changes. Several authors considered processes to furnish stored magnetic field energy in the solar atmosphere for dissipative processes which are characterized by departures from a simple potential field which represents the minimum-energy state without extractable energy. Such a storage can be established by a twisting of magnetic field lines which is described by a variation of the parameter α in Equation (V.12) and the presence of field-aligned currents

$$\mathbf{j} = \frac{c}{4\pi} \operatorname{curl} \mathbf{B}. \tag{V.13}$$

A first proposal for a detailed mechanism of such a build up of flare energy by the rotational motion of sunspots was made by Stenflo (1969) essentially based on earlier statements of Alfvén (1963), who pointed out that field-aligned currents should be produced by photospheric motions if the integral along a magnetic filament F is

$$\int_F (\mathbf{v} \times \mathbf{B}) \cdot \mathrm{d}\mathbf{s} \neq 0. \tag{V.14}$$

As mentioned by Sturrock (1966) the non-current-free state does not correspond to minimum energy conditions thus leading to the principal possibility of the development of instabilities and energy dissipation.

Almost similarily the flare build up by sheared magnetic fields was considered by Nakagawa and co-workers (Nakagawa *et al.*, 1971; Raadu and Nakagawa, 1971; Nakagawa *et al.*, 1973; Tanaka and Nakagawa, 1973).

Another approach to the question of flare energy build-up processes by force-free magnetic fields was proposed by Low (1973). Here the change of α was obtained by the assumption of the action of a resistive magnetic field diffusion in a passive medium, i.e. a tenuous compressible medium being free to move to accomodate the changing magnetic field configuration. Such a situation leads spontaneously to steep magnetic field gradients and instability. A generalization of this discussion on the basis of the existence properties of the corresponding set of equations was made by Birn *et al.* (1977).

5.6.3. Energy Release in the Hydrodynamic Flare Stage

a *Current-Sheet Models* (*General*)

Proposing an origin of the flare energy in a dissipation of magnetic fields, a reconnection of field lines is expected near neutral points of the magnetic field forming a current sheet. Originally two basic types of a macroscopic field configuration were suggested as being favorable for a magnetic field annihilation: The classical current sheet near an X- or Y-shaped neutral point (Sweet, 1958) and twisted magnetic fields (Gold and Hoyle, 1960).

Starting from these model assumptions a great number of variants have been attempted. As an example Barnes and Sturrock (1972) presented a model demonstrating how a current sheet might be built up from a force-free magnetic field configuration.

The hitherto most detailed concrete realizations of magnetic field annihilation processes go back to Syrovatskij's mechanism of dynamic field dissipation (Syrovatskij, 1966a) and Petschek's mechanism invoking the action of a shock wave (Petschek, 1964) which will be considered in the following subsections. Reviews on these matters were given by Vasyliunas (1975) and Syrovatskij (1975).

In summary, the main features invoked by hydrodynamic models of a magnetic field annihilation are: (1) Approaching magnetic field patterns or twisting of fields; (2) existence of current sheets near a neutral line of magnetic fields; and (3) action of hydromagnetic waves.

But it must be remarked, that the above mentioned macroscopic, hydrodynamic models can never be exhaustive in describing all features involved in the flare phenomenon. Hydrodynamic processes refer rather to a special group of flare processes realized in the hydrodynamic part or stage of a flare. Microscopic processes are likely to be involved forming a kinetic stage of the flare development (cf. Section 5.6.4). In order to make a magnetic field 'digestible' for an annihilation process a shift from larger- into smaller-scale processes is anticipated, i.e. a reduction to smaller structure elements coming into the range of microinstabilities.

b. *Dynamic Magnetic Field Dissipation*

The central point of any flare theory invoking an active role of the magnetic field is the finding of a real mechanism for a transformation of the magnetic energy into kinetic and wave energies comprising the production of accelerated particles, macroscopic plasma motions, heat, and different wave modes.

The basic principles of a magnetic field reconnection (annihilation) have already been established by Dungey (1953), Sweet (1958), and Parker (1963). In the picture of a two-dimensional representation magnetic field lines of opposite direction are assumed to be carried together in the vicinity of an X-shaped neutral point by a plasma flow of a velocity v_p (Figure V.8). There arise a diffusion region and current sheet where the magnetic flux is assumed to be annihilated partly through Ohmic dissipation

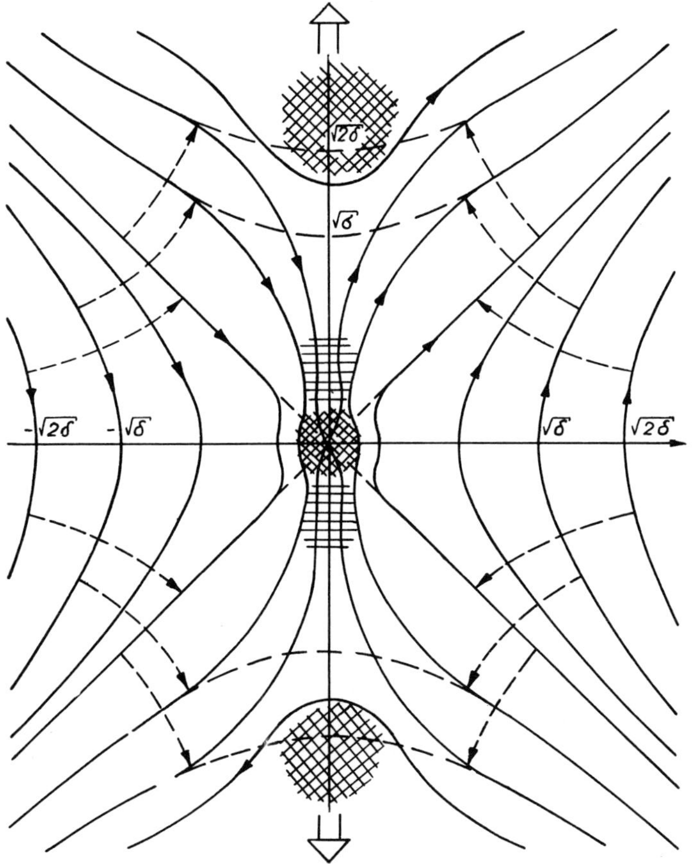

Fig. V.8. Idealized picture of the magnetic field configuration near an X-type neutral point undergoing a displacement δ leading to a MHD plasma flow (after Syrovatskij, 1966).

(heat) and partly through acceleration (kinetic energy of the plasma). Considerable progress in this question was achieved by Syrovatskij (1966a, ff.) who presented a mechanism of collisionless dynamic field dissipation directly transforming the magnetic energy into kinetic energy of accelerated particles in the vicinity of a magnetic neutral line. Considering the MHD flow near such a neutral line represented as a neutral line between two parallel currents, solutions are obtained if the currents change, corresponding to approaching magnetic field structure by small displacements δ. The latter may be realized by new magnetic fluxes emerging into a previously existing field structure of opposite polarity. Under special conditions the solution has the form of a converging cylindric wave leading to a cumulative effect with a strong increase of the field gradients towards the neutral line. The ratio of the electric current density $\mathbf{j}$ to the plasma density N grows up to high values. This indicates that the MHD condition of frozen-in magnetic fields breaks down and the magnetic field changes into an inductive electric field which directly accelerates charged particles.

The width of the current sheet increases with time up to a certain value. Plasma is squeezed out of the sheet at the edges and becomes compressed in two ropes along the neutral sheet furnished by a corresponding plasma flow towards the sheet from the outer regions.

In the above described way two effects are the preconditions necessary for a proper action of the mechanism: A strong compression of the magnetic field and an extreme rarefaction of the plasma near the neutral sheet. More detailed numerical computations revealed a number of difficulties for meeting favorable conditions, which, however, can clearly be overcome (Anzer, 1973; Syrovatskij, 1975). A special point requiring further consideration is the development of plasma turbulence in the neutral sheet.

In order to explain the impulsive flare phase the process of current interruption similar to the proposal of Alfvén and Carlquist (1967) can be adopted. Such a current-sheet interruption may arise if the ratio j/N exceeds a certain critical value necessary to maintain the current. Then plasma instabilities may set in and an additional effective acceleration is expected to occur. Though there are theoretical objections against this process (Smith and Priest, 1972) experimental results seem to support the possibility of such a process. However, there are still a number of open points in the whole mechanism. Some of those questions are the following: (1) Action of plasma instabilities and quantitative evaluation of the involved nonlinear processes; (2) more exact determination of the cause, topology, and temporal development of magnetic field changes (both theoretically and experimentally); and (3) comparison with real flare observations, e.g. explanation of flare loops and the morphology of different flare-burst compenents, etc.

c. *The Petschek Mechanism*

An alternative model of the processes taking place at the vicinity of a neutral sheet was proposed by Petschek (1964). Petschek's model was the first approach with a sufficiently fast magnetic field annihilation as required for solar flare processes. The model invokes the action of a MHD slow-mode shock wave going out from the diffusion region leading to a maximum reconnection rate. Since the model assumes a plasma flow into the field reversal region, the magnetic field lines are refracted at the wave front toward the wave-normal direction (Figure V.9). This reduces the magnetic field and the plasma flow is accelerated along the wave front. In order to obtain a stationary state the wave propagation must be balanced by the plasma flow. It is important to note that the wave velocity and plasma-flow speed are independent of the resistivity and can be much larger than the speed for magnetic field diffusion. The dominating influence of the diffusion is expected to be restricted to a relatively small region near the neutral line.

The proposed process of the magnetic field reconnection (merging) can be interpreted simply as a collision of two plasma flows (jets) carrying oppositely directed magnetic fields (Vasiliunas, 1975). They give rise to slow shocks which remain attached to the diffusion region since they cannot propagate perpendicular to the magnetic field.

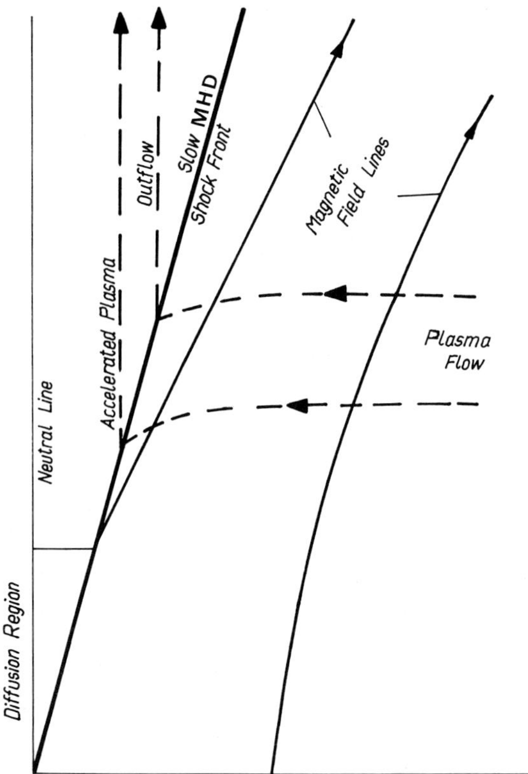

Fig. V.9. Scheme of the magnetic field lines and plasma flow in the first quadrant for Petschek's model of fast reconnection. The outward propagation of the shock front is thought to be compensated by an equal (reverse) plasma-flow velocity.

Based on this general principle a number of modifications and extensions have been made e.g. allowing for rotational discontinuities of the magnetic field (Kantrowitz and Petschek, 1966; Petschek and Thorne, 1967) and alternative solutions due to changed configurations (Sonnenrup, 1970; Yeh and Axford, 1970; Yeh and Dryer, 1973; Cowley, 1974; Priest and Cowley, 1975; Priest and Soward, 1975).

Comparing Petschek's model with Syrovatskij's model, several differences can be noted (cf. also Syrovatskij, 1975):

(a) In the center of Petschek's diffusion region the current density is a minimum compared with the boundary region. On the other hand, in the center of the current sheet of Syrovatskij's model the current density becomes a maximum.

(b) In Petschek's model the width of the diffusion region decreases with increasing conductivity, whereas in the case of a neutral current sheet it should increase.

(c) In Petschek's model the parameter $P/(H^2/(8\pi))$ must be sufficiently high, but in the theory of neutral sheets it should be $\ll 1$.

Further detailed differences in different models concern the special form of the adopted magnetic field model (e.g. homogeneity) and the resulting plasma flow (e.g. stationarity).

Until now different laboratory experiments seem to support (to a certain extent) the dynamic dissipation and also the field-line merging process by a Petschek wave. It appears, therefore, that a synthesis of both mechanisms cannot be totally excluded as appropriate for a wide range of realizations.

5.6.4. Kinetic Stages of the Flare–Energy Release

a. *Analogies to Laboratory Experiments*

It was already mentioned in Section 5.6.3.a that the hydrodynamic stage of the flare seems to be insufficient to describe the *totality* of all flare phenomena and processes. Analogies to laboratory experiments may help to discern what may happen in the kinetic stage of a flare (Baum and Bratenahl, 1974).

In experiments on the dense plasma focus a (z-) pinch discharge can be referred to as the magnetohydrodynamic stage (cf. e.g. Maisonnier *et al.*, 1971; Newman and Petrosian, 1975). From the pinch discharge a small X-ray and neutron burst can be observed. But this pinch is only a first and very transient stage before the focus phase sets in. The plasma column resulting from the z-pinch is immediately disrupted by MHD instabilities. There follows a characteristic interruption of the X-ray and neutron emissions called the *dark pause* (of about 50 ms duration in this particular experiment), during which all energy is stored internally in the magnetic field formerly ordered but now 'mixed' with the plasma by the action of turbulence resulting from the instability. The turbulent interaction of the magnetic field and the plasma produces an acceleration of high-energy particles, which is displayed by the laboratory experiments (Johnson, 1974; Newman and Petrosian, 1975; Gribkov, 1975). As a consequence of that acceleration a second (strong) X-ray burst and neutron burst is observed. Beside these emissions an enhanced level of plasma waves can be observed at the fundamental and first harmonic of the plasma frequency. Finally a thermalization of the accelerated particles follows, resulting in a heating of the plasma.

As supported by the laboratory experiments the solar flare phenomenon also comprises a number of consecutive processes and developing instabilities, which correspond to different flare stages and components. Regarding the microinstabilities which are presumably forming an essential part in the chain of consecutive processes, the main work has yet to be done.

b. *Deficiencies of Previous Flare Theories*

The main difficulties for a quantitative interpretation of the essential parts of the flare process consist of the need of a highly nonlinear mathematical description. There is a considerable number of spirited attempts to tackle the problem; in general, however, the inclusion of the kinetic stage of a flare seems not yet sufficiently treated in contrast to the fruitful approaches by the hydrodynamic models.

On the other hand, the facilities for checking special models of the Sun are rather limited since a direct probing is not possible. But also indirect methods are still

TABLE V.7

Survey of flare theories

Approach	References
First beginnings of a 'discharge theory'; Invocation of a current sheet near a neutral line and subsequent particle acceleration	Giovanelly (1946, 1947, 1948), Dungey (1953, 1958). Sweet (1958, 1969), Parker (1963), Takakura (1961a).
Reduction of the dissipation time (electric conductivity) e.g. by – finite-conductivity instability – tearing-mode instability – plasma waves – filamentary structure (skin effect in twisted magnetic fields) – plasma turbulence – neutral atoms (and anihilation of twisted magnetic fields) – tensor character of the conductivity and boundary effects – current instability in reconnecting current sheets	 Jaggi (1963, 1965), Furth *et al.* (1963), Scarf *et al.* (1965), Wentzel (1963, 1964), Romanchuk and Krivodubskij (1974), Gold and Hoyle (1960), Piddington (1967), Smith (1976).
Pinch effect in neutral points of the magnetic field	Severnij (1958a, b), Severnij and Shabanskij (1960, 1961), Syrovatskij (1962).
Action of plasma turbulence + tearing mode	Friedman and Hamburger (1969), Coppi and Friedland (1971), Parker (1973a, b), Coppi (1975).
Accumulation of subphotospheric MHD waves; Thermal run-away as trigger mechanism	Pneuman (1967), Warwick (1967). Kahler and Kreplin (1970).
Flare-energy storage by kinetic energy of particles continuously accelerated in active regions and inhibited outflow of solar wind; Opening of a magnetic bottle in the high corona	Elliot (1969, 1964), Warwick (1962), Carmichael (1964). Schatzman (1966, 1967).
Resistive tearing-mode instability in Y-shaped neutral points	Sturrock (1964, 1966, 1968), Strauss and Papagiannis (1971).
Storage and release of gravitational potential energy	Sturrock and Coppi (1964).
Mechanism of current interruption; Unstable arch model	Alfvén (1963), Jacobson and Carlquist (1964), Alfvén and Carlquist (1967), Carlquist (1969), Smith and Priest (1972). Spicer (1977a, b).
Enhanced magnetic-field annihilation (field-line 'reconnection') in different modifications by MHD shock waves and a plasma flow toward the neutral line	Petschek (1964), Kantrowitz and Petschek (1966), Green and Sweet (1967), Petschek and Thorne (1967), Sonnenrup (1970), Yeh and Axford (1970), Cowley (1974), Vasyliunas (1975), Priest and Cowley (1975), Priest and Soward (1975), Soward and Priest (1977).

TABLE V.7 (*continued*)
Survey of flare theories

Approach	References
Dynamic dissipation of magnetic fields near neutral current sheets due to approaching magnetic field patterns	Syrovatskij (1966a, b, 1967, 1968, 1969, 1972, 1974, 1975), Křivský (1968), Anzer (1973), Syrovatskij and Shmeleva (1972), Gerlakh and Syrovatskij (1974), Bulanov and Syrovatskij (1974), Somov and Syrovatskij (1974, 1976), Somov (1976), Somov *et al.* (1977).
Nuclear reactions in neutron flares	Lingenfelter and Ramaty (1967).
Excess heating of nonlinear Alfvén waves;	Uchida and Kaburaki (1974), Wentzel (1974).
Action of short-lived magnetohydrodynamic vortices from the solar interior;	Nakagawa *et al.* (1973), Komelkov (1962).
Subphotospheric magnetic field twist	Piddington (1973, 1974).
Photospheric dynamo action with energy store in Hall currents;	Sen and White (1972).
Energy storage in sheared force-free magnetic fields;	Zirin and Tanaka (1973), Tanaka and Nakagawa (1973).
Inclusion of type III burst acceleration by collisionless tearing-mode instability in a neutral current sheet	Priest and Heyvaerts (1974).
Evaporating hot gas from deeper levels	Sturrock (1973), Hudson (1973), Hirayama (1974).
Model of filament destabilization	Pustilnik (1973).
Invocation of hydrodynamic and kinetic stages in analogy to dense plasma focus experiments;	Kleczek *et al.* (1976).
Laboratory experiments simulating flare phenomena;	Baum and Bratenahl (1974), Elton and Lie (1972).
Experimental evidence of current penetration and plasma disruption	Stix (1976).
Action of an 'antidynamo' (hydrodynamic flow with $N\rho v^2/2 > H^2/8\pi$);	Vainstejn (1965).
Instability of an emerging magnetic flux region;	Vandakurov (1974).
Unspecified point explosion with subsequent blast wave;	Vesecky and Meadows (1969), Gusejnov (1961).
Resistive diffusion of force-free magnetic fields;	Low (1973).
Build-up of flare energy by photospheric motions;	Stenflo (1969), Barnes and Sturrock (1972).
Quasi-stationary electric current + development of kinetic instabilities;	Vilkovskij (1973).
Infall–impact model of prominence material with subsequent shock formation	Hyder (1967), Hyder *et al.* (1973).

TABLE V.8

Development of flare processes

Stage	Energy	Process Involved	Phenomenon
'Quiet' Sun	(Nuclear)		(Hydrogen burning in the Sun's interior)
Active region build-up	kinetic (turbulence),	convection and differential rotation,	development of active regions
	magnetic subphotospheric currents	penetration of magnetic field into plasma	
Flare build-up	magnetic (currents above the photosphere)	photospheric motions (rotational)	twisting of magnetic fields
Flare triggering		thermal instability, shock waves, etc.	
Flare release			
a) MHD stage	quasi-thermal	macroscopic pinch-like discharge, MHD instability	gradual flare onset (microwave and soft X-ray burst),
b) 'kinetic' stages	nonthermal kinetic,	acceleration of energetic particles, turbulent field-plasma interaction	impulsive flare (Hα flash, hard X-ray and type III radio burst),
	plasma turbulence,	micro-instabilities	explosive phase (type IV radio burst, proton flare),
	nonthermal/thermal	?	post-flare phase (noise storm, soft X-ray recovery)

incomplete. Possibly a good deal of our unsatisfactory knowledge about solar flare processes is due to the lack of flare pictures in the microwave region providing sufficient temporal, spatial, and spectral resolution. It would be a unique opportunity to observe magnetic-field controlled radiation from the central regions of the impulsive and explosive flare phases.

We cannot delve into all details of special flare theories since the field is rather extended now and already beyond the scope of solely radio physics. Instead of this, we try to summarize in Table V.7 concisely a number of approaches, which may help to enter into the special literature.

5.6.5. Summary and Prospects

Now we shall briefly recapitulate some major points concerning the solar flare phenomenon which comprises also the main part of solar radio astronomy. With regard to the different components of solar radio-burst emissions we have distinguished the following main phases: Pre-heating, impulsive phase, explosive phase, and post-explosive phase.

In this way it can be concluded that solar flares consist of a series of consecutive physical processes which, only in very gross structures, can be described by the MHD approximation. It is to be expected that after a MHD-determined instability a decay of the Maxwellian distribution follows and a development of smaller-scale instabilities accessible to kinetic theory sets in. Obviously the supposed sequence of instabilities can be quite different in different cases as individual forms of the development of particular flare events show. The later stages, in particular, may require a triggering of instabilities in greater heights of the solar atmosphere. Table V.8 summarizes in a very tentative form a rough scheme of anticipated processes. It is to be anticipated that in due time details of this picture will become successively clearer.

LITERATURE

A. Textbooks

1. RADIO ASTRONOMY

Christiansen, W. N. and Högbom, J. A.: 1969, *Radiotelescopes*, Cambridge Univ. Press.

Elgarøy, Ø.: 1977, *Solar Noise Storms*, Pergamon Press.

Hey, J. S.: 1973, *The Evolution of Radio Astronomy*, P. Elek Science Ltd., London.

Jennison, R. C.: 1966, *Introduction to Radio Astronomy*, Newnes, London.

Kraus, J. D.: 1966, *Radio Astronomy*, McGraw Hill Co., New York, St. Louis, San Francisco, Toronto, London, Sydney.

Krüger, A.: 1972,* *Physics of Solar Continuum Radio Bursts*, Akademie-Verlag, Berlin.

* A list of misprints is available on request to the author.

Kundu, M. R.: 1965, *Solar Radio Astronomy*, Interscience Publ., New York, London, Sydney (Preprint Univ. Michigan 1964).

Pacholczyk, A. G.: 1970, *Radio Astrophysics*, Freeman and Co., San Francisco.

Pawsey, J. L. and Bracewell, R. N.: 1955, *Radio Astronomy*, Clarendon Press, Oxford.

Schelkunoff, S. A. and Friis, H. T.: 1952, *Antennas: Theory and Practice*, John Wiley and Sons, Inc. New York.

Smith, A. G.: 1967, *Radio Exploration of the Sun*, D. van Nostrand Co., Princeton, Toronto, London.

Zheleznyakov, V. V.: 1964: *Radio Emission of the Sun and Planets* (in russ.), Izdat. 'Nauka', Moscow (engl. transl. Pergamon Press, 1970).

2. SOLAR AND SPACE PHYSICS

Athay, R. G.: 1976, *The Solar Chromosphere and Corona-Quiet Sun*, D. Reidel Publ. Co., Dordrecht, Boston, London.

Billings, D. E.: 1966, *A Guide to the Solar Corona*, Academic Press, New York.

Brandt, J. C.: 1970, *Introduction to the Solar Wind*, W. H. Freeman and Co., San Francisco.

Bray, R. J. and Loughhead, R. E.: 1964, *Sunspots*, Chapman and Hall Ltd., London.

Bray, R. J. and Loughhead, R. E.: 1967, *The Solar Granulation*, Chapman and Hall Ltd., London.

Bray, R. J. and Loughhead, R. E.: 1974, *The Solar Chromosphere*, Chapman and Hall Ltd., London.

De Jager, C.: 1959, 'Structure and Dynamic of the Solar Atmosphere', in S. Flügge (ed.), *Encyclopedia of Physics*, Vol. 52, p. 80, Springer-Verlag Berlin, Göttingen, Heidelberg.

Gibson, E. G.: 1973, *The Quiet Sun*, NASA, Washington.

Hundhausen, A. J.: 1972, *Coronal Expansion and Solar Wind*. Springer-Verlag Berlin, Heidelberg, New York.

Makarova, E. A. and Kharitonov, A. V.: 1972, *Distribution of Energy in the Solar Spectrum and the Solar Constant* (in russ.), Izdat. 'Nauka', Moscow.

McIntosh, P. S. and Dryer, M. (ed.): 1972, *Solar Activity Observations and Predictions*, Mass. Inst. Technology.

Parker, E. N.: 1963, *Interplanetary Dynamical Processes*, Interscience Publ., New York.

Robinson, N.: 1966, *Solar Radiation*, Elsevier Publ. Co., Amsterdam, London, New York.

Roederer, J. G.: 1970, *Dynamics of Geomagnetically Trapped Radiation*, Physics and Chemistry in Space, Vol. 2, Springer-Verlag, Heidelberg, New York.

Rossi, B. and Olbert, S.: 1970, *Introduction to the Physics of Space*, McGraw-Hill Book Co., New York, St. Louis, San Francisco, London, Sydney, Toronto, Mexico, Panama.

Shklovskij, I. S.: 1962, *Physics of the Solar Corona* (in russ.), 2nd ed., Gos. Izdat. Fiz.-Mat. Lit., Moscow.

Smith, H. J. and Smith, E. v. P.: 1963, *Solar Flares*, MacMillan Co., New York, London.
Švestka, Z.: 1976, *Solar Flares*, D. Reidel Publ. Co., Dordrecht.
Tandberg-Hanssen, E.: 1967, *Solar Activity*, Blaisdell Publ. Co., Waltham, Mass., Toronto, London.
Tandberg-Hanssen, E.: 1974, *Solar Prominences*, D. Reidel Publ. Co., Dordrecht.
Thomas, R. N. and Athay, R. G.: 1961, *Physics of the Solar Chromosphere*, Interscience Publ., New York, London.
Unsöld, A.: 1955, *Physik der Sternatmosphären*, Springer-Verlag, Berlin, Göttingen, Heidelberg.
Vitinskij, Yu. I.: 1973, *Cyclicity and Prognoses of Solar Activity* (in russ.), Izdat. 'Nauka', Leningrad.
Waldmeier, M.: 1955, *Ergebnisse und Probleme der Sonnenforschung*, Akad. Verlagsgesellschaft Geest und Portig K.-G., Leipzig.
Zirin, H.: 1966, *The Solar Atmosphere*, Blaisdell Publ. Co., Waltham, Mass., Toronto, London.

3. PLASMA PHYSICS AND THEORY

Adams, R. N. and Denman, E. D.: 1966, *Wave Propagation and Turbulent Media*, Amer. Elsevier Publ. Co., New York.
Akhiezer, A. I., Akhiezer, I. A., Polovin, R. V., Sitenko, A. G., and Stepanov, K. N.: 1974, *Plasma Electrodynamics* (in russ.), Izdat. 'Nauka', Moscow.
Alfvén, H. and Fälthammar, C.-G.: 1963, *Cosmical Electrodynamics, Fundamental Principles*, Clarendon Press, Oxford (2nd ed.), (First ed.: Alfvén, H.: 1950, Oxford Press, Oxford).
Allis, W. P., Buchsbaum, S. J., and Bers, A.: 1963, *Waves in Anisotropic Plasmas*, M.I.T. Press, Cambridge, Mass.
Bekefi, G.: 1966, *Radiation Processes in Plasmas*, John Wiley and Sons, Inc., New York, London, Sydney.
Bogolyubov, N. N.: 1946, *Problems of a Dynamical Theory in Statistical Physics* (in russ.), Moscow (engl. tansl. in J. de Boer and G. E. Uhlenbeck (ed.), *Studies in Statistical Mechanics*, Vol. 1, North-Holland, Amsterdam 1962).
Briggs, R. J.: 1964, *Electron-Stream Interaction with Plasmas*, M.I.T. Press, Cambridge, Mass.
Budden, K. G.: 1961, *Radio Waves in the Ionosphere*, Univ. Press Cambridge.
Cap, F.: 1970, 1972, *Einführung in die Plasmaphysik I–III*, Akademie-Verlag Berlin, Pergamon Press Oxford, Vieweg & Sohn, Braunschweig.
Cap, F.: 1976, *Handbook on Plasma Instabilities*, Vol. 1 and 2, Academic Press, New York, San Francisco, London.
Chandrasekhar, S.: 1960, *Radiative Transfer*, Dover Publ., New York.
Davidson, R. C.: 1972, *Methods in Nonlinear Plasma Theory*, Academic Press, New York, London.
Denisse, J. F. and Delcroix, J. L.: 1961, *Théorie des ondes dans les plasmas*, Dunod, Paris.
Drummond, J. E.: 1961, *Plasma Physics*, McGraw-Hill Book Co., New York, Toronto, London.
Dungey, J. W.: 1958, *Cosmic Electrodynamics*, Univ. Press, Cambridge.
Frank-Kamenetskij, D. A.: 1964, *Lectures on Plasma Physics* (in russ.), Atomizdat., Moscow.
Ginzburg, V. L.: 1960, *Propagation of Electromagnetic Waves in Plasma* (in russ.), Gos. Izdat. Fiz.-Mat. Lit., Moscow (2nd ed. 1967) (engl. transl. Gordon and Breach, New York/North-Holland Publ. Co., Amsterdam, 1961).
Ginzburg, V. L. and Syrovatskij, S. I.: 1963, *Origin of Cosmic Rays* (in russ.), Izdat. Akad. Nauk, Moscow.
Hasegawa, A.: 1975, *Plasma Instabilities and Nonlinear Effects*, Springer-Verlag, Berlin, Heidelberg, New York.
Heald, M. A. and Wharton, C. B.: 1965, *Plasma Diagnostics with Microwaves*, John Wiley and Sons, Inc., New York, London, Sydney.
Heitler, J. W.: 1954, *The Quantum Theory of Radiation* (2nd ed.) Clarendon Press, Oxford.
Ivanenko, D. and Sokolov, A.: 1949, *Classical Theory of Fields* (in russ.), Gos. Izdat. Tekh. Teor. Lit. Moscow, Leningrad.
Jancel, R. and Kahan, T.: 1963, *Électrodynamique des plasmas*, Dunod, Paris (engl. transl. John Wiley and Sons, Inc., London, New York, Sydney 1966).
Jelly, J. V.: 1958, *Čerenkov Radiation*, Pergamon Press, London.
Kadomtsev, B. B.: 1976, *Collective Phenomena in Plasma* (in russ.), Izdat 'Nauka', Moscow.
Kaplan, S. A. and Tsytovich, V. H.: 1972, *Plasma Astrophysics* (in russ.), Izdat 'Nauka', Moscow (engl. transl. Pergamon Press, 1973).
Karpman, V. I.: 1973, *Nonlinear Waves in Dispersion Media* (in russ.), Izdat. 'Nauka', Moscow.
Klimontovich, Yu. L.: 1964, *The Statistical Theory of Non-Equilibrium Processes in a Plasma* (in russ.), Izdat. MGU, Moscow (engl. transl. Pergamon Press 1967).

Krall, N. A. and Trivelpiece, A. W.: 1973, *Principles of Plasma Physics*, McGraw-Hill Book Co., New York.
Landau, L. D. and Lifshits, E. M.: 1962, *Field Theory* (4th ed., in russ.). Gos. Izdat Fiz.-Mat. Lit. Moscow 1962, engl. transl.: *The Classical Theory of Fields*, Addison-Wesley, Reading, Mass., 1951.
Lehnert, B.: 1964, *Dynamics of Charged Particles*, North-Holland Publ. Co., Amsterdam.
Mikhailovskij, A. B.: 1970/71, *Theory of Plasma Instabilities* (in russ.), Vol. I and II, Atomizdat, Moscow.
Montgomery, D. C. and Tidman, D. A.: 1964, *Plasma Kinetic Theory*, McGraw-Hill Book Co., New York.
Panofsky, W. K. H. and Phillips, M.: 1955, *Classical Electricity and Magnetism*, Addison-Wesley Publ. Co., Reading (Mass.).
Piddington, J. H.: 1969, *Cosmic Electrodynamics*, John Wiley and Sons, Inc., New York.
Pikelner, S. B.: 1961, *Fundamentals of Cosmical Electrodynamics* (in russ.), Goz. Izdat. Fiz.-Mat. Lit., Moscow.
Ratcliffe, J. A.: 1959, *The Magneto-Ionic Theory and Its Applications to the Ionosphere*, Univ. Press, Cambridge.
Sagdeev, R. Z. and Galeev, A. A.: 1969, *Nonlinear Plasma Theory* (revised and edited by T. M. O'Neil and D. C. Book), W. A. Benjamin, Inc., New York, Amsterdam.
Sears, F. W.: 1953, *Thermodynamics, The Kinetic Theory of Gases, and Statistical Mechanics* (2nd ed.), Addison-Wesley Publ. Co., Reading, Mass.
Sokolov, A. A. and Ternov, I. M. (ed.): 1966, *Synchroton Radiation* (in russ.), Izdat. 'Nauka', Moscow.
Spitzer, L., Jr.: 1962, *Physics of Fully Ionized Gases*, 2nd ed., Interscience Publishers, New York, London.
Stix, T. H.: 1962, *The Theory of Plasma Waves*, McGraw-Hill Book Co., New York, San Francisco, Toronto, London.
Tsytovich, V. N.: 1967, *Nonlinear Effects in Plasma* (in russ.), Izdat. 'Nauka', Moscow (engl. transl. Plenum Press, New York, London, 1970).
Tsytovich, V. N.: 1971, *Theory of Turbulent Plasma* (in russ.), Atomizdat., Moscow (engl. transl. Plenum Press, New York, London, 1976); cf. also *An Introduction to the Theory of Plasma Turbulence*, Pergamon Press, Oxford, New York, 1972.
Van Kampen, N. G. and Felderhof, B. U.: 1967, *Theoretical Methods in Plasma Physics*, North-Holland Publ. Co., Amsterdam.
Wu. T.-Y.: 1966, *Kinetic Equations of Gases and Plasmas*, Addison-Wesley Publ. Co., Reading, Mass., Palo Alto, London, Don Mills, Ontario.
Zheleznyakov, V. V.: 1977, *Electromagnetic Waves in Cosmical Plasmas*, Generation and Propagation (in russ.), Izdat. 'Nauka', Moscow.

4. SYMPOSIA ETC.

Athay, R. G. (ed.): 1974, *Chromospheric Fine Structure*, Proc. of IAU Symposium No. 56 held at Surfer's Paradise, Queensland, Australia 1973, D. Reidel Publ. Co., Dordrecht, Boston.
Athay, R. G. and Newkirk, G. Jr. (ed.): 1969, *Chromosphere-Corona Transition Region*, Manuscripts presented at a Conference held in Boulder, Colo., High Altitude Observatory, Boulder.
Bracewell, R. N. (ed.): 1959, *Paris Symposium on Radio Astronomy*, Stanford Univ. Press, Stanford, Calif.
De Jager, C. (ed.): 1965, *The Solar Spectrum*, Proc. of the Symposium Utrecht 1963, D. Reidel Publ. Co., Dordrecht.
De Jager, C. and Švestka, Z.: 1969, *Solar Flares and Space Research*, Proc. of the Symposium held on the Occasion of the 11th Plenary Meeting of COSPAR, Tokyo 1968, North-Holland Publ. Co., Amsterdam.
Dyer, E. R. (ed.): 1972, *Solar-Terrestrial Physics*, Proc. of the Int. Symposium in Leningrad 1970, D. Reidel Publ. Co., Dordrecht.
Emming, J. G. (ed.): 1967, *Electromagnetic Radiation in Space*, Proc. of the Third ESRO Summer School in Space Physics, Alpbach (Austria) 1965, D. Reidel Publ. Co., Dordrecht.
Evans, J. W. (ed.): 1963, *The Solar Corona*, IAU Symposium No. 16, Cloudcroft 1961, Academic Press, New York, London.
Gelfreikh, G. B. and Livshits, M. A. (ed.): 1972, *Radioastronomical Observations of the Solar Eclipse of 20th May, 1966* (in russ.), Izdat. 'Nauka', Moscow.
Hess, W. N. (ed.): 1964, *AAS-NASA Symposium on the Physics of Solar Flares*, Proc. of a Symposium held at Greenbelt, Md. 1963, NASA Washington.
Howard, R. (ed.): 1971, *Solar Magnetic Fields*, IAU Symposium No. 43, Paris 1970, D. Reidel Publ. Co., Dordrecht.
Hundhausen, A. J. and Newkirk, G., Jr.: 1974, *Flare-Produced Shock Waves in the Corona and in Inter-*

planetary Space, Manuscripts presented at a Conference held in Boulder, 1972, High Altitude Observatory/NCAR, Boulder.
Kadomtsev, B. B., Rosenbluth, M. N., and Thompson, W. B. (ed.): 1965, 'Plasma Physics', *Lectures presented at the Seminar on Plasma Physics Organized by and Held at the Int. Centre for Theoret. Physics*, Trieste 1964, Int. Atomic Energy Agency, Vienna.
Kane, S. R. (ed.): 1975, *Solar Gamma-, X, and EUV Radiation*, IAU Symposium No. 68, D. Reidel Publ. Co., Dordrecht, Boston, London.
Kiepenheuer, K. O. (ed.): 1966, *The Fine Structure of the Solar Atmosphere*, Colloqu. Anacapri June 6–8, 1966, Franz Steiner Verl., Wiesbaden.
Kiepenheuer, K. O.: 1968, *Structure and Development of Solar Active Regions*, IAU Symposium No. 35, Budapest 1967, D. Reidel Publ. Co., Dordrecht, Boston, London.
Lüst, R. (ed.): 1965, *Stellar and Solar Magnetic Fields*, IAU Symposium No. 22, Munich 1963, North-Holland Publ. Co., Amsterdam.
Mackin, R. J. and Neugebauer, M. (ed.): 1966, *The Solar Wind*, Proc. of a Conference held at Caltech, Pasadena, April 1–4, 1964, Pergamon Press, Oxford, London, Edinburgh, New York, Toronto, Paris, Frankfurt.
Macris, C. J. (ed.): 1971, *Physics of the Solar Corona*, Proc. of NATO Advanced Study Inst. on Physics of the Solar Corona, Athens 1970, D. Reidel Publ. Co., Dordrecht, Boston, London.
Mangeney, A. (ed.): 1972, *Proc. Summer School on Plasma Physics and Solar Radio Astronomy*, Meudon, France.
McCormac, B. M. (ed.): 1970, *Particles and Fields in the Magnetosphere*, Proc. of a Symposium at Santa Barbara, Calif., 1969, D. Reidel Publ. Co., Dordrecht, Boston, London.
Newkirk, G., Jr. (ed.): 1974, *Coronal Disturbances*, IAU Symposium No. 57 held at Surfer's Paradise, Queensland, Australia, 1973, D. Reidel Publ. Co., Dordrecht, Boston, London.
Ortner, J. and Maseland, H. (ed.): 1965, *Introduction to Solar Terrestrial Relations*, Proc. of the Summer School, Alpach (Austria) 1963, D. Reidel Publ. Co., Dordrecht, Boston, London.
Righini, G. (ed.): 1960, *Radioastronomia solare, Rendiconti della Scuola Internazionale di Fisica* 'Enrico Fermi', Corso XII, Varenna 1959, N. Zanichelli: Ed., Bologna.
Rosenbluth, M. N. (ed.): 1964, *Teoria del plasmi, Rendiconti della Scuola Internazionale di Fisica* 'Enrico Fermi' XXV Corso, Varenna 1962, Academic Press, New York, London.
Stickland, A. C. (Gen. ed.): 1968, 69, *Annals of the IQSY*, Vol. 1–7, M.I.T. Press, Cambridge, Mass.
Wentzel, D. G. and Tidman, D. A. (ed.): 1969, *Plasma Instabilities in Astrophysics*, Proc. of a Conference held on Monterey Peninsula, Calif. 1968, Gordon and Breach, New York, London, Paris.

5. TABLES ETC.

Abramowitz, M. and Stegun, I. A.: 1965, *Handbook of Mathematical Functions*, Dover, Publ., Inc., New York.
Allen, C. W.: 1973, *Astrophysical Quantities* (3rd ed.), Athlone Press, London.
Hellwege, K.-H. (ed.): 1965, *Landoldt-Börnstein, Neue Serie, Zahlenwerte aus Naturwissenschaften und Technik*, Gruppe VI, Bd. 1 Astronomie und Astrophysik, Springer-Verlag, Berlin, Heidelberg, New York.
Švestka, Z. and Simon, P. (ed.): 1975, *Catalogue of Solar Particle Events*, D. Reidel Publ. Co., Dordrecht.

B. Single publications*

Abrami, A.: 1968, *Mem. Soz. Astron. Ital.* **39**, 371.
Abrami, A.: 1970a, *Solar Phys.* **11**, 104.
Abrami, A.: 1970b, Report UAG-8, WDC A, Boulder, p. 163.
Abrami, A.: 1972a, *Nature Physical Science* **238**, 25.
Abrami, A.: 1972b, in A. Mangeney (ed.), *Plasma Physics and Solar Radio Astronomy*, Meudon, p. 233.
Abranin, E. P., Bazelyan, L. L., Goncharov, N. Yu., Zaitsev, V. V., Zinichev, V. A., Levin, B. N., Rapaport, V. D., and Tsybko, Ya. G.: 1977, *Astron. Zh.* **54**, 146.
Achong, A.: 1974, *Solar Phys.* **37**, 477.
Acton, L. W.: 1968, *Astrophys. J.* **152**, 305.
Akabane, K.: 1958, *Ann. Tokyo Astron. Obs.*, Ser. 2, **6**, 57.
Akabane, K. and Cohen, M. H.: 1961, *Astrophys. J.* **133**, 258.

* Review of literature terminates at about September 1977.

Akabane, K., Nakasima, H., Ohki, K., Moriyama, F., and Miyaji, T.: 1973, *Solar Phys.* **33**, 431.
Akasofu, S.-I. and Yoshida, S.: 1967, *Planet. Space Sci.* **15**, 942.
Akinyan, S. T.: 1972, *Astron. Zh.* **49**, 579.
Akinyan, S. T. and Chertok, I. M.: 1977, *Geomagnetizm i Aeronomiya* **17**, 596.
Akinyan, S. T., Fomichev, V. V., and Chertok, I. M.: 1977, *Geomagnetizm i Aeronomiya* **17**, 10 and 177.
Akinyan, S. T., Iskov, V. V., Mogilevskij, E. I., Böhme, A., Fürstenberg, F., and Krüger, A.: 1976, *Contr. Astron. Obs. Skalnate Pleso* **6**, 35.
Akinyan, S. T., Mogilevskij, E. I., Böhme, A., and Krüger, A.: 1971, *Solar Phys.* **20**, 112.
Alexander, J. K., Malitson, H. H., and Stone, R. G.: 1969, *Solar Phys.* **8**, 388.
Alfvén, H.: 1947, *Monthly Notices Roy. Astron. Soc.* **107**, 211.
Alfvén, H.: 1963, in J. W. Evans (ed.), *The Solar Corona*, IAU Symposium No. 16., Academic Press, p. 35.
Alfvén, H. and Carlquist, P.: 1967, *Solar Phys.* **1**, 220.
Alfvén, H. and Herlofson, N.: 1950, *Phys. Rev.* **78**, 616.
Alissandrakis, C. E. and Kundu, M. R.: 1975, *Solar Phys.* **41**, 119.
Allen, C. W.: 1965, *Space Sci. Rev.* **4**, 91.
Alon, I., Arsac, J., and Steinberg, J. L.: 1953, *Compt. Rend. Acad. Sci. Paris*, **237**, 300.
Altschuler, M. D. and Newkirk, G.: 1969, *Solar Phys.* **9**, 131.
Altschuler, M. D., Smith, D. F., Swarztrauber, P. N., and Priest, E. R.: 1973, *Solar Phys.* **32**, 153.
Altschuler, M. D., Trotter, D. E., and Orrall, F. Q.: 1972, *Solar Phys.* **26**, 354.
Alurkar, S. K. and Bhonsle, R. V.: 1969, *Solar Phys.* **9**, 198.
Alvarez, H. and Haddock, F. T.: 1973a, *Solar Phys.* **29**, 197.
Alvarez, H. and Haddock, F. T.: 1973b, *Solar Phys.* **30**, 175.
Alvarez, H., Haddock, F., and Lin, R. P.: 1972, *Solar Phys.* **26**, 468.
Alvarez, H., Haddock, F. T., and Potter, W. H.: 1973, *Solar Phys.* **31**, 493.
Alvarez, H., Haddock, F. T., and Potter, W. H.: 1974, *Solar Phys.* **34**, 413.
Alvarez, H., Lin, R. P., and Bame, S. J.: 1975, *Solar Phys.* **44**, 485.
Ambrož, P., Bumba, V., Howard, R., and Sykora, J.: 1971, in R. Howard (ed.), *Solar Magnetic Fields*, IAU Symposium No. 43, D. Reidel Publ. Co., p. 696.
Anderson, K. A.: 1969, *Solar Phys.* **6**, 111.
Anderson, K. A.: 1972, *Solar Phys.* **27**, 442.
Anderson, K. A. and Winckler, J. R.: 1962, *J. Geophys. Res.* **67**, 4103.
Antalova, A.: 1967, *Bull. Astron. Inst. Czech.* **18**, 61.
Anzer, U.: 1973, *Solar Phys.* **30**, 459.
Appleton, E. V. and Hey, J. S.: 1946, *Phil. Mag.*, Ser. 7, **37**, 73.
Arisawa, M.: 1971, *Proc. Res. Inst. Atmospherics, Nagoya Univ.* **18**, 89.
Arnoldy, R. L., Kane, S. R., and Winckler, J. R.: 1968, *Astrophys. J.* **151**, 711.
Aubier, M. G.: 1974, *Astron. Astrophys.* **32**, 141.
Aubier, M. and Boischot, A.: 1972, *Astron. Astrophys.* **19**, 343.
Aubier, M., Boischot, A., Daigne, G., Leblanc, Y., Lecacheux, A., de la Noë, J., Møller-Pedersen, B., and Rosolen, C.: 1975, *Oss. e Mem. Oss. Arcetri* **105**, 110.
Aubier, M. and de La Noë, J.: 1971, *Astron. Astrophys.* **12**, 491.
Aubier, M., Leblanc, Y., and Boischot, A.: 1971, *Astron. Astrophys.* **12**, 435.
Aurass, H., Böhme, A., and Krüger, A.: 1976, *Phys. Solariterr., Potsdam*, No. **2**, 71.
Avery, L. W.: 1976, *Solar Phys.* **49**, 141.
Avery, L. W., Feldman, P. A., Gaizaukas, V., Roy, J.-René, and Wolfson, C. J.: 1977, *Astron. Astrophys.* **56**, 327.
Avignon, Y., Lantos, P., Palagi, F., and Patriarchi, P.: 1975, *Solar Phys.* **45**, 141.
Avignon, Y., Martres, M. J., and Pick, M.: 1966, *Ann. Astrophys.* **29**, 33.
Axisa, F.: 1974, *Solar Phys.* **35**, 207.
Axisa, F., Avignon, Y., Martres, M. J., Pick, M., and Simon, P.: 1971, *Solar Phys.* **19**, 110.
Axisa, F., Martres, M. J., Pick, M., and Soru-Escaut, I.: 1973, *Solar Phys.* **29**, 163.
Babcock, H. D.: 1959, *Astrophys. J.* **130**, 364.
Bachurin, A. F., Dvoryashin, A. S., and Eryushev, N. N.: 1976, *Izv. Krymsk. Astrofiz. Obs.* **55**, 70.
Bachurin, A. F., Eryushev, N. N., and Tsvetkov, L. I.: 1974, *Izv. Krymsk. Astrofiz. Obs.* **50**, 180.
Bai, T. and Ramaty, R.: 1976, *Solar Phys.* **49**, 343.
Bailey, D. K.: 1964, *Planet. Space Sci.* **12**, 495.

Baldwin, D. E., Bernstein, I. B., and Weenink, M. P. H.: 1969, in A. Simon and W. B. Thompson (ed.) *Advances in Plasma Physics*, Vol. 3, Interscience Publ., New York, London, Sydney, Toronto, p. 1.

Baldwin, D. E.: 1974, *Phys. Letters* **12**, 202.

Barnes, C. W. and Sturrock, P. A.: 1972, *Astrophys. J.* **174**, 659.

Barrow, C. H., and Achong, A.: 1975, *Solar Phys.* **45**, 459 (and 467).

Basu, D.: 1966, *Nature* **212**, 804.

Baum, P. J. and Bratenahl, A.: 1974, *J. Plasma Phys.* **11**, 93.

Baumback, M. M., Kurth, W. S., and Gurnett, D. A.: 1976, *Solar Phys.* **48**, 361.

Bazelyan, L. L., Braude, S. Ya., and Men, A. V.: 1970, *Astron. Zh.* **47**, 188; *Soviet Astron.* **14**, 153.

Bazelyan, L. L., Goncharov, N. Yu., Zaitsev, V. V., Zinichev, V. A. Rapoport, V. D., and Tsybko, Ya. G.: 1974, *Solar Phys.* **39**, 213 and 223.

Bazelyan, L. L., Goncharov, N. Yu., Zaitsev, V. V., Zinichev, V. A., Rapoport, V. D., and Tsybko, Ya. G.: 1977, *Solar Phys.* **52**, 141.

Beckmann, J. E.: 1968, *Nature* **220**, 52.

Beigman, I. L.: 1974, P. N. Lebedev Phys. Inst. Moscow, Preprint No. 68.

Bekefi, G., Hirshfield, J. L., and Brown, S. C.: 1961, *Phys. Rev.* **122**, 1037.

Belcher, J. W.: 1971, *Astrophys. J.* **168**, 509.

Belcher, J. W. and Davies, L., Jr.: 1971, *J. Geophys. Res.* **76**, 3534.

Bell, M. B.: 1972, *Solar Phys.* **27**, 137.

Bell, M. B., Covington, A. E., and Kennedy, W. A.: 1973, *Solar Phys.* **28**, 123.

Benz, A. O.: 1977, *Astrophys. J.* **211**, 270.

Benz, A. O., Berney, M., and Santin, P.: 1977, *Astron. Astrophys.* **56**, 123.

Benz, A. O. and Gold, T.: 1971, *Solar Phys.* **21**, 157.

Benz, A. O. and Kuijpers, J.: 1976, *Solar Phys.* **46**, 275.

Benz, A. O. and Tarnstrom, G. L.: 1976, *Astrophys. J.* **204**, 597.

Berger, P. S. and Simon, M.: 1972, *Astrophys. J.* **171**, 191.

Bhonsle, R. V. and Mattoo, S. K.: 1974, *Astron. Astrophys.* **30**, 301.

Bhonsle, R. V. and McNarry, L. R.: 1964, *Astrophys. J.* **139**, 1312.

Billings, D. E. and Roberts, W. O.: 1964, *Astrophys. Norv.* **9**, 141.

Birn, J., Goldstein, H., and Schindler, K.: 1977, Preprint submitted to *Solar Phys.*

Biswas, S. and Fichtel, C. E.: 1965, *Space Sci. Rev.* **4**, 705.

Blum, E. J.: 1974, in E. Schanda (ed.), *Proc. of the 4th Meeting of CESRA*, Inst. Appl. Phys., Univ., Berne, p. 31.

Bocchia, R. and Poumeyrol, F.: 1974, *Solar Phys.* **38**, 193.

Böhme, A.: 1962, *Gerlands Beitr. Geophys.* **71**, 211.

Böhme, A.: 1964a, *Beitr. Plasmaphys.* **4**, 137.

Böhme, A.: 1964b, *Mber. DAW Berlin* **6**, 1.

Böhme, A.: 1968, in K. O. Kiepenheuer (ed.), *Structure and Development of Solar Regions*, IAU Symposium No. 35, D. Reidel Publ. Co., p. 570.

Böhme, A.: 1969, *Gerlands Beitr. Geophys.* **3**, 193 and **4**, 283.

Böhme, A.: 1971, Doctoral Thesis, DAW Berlin.

Böhme, A.: 1972, *Solar Phys.* **24**, 457.

Böhme, A. and Krüger, A.: 1969, *Geodät. Geophys. Veröff.* Potsdam II, **10**, 49 and 59, and *Solnechno-Zemnaya Fizika, Moscow*, **1**, 19.

Böhme, A. and Krüger, A.: 1973, US Dept. Commerce, Spec. Report UAG-28, Boulder, Part I, p. 260.

Böhme, A., Fürstenberg, F., and Krüger, A.: 1974, *Solar Phys.* **39**, 207.

Böhme, A., Fürstenberg, F., Hildebrandt. J., Hoyng, P., Krüger, A., Saal, O., and Stevens, G. A.: 1976, *HHI-STP-Report* No.6, Berlin.

Böhme, A., Fürstenberg, F., Hildebrandt, J., Saal O., Krüger, A., Hoyng, P., and Stevens, G. A.: 1977, *Solar Phys.* **53**, 139.

Bohm, D. and Gross, E. P.: 1949, *Phys. Rev.* **75**, 1851 and 1864.

Boischot, A.: 1957, *Compt. Rend. Acad. Sci. Paris* **244**, 1326.

Boischot, A.: 1958, *Ann. Astrophys.* **21**, 273.

Boischot, A.: 1959, in R. N. Bracewell (ed.), *Paris Symposium on Radio Astronomy*, Stanford Univ. Press, p. 186.

Boischot, A.: 1967, *Ann. Astrophys.* **30**, 85.

Boischot, A.: 1974a, in G. Newkirk, Jr. (ed.), *Coronal Disturbances*, IAU Symposium No. 57, D. Reidel Publ. Co., p. 423.
Boischot, A.: 1974b, *Astron. Astrophys.* **33**, 379.
Boischot, A. and Clavelier, B.: 1967, *Astrophys. Letters* **1**, 7.
Boischot, A. and Clavelier, B.: 1968, *Ann. Astrophys.* **31**, 445 and 531.
Boischot, A., Clavelier, B., Mangeney, A., and Lacombe, C.: 1967, *Comptes Rendus Acad. Sci. Paris* **265**, 1151.
Boischot, A. and Daigne, G.: 1968, *Ann. Astrophys.* **31**, 531.
Boischot, A., de la Noë, J., du Chaffaut, M., and Rosolen, C.: 1971, *Comptes Rendus Acad. Sci. Paris* **272**, 166.
Boischot, A., de la Noë, J., and Møller-Pedersen, B.: 1970, *Astron. Astrophys.* **4**, 159.
Boischot, A. and Denisse, J. F.: 1957, *Comptes Rendus Acad. Sci. Paris* **245**, 2194.
Boischot, A. and Fokker, A. D.: 1959, in R. N. Bracewell (ed.), *Paris Symposium on Radio Astronomy*, Stanford Univ. Press, p. 263.
Boischot, A., Haddock, F. T. and Maxwell, A.: 1960, *Ann. Astrophys.* **23**, 478.
Boischot, A. and Lecacheux, A.: 1975, *Astron. Astrophys.* **40**, 55.
Boischot, A. and Pick, M.: 1962, *J. Phys. Soc. Japan* **17**, Suppl. A–2, 203.
Borovik, V. N., Gelfreikh, T. B., Krüger, A., and Peterova, N. G.: 1968, *Solnechnye Dannye, Leningrad* No. 9, 101.
Bougeret, J. L.: 1972, in J. Poumeyrol (ed.), *Proc. of the 3rd Meeting of CESRA*, Obs. Univ. Bordeaux, Floriac, p. 182.
Bougeret, J. L.: 1973, *Astron. Astrophys.* **24**, 53.
Bougeret, J. L.: 1977, *Astron. Astrophys.* **60**, 131.
Bougeret, J. L., Caroubalos, C., Mercier, C., and Pick, M.: 1970, *Astron. Astrophys.* **6**, 406.
Bracewell, R. N.: 1962, in S. Flügge (ed.), *Handbuch der Physik*, Springer-Verlag **54**, 42.
Bracewell, R. N. and Swarup, G.: 1961, *I.R.E. Trans. Antennas Propagation* **AP–9**, 22.
Bradford, H. M. and Hughes, V. A.: 1974, *Astron. Astrophys.* **31**, 419.
Brambilla, M.: 1970, *Plasma Phys.* **12**, 903.
Braude, S. Ya., Bruk, Yu. M., Melyanovskij, P. A., Men, A. V., Sodin, L. G., and Sharykin, N. K.: 1971, Preprint Nr. 7, Inst. Radiophys. Electr. Charkov.
Braude, S. Ya., Lebedeva, O. M., Megn, A. V., Ryabov, B. P., and Zhouk, I. N.: 1969, *Monthly Notices Roy. Astron. Soc.* **143**, 289.
Bregman, J. D. and Felli, M.: 1976, *Astron. Astrophys.* **46**, 41.
Brice, N.: 1964, *J. Geophys. Res.* **69**, 4515.
Brown, J. C., McClymont, A. N., and McLean, I .S.: 1974, *Nature* **247**, 448.
Brown, J. C. and Melrose, D. B.: 1977, *Solar Phys.* **52**, 117.
Bruzek, A.: 1967, in J. N. Xanthakis (ed.), *Solar Physics*, Interscience Publ., p. 399.
Bruzek, A.: 1974, in G. Newkirk (ed.), *Coronal Disturbances*, IAU Symposium No. 54, D. Reidel Publ. Co., p. 323.
Bryant, D. A., Cline, T. L., Desai, V. D., and McDonald, F. B.: 1965, *Phys. Rev. Letters* **14**, 481.
Buhl, D. and Tlamicha, A.: 1968, *Astrophys. J. Letters* **153**, L189.
Buhl, D. and Tlamicha, A.: 1970, *Astron. Astrophys.* **5**, 102.
Bulanov, S. V. and Syrovatskij, S. I.: 1974, *Trudy Fiz. Inst. Lebedeva* **74**, 88.
Bumba, V.: 1972, in C. P. Sonett, P. J. Coleman, Jr., and J. M. Wilcox (ed.), *Solar Wind*, NASA SP–308, p. 31.
Bumba, V. and Howard, R.: 1965, *Astrophys. J.* **142**, 796.
Bumba, V., and Howard, R.: 1969, in C. de Jager and Z. Švestka (ed.), *Solar Flares and Space Research*, North-Holland Publ. Co., p. 387.
Bumba, V., Howard, R., Kopecký, M., and Kuklin, G. V.: 1969, *Bull. Astron. Inst. Czech.* **20**, 18.
Burke, B. F. and Franklin, K. L.: 1955, *J. Geophys. Res.* **60**, 213.
Burkhardt, G. and Schlüter, A.: 1949, *Z. Astrophys.* **26**, 295.
Burlaga, L. F.: 1969, *Solar Phys.* **7**, 57.
Burlaga, L. F.: 1971, *Space Sci. Rev.* **12**, 600.
Burlaga, L. F. and Ness, N. F.: 1968, *Can. J. Phys.* **46**, 5962.
Burlaga, L. F. and Ogilvie, K. W.: 1969, *J. Geophys. Res.* **74**, 2815.
Butz, M., Fürst, E., Hirth, W., and Kundu, M. R.: 1975, *Solar Phys.* **45**, 125.

Canuto, V. and Chiu, H. Y.: 1971, *Space Sci. Rev.* **12**, 3.
Carlquist, P.: 1969, *Solar Phys.* **7**, 377.
Carmichael, H.: 1962, *Space Sci. Rev.* **1**, 28.
Carmichael, H.: 1964, in N. Ness (ed.), *AAS-NASA Symposium on the Physics of Solar Flares*, NASA, Washington, p. 451.
Caroubalos, C., Aubier, M., Leblanc, Y., and Steinberg, J. L.: 1972, *Astron. Astrophys.* **16**, 374.
Caroubalos, C., Heyvaerts, J., Pick, M., and Trottet, G.: 1974a, *Solar Phys.* **37**, 205.
Caroubalos, C., Pick, M., Rosenberg, H., and Slottje, C.: 1973, *Solar Phys.* **30**, 473.
Caroubalos, C., Poquérusse, M., and Steinberg, J. L.: 1974b, *Astron. Astrophys.* **32**, 255.
Caroubalos, C. and Steinberg, J. L.: 1974, *Astron. Astrophys.* **32**, 245.
Castelli, J. P.: 1970, Preprint AFCRL Bedford, Mass.
Castelli, J. P. and Aarons, J.: 1965, in J. Aarons (ed.), *Solar System Radio Astronomy*, Plenum Press, New York, p. 49.
Castelli, J. P., Aarons, J., and Michael, G. A.: 1967, *J. Geophys. Res.* **72**, 5491.
Castelli, J. P., Guidice, D. A., Forrest, D. J., and Babcock, R. R.: 1974, *J. Geophys. Res.* **79**, 889.
Castelli, J. P. and Richards, D. W.: 1971, *J. Geophys. Res.* **76**, 8409.
Ceballos, J. C. and Lantos, P.: 1972, *Solar Phys.* **22**, 142.
Chambe, G. and Lantos, P.: 1971, *Solar Phys.* **17**, 97.
Chambe, G. and Sain, A.: 1969, *Astron. Astrophys.* **2**, 133.
Cheng, C. C.: 1972a, *Solar Phys.* **22**, 178.
Cheng, C. C.: 1972b, *Space Sci. Rev.* **13**, 3.
Chernov, G. P.: 1973, *Astron. Zh.* **50**, 1254.
Chernov, G. P.: 1976a, *Astron. Zh.* **53**, 798 and 1027.
Chernov, G. P.: 1976b, in *Fisika solnechnoj aktivnosti*, Izdat. 'Nauka', Moscow, p. 112.
Chernov, G. P.: 1977, *Astron. Zh.* **54**, 1081.
Chernov, G. P., Chertok, T. M., Fomichev, V. V., and Markeev, A. K.: 1972, *Solar Phys.* **24**, 215.
Chernov, G. P., Korolev, O. S., and Markeev, A. K.: 1975, *Solar Phys.* **44**, 435.
Chertok, I. M., Fomichev, V. V., Krüger, A., and Willimczik, W.: 1972, *Solar Phys.* **25**, 452.
Chertok, I. M., and Krüger, A.: 1973, *Astron. Nachr.* **294**, 241.
Chin, Y. C., Lusignan, B. B., and Fung, P. C. W.: 1971, *Solar Phys.* **16**, 135.
Chiu, Y. T.: 1970, *Solar Phys.* **13**, 420.
Chiuderi, C., Giachetti, R., and Rosenberg, H.: 1973, *Solar Phys.* **33**, 225.
Chiuderi, C. and Giovanardi, C.: 1975, *Solar Phys.* **41**, 35.
Chiuderi, C., Chiuderi Drago, F., and Noci, G.: 1971, *Solar Phys.* **17**, 369.
Chiuderi, C., Chiuderi Drago, F., and Noci, G.: 1972, *Solar Phys.* **26**, 343.
Chiuderi Drago, F., Fürst, E., Hirth, W., and Lantos, P.: 1975, *Astron. Astrophys.* **39**, 429.
Chiuderi Drago, F. and Noci, G.: 1976, *Astron. Astrophys.* **48**, 367.
Chiuderi Drago, F. and Patriarchi, P.: 1974, *Solar Phys.* **37**, 403.
Christiansen, W. N.: 1966, in E. Herbays (ed.), *Progress in Radio Science*, Vol. V, Elsevier Publ. Co., Amsterdam, London, New York, p. 45.
Christiansen, W. N. and Hindman, J. V.: 1951, *Nature* **167**, 635.
Christiansen, W. N. and Mathewson, D. S.: 1958, *Proc. Inst. Radio Engrs.* **46**, 127.
Christiansen, W. N. and Warburton, J. A.: 1953, *Australian J. Phys.* **6**, 262.
Chubb, T. A.: 1972, in E. R. Dyer (ed.), *Solar-Terrestrial Physics/1970*, Part I (ed. by C. de Jager), D. Reidel Publ. Co., p. 99.
Chubb, T. A., Kreplin, R. W., and Friedman, H.: 1966, *J. Geophys. Res.* **71**, 3611.
Chupp, E. L.: 1971, *Space Sci. Rev.* **12**, 486.
Chupp, E. L., Forrest, D. J., Higbie, P. R., Suri, A. N., Tsai, C., and Dunphy, P. P.: 1973, *Nature* **241**, 333.
Clavelier, B.: 1967, *Ann. Astrophys.* **30**, 895.
Clavelier, B., Jarry, M. F., and Pick, M.: 1968, *Ann. Astrophys.* **31**, 523.
Cliver, E. W., Hurst, M. D., Wefer, F. L., and Bleiweiss, M. P.: 1976, *Solar Phys.* **48**, 307.
Coates, R. J.: 1961, *Astrophys. J.* **133**, 723.
Coates, R. J., Gibson. J. E., and Hagen, J. P.: 1958, *Astrophys. J.* **128**, 406.
Cogdell, J. R.: 1972, *Solar Phys.* **22**, 147.
Cohen, M. H.: 1958, *Proc. Inst. Radio Engrs.* **46**, 172.

Cohen, M. H.: 1960, *Astrophys. J.* **131**, 664.
Cohen, M. H.: 1961a, *Astrophys. J.* **133**, 978.
Cohen, M. H.: 1961b, *Phys. Rev.* **123**, 711.
Cohen, M. H. and Dwarkin, M.: 1961, *J. Geophys. Res.* **66**, 411.
Cole, T. W.: 1973, *Astrophys. Letters* **15**, 59.
Cole, T. W. and Ables, J. G.: 1974, *Astron. Astrophys.* **34**, 149.
Coleman, P. J., Jr.: 1967, *Planetary Space Sci.* **15**, 953.
Coppi, B.: 1964, *Ann. Phys.* **30**, 178.
Coppi, B.: 1975, *Astrophys. J.* **195**, 545.
Coppi, B., and Friedland, A. B.: 1971, *Astrophys. J.* **169**, 379.
Covington, A. E.: 1947, *Nature* **159**, 405.
Covington, A. E.: 1951, *J. Roy. Astron. Soc. Can.* **45**, 49.
Covington, A. E.: 1960, *J. Roy. Astron. Soc. Can.* **54**, 17.
Covington, A. E.: 1967, *J. Roy. Astron. Soc. Can.* **61**, 314.
Covington, A. E.: 1969, *J. Roy. Astron. Soc. Can.* **63**, 125.
Covington, A. E.: 1973, *Solar Phys.* **33**, 439.
Covington, A. E.: 1974, *J. Roy. Astron. Soc. Can.* **68**, 31.
Covington, A. E. and Harvey, G. A.: 1961, *Nature* **192**, 152.
Covington, A. E., Harvey, G. A., and Dodson, H. W.: 1962, *Astrophys. J.* **135**, 531.
Covington, A. E., Legg, T. H., and Bell, M. B.: 1967, *Solar Phys.* **1**, 465.
Cowley, S. W. H.: 1971, *Cosmic Electrodynamics* **2**, 90.
Cowley, S. W. H.: 1974, *J. Plasma Phys.* **12**, 319 and 341.
Crawford, F. W.: 1965a, *Nucl. Fusion* **5**, 73.
Crawford, F. W.: 1965b, *Radio Science* **69**, 709.
Crawford, F. W.: 1968, in *A Survey of Phenomena in Ionized Gases*, Int. Atomic Energy Agency, Vienna.
Croom, D. L.: 1970a, *Solar Phys.* **15**, 414.
Croom, D. L.: 1970b, *Astrophys. Letters* **7**, 133.
Croom, D. L.: 1970c, *J. Geophys. Res.* **75**, 6940.
Croom, D. L.: 1971a, *Solar Phys.* **19**, 152.
Croom, D. L.: 1971b, *Solar Phys.* **19**, 171.
Croom, D. L.: 1971c, *Nature Phys. Sci.* **229**, 142.
Croom, D. L. and Powell, R. J.: 1969, *Nature* **221**, 945.
Croom, D. L. and Powell, R. J.: 1971, *Solar Phys.* **20**, 136.
Culhane, J. L. and Acton, L. W.: 1974, *Ann. Rev. Astron. Astrophys.* **12**, 359.
Daene, H.: 1965, *Mber. DAW Berlin* **7**, 111.
Daene, H.: 1967, *Astron. Nachr.* **289**, 279.
Daene, H. and Fomichev, V. V.: 1971, HHI Suppl. Ser. Solar Data II.6, 241.
Daene, H., Křivský, L., and Krüger, A.: 1966, *Mber. DAW Berlin* **8**, 807.
Daene, H. and Krüger, A.: 1966a, *Astron. Nachr.* **289**, 105 and 117.
Daene, H. and Krüger, A.: 1966b, *Mber. DAW Berlin* **8**, 47.
Daene, H., Scholz, D., and Voigt, W.: 1970, HHI Suppl. Ser. Solar Data II.1, 2 p. 1 and 131.
Daene, H. and Voigt, W.: 1965, Publ. Astrophys. Obs. Potsdam No. 105.
Daigne, G.: 1968, *Nature* **220**, 5167.
Daigne, G.: 1975a, *Astron. Astrophys.* **38**, 141.
Daigne, G.: 1975b, *Astron. Astrophys.* **42**, 71.
Daigne, G. and Møller-Pedersen, B.: 1974, *Astron. Astrophys.* **37**, 355.
Daigne, G., Oudet, C., and Vinokur, M.: 1967, *Compt. Rendus Acad. Sci. Paris* **265**, 949.
D'Angelo, N.: 1972, in A. Mangeney (ed.), *Plasma Physics and Solar Radio Astronomy*, Meudon, p. 14.
Das Gupta, M. K. and Basu, D.: 1965, *J. Atmospheric Terrest. Phys.* **27**, 1029.
Das Gupta, M. K., Chattopadhyay, T., and Sarkar, S. K.: 1977, *Solar Phys.* **51**, 409.
Das Gupta, M. K. and Sarkar, S. K.: 1971, *Solar Phys.* **18**, 276.
Das Gupta, M. K. and Sarkar, S. K.: 1972, *Solar Phys.* **26**, 378.
Datlowe, D. W., Hudson, H. S., and Peterson, L. E.: 1974, *Solar Phys.* **35**, 193.
Davis, W. D.: 1977, *Solar Phys.* **54**, 139.
De Feiter, L. D.: 1972, *Space Sci. Rev.* **13**, 827.

De Feiter, L. D.: 1974, in: M. J. Rycroft and R. D. Reasenberg (ed.), *Space Research XIV*, Akademie Verlag, Berlin p. 413.
De Feiter, L. D.: 1975, *Space Sci. Rev.* **17**, 181.
De Feiter, L. D.: 1976, *Solar Phys.* **47**, 15.
De Feiter, L. D. and de Jager, C.: 1973, *Solar Phys.* **28**, 183.
De Groot, T.: 1966, *Rech. Astron. Obs. Utrecht* **18**, 1.
De Groot, T.: 1970, *Solar Phys.* **14**, 176.
De Groot, T., Loonen, J., and Slottje, C.: 1976, *Solar Phys.* **49**, 321.
De Jager, C.: 1955, *Ann. Geophys.* **11**, 330.
De Jager, C.: 1960, in H. Kallmann Bijl (ed.), *Space Research I*, North-Holland Publ. Co., p. 628.
De Jager, C.: 1962, *Space Sci. Rev.* **1**, 487.
De Jager, C.: 1965a, in J. Ortner and H. Maseland (ed.), *Introduction to Solar Terrestrial Relations*, D. Reidel Publ. Co., p. 13.
De Jager, C.: 1965b, *Ann. Astrophys.* **28**, 263.
De Jager, C.: 1967, *Solar Phys.* **2**, 327.
De Jager, C.: 1969, in C. de Jager and Z. Švestka (ed.), *Solar Flares and Space Research*, North-Holland Publ. Co., p. 1.
De Jager, C.: 1972, in C. De Jager (ed.), *Solar-Terrestrial Physics*, D. Reidel Publ. Co., Part I, p. 1.
De Jager, C.: 1975, *Solar Phys.* **40**, 133.
De Jager, C. and Kundu, M. R.: 1963, in W. Priester (ed.), *Space Research III*, North-Holland Publ. Co., p. 836.
De Jager, C. and van't Veer, F.: 1958, *Rech. Astron. Obs. Utrecht* **15**, 1.
Denisse, J. F.: 1950, *J. Phys. Radium* **11**, 164.
Denisse, J. F.: 1960, *Inf. Bull. Solar Radio Obs. Europe* **4**, 3.
Denisse, J. F. and Kundu, M. R.: 1957, *Comptes Rendus Acad. Sci. Paris*, **244**, 45.
de la Noë, J.: 1974, *Solar Phys.* **37**, 225.
de la Noë, J.: 1975, *Astron. Astrophys.* **43**, 201.
de la Noë, J. and Boischot, A.: 1972, *Astron. Astrophys.* **20**, 55.
de la Noë, J. and Møller-Pedersen, B.: 1971, *Astron. Astrophys.* **12**, 371.
de la Noë, J., Møller-Pedersen, B., and Boischot, A.: 1976, *Solar Phys.* **46**, 505.
Dicke, R. H.: 1946, *Rev. Sci. Inst.* **17**, 268.
Dittmer, P. H.: 1975, *Solar Phys.* **41**, 227.
Dodge, J. C.: 1975, *Solar Phys.* **42**, 445.
Dodson, H. W. and Hedeman, E. R.: 1960, *Astron. J.* **65**, 51.
Dodson, H. W. and Hedeman, E. R.: 1964, in W. Hess (ed.), *AAS-NASA Symposium on Physics of Solar Flares*, NASA, Washington, p. 15.
Dodson, H. W. and Hedeman, E. R.: 1966, *Astrophys. J.* **145**, 224.
Dodson, H. W. and Hedeman, E. R.: 1968, in K. O. Kiepenheuer (ed.), *Structure and Development of Solar Active Regions*, IAU Symposium No. 35, D. Reidel Publ. Co., p. 56.
Dodson, H. W., Hedeman, E. R., and Covington, A. E.: 1954, *Astrophys. J.* **119**, 541.
Dodson, H. W., Hedeman, E. R., and Owren, L.: 1953, *Astrophys. J.* **118**, 169.
Dollfus, A. and Martres, M. J.: 1977, *Solar Phys.* **53**, 449.
Doschek, G. A.: 1972, *Space Sci. Rev.* **13**, 765.
Drake, F. D.: 1962, *Nature* **195**, 893 and 894.
Drake, J. F.: 1971, *Solar Phys.* **16**, 152.
Dravskikh, A. F.: 1960, *Izv. Glav. Astron. Obs. Pulkovo* **21**, No. 164, 128.
Dröge, F.: 1967, *Z. Astrophys.* **66**, 200.
Dröge, F.: 1977, *Astron. Astrophys.* **57**, 285.
Drummond, W. E. and Rosenbluth, M. N.: 1960, *Phys. Fluids* **3**, 45.
Dryer, M.: 1974, *Space Sci. Rev.* **15**, 403.
Dulk, G. A.: 1970a, *Proc. Astron. Soc. Australia* **1**, 308.
Dulk, G. A.: 1970b, *Proc. Astron. Soc. Australia* **1**, 372.
Dulk, G. A.: 1971a, *Australian J. Phys.* **24**, 177.
Dulk, G. A.: 1971b, *Australian J. Phys.* **24**, 217.
Dulk, G. A.: 1973, *Solar Phys.* **32**, 491.
Dulk, G. A. and Altschuler, M. D.: 1971, *Solar Phys.* **20**, 438.

Dulk, G. A. and Nelson, G. J.: 1973, *Proc. Astron. Soc. Australia* **2**, 211.
Dulk, G. A. and Sheridan, K. V.: 1974, *Solar Phys.* **36**, 191.
Dulk, G. A., Sheridan, K. V., Smerd, S. F., and Withbroe, G. L.: 1977, *Solar Phys.* **52**, 349.
Dulk, G. A. and Smerd, S. F.: 1971, *Australian J. Phys.* **24**, 185.
Dulk, G. A., Stewart, R. T., Black, H. C., and Johns, I. A.: 1971, *Australian J. Phys.* **24**, 239.
Duncan, R. A.: 1974, *Proc. Astron. Soc. Australia* **2**, 255.
Dunckel, N.: 1974, NASA Technical Report No. 3469-2.
Dunckel, N. and Helliwell, R. A.: 1969, *J. Geophys. Res.* **74**, 6371.
Dunckel, N. and Helliwell, R. A.: 1972, *Solar Phys.* **25**, 197.
Dunckel, N., Helliwell, R. A., and Vesecky, J.: 1972, *Solar Phys.* **25**, 197.
Dungey, J. W.: 1953, *Phil. Mag.*, Ser. 7, **44**, 725.
Dungey, J. W.: 1958, in B. Lehnert (ed.), *Electromagnetic Phenomena in Cosmical Physics*, IAU Symposium No. 6, University Press Cambridge, p. 135.
Dungey, J. W.: 1975, *Space Sci. Rev.* **17**, 173.
Dupree, A. K.: 1968, *Astrophys. J. Letters* **152**, L125.
Dupree, T. H.: 1963, *Phys. Fluids* **6**, 1714.
Dupree, T. H.: 1964, *Phys. Fluids* **7**, 923.
Durasova, M. S., Kobrin, M. M., and Yudin, O. I.: 1971, *Nature* **229**, 83.
Earl, J. A.: 1961, *Phys. Rev. Letters* **6**, 125.
Eckhoff, H. K.: 1966, Report No. 18, Inst. Theoret. Astrophys., Blindern-Oslo.
Eddy, J. A.: 1976a, *Science* **192**, 1189.
Eddy, J. A.: 1976b, Paper presented at the AGU/International Symposium on Solar-Terrestrial Physics, Boulder, Colo., June 1976.
Edelson, S., Mayfield, E. B., and Shimabukuro, F. I.: 1971, *Nature Phys. Sci.* **232**, 82.
Efanov, V. A., Kislyakov, A. G., and Moiseev, I. G.: 1972, *Solar Phys.* **24**, 142.
Efanov, V. A., Kislyakov, A. G., Moiseev, I. G., and Naumov, A. I.: 1969, *Solar Phys.* **8**, 331.
Efanov, V. A. and Moiseev, I. G.: 1973, *Izv. Krymsk. Astrofiz. Obs.* **47**, 59.
Eidman, V. J.: 1958, *Zh. Eksperim. Teor. Fiz.* **34**, 131.
Eidman, V. J.: 1959, *Zh. Experim. Teor. Fiz.* **36**, 1335.
Einstein, A.: 1917, *Phys. Z.* **18**, 121.
Elgarøy, Ø.: 1957, *Nature* **180**, 862.
Elgarøy, Ø.: 1959, *Nature* **184**, 887.
Elgarøy, Ø.: 1961, *Astrophys. Norv.* **7**, 123.
Elgarøy, Ø.: 1968, *Astrophys. Letters* **3**, 39.
Elgarøy, Ø. and Eckhoff, H. K.: 1966, *Astrophys. Norv.* **10**, 127.
Elgarøy, Ø. and Lyngstad, E.: 1972, *Astron. Astrophys.* **16**, 1.
Elgarøy, Ø. and Sveen, O. P.: 1973, *Solar Phys.* **32**, 231.
Elgarøy, Ø. and Ugland, O.: 1970, *Astron. Astrophys.* **5**, 372.
Eliseeva, L. A. and Yurovskaya, L. I.: 1974, *Izv. Krymsk. Astrofiz. Obs.* **49**, 42.
Elliot, H.: 1964, *Planet. Space Sci.* **12**, 657.
Elliot, H.: 1969, in C. de Jager and Z. Svestka (ed.), *Solar Flares and Space Research*, North-Holland Publ. Co., Amsterdam. p. 356.
Ellis, G. R. A.: 1964, *Australian J. Phys.* **17**, 63.
Ellis, G. R. A.: 1969, *Australian J. Phys.* **22**, 177.
Ellis, G. R. A.: 1972, *Proc. Astron. Soc. Australia* **2**, 135.
Ellis, G. R. A.: 1973, *Australian J. Phys.* **26**, 253.
Ellis, G. R. A. and McCulloch, P. M.: 1966, *Nature* **211**, 1070.
Ellis, G. R. A. and McCulloch, P. M.: 1967, *Australian J. Phys.* **20**, 583.
Ellison, M. A.: 1963, *Quart. J. Roy. Astron. Soc.* **4**, 62.
Ellison, M. A. and Reid, J. H.: 1964, in H. Odishaw (ed.), *Research in Geophysics*, Vol. 1, M.I.T. Press, p. 43.
El-Raey, M. and Amer, R.: 1975, *Solar Phys.* **45**, 533.
El-Raey, M. and Scherrer, P.: 1973, *Solar Phys.* **30**, 149.
Elton, R. C. and Lie, T. N.: 1972, *Space Sci. Rev.* **13**, 747.
Elwert, G.: 1948, *Z. Naturforsch.* **3a**, 477.
Elwert, G.: 1954, *Z. Naturforsch.* **9a**, 637.
Elwert, G.: 1961, *J. Geophys. Res.* **66**, 391.

Elwert, G. and Haug, E.: 1971, *Solar Phys.* **20**, 413.
Elzner, L. R.: 1976, *Astron. Astrophys.* **47**, 9.
Énomé, S.: 1964, *Publ. Astron. Soc. Japan* **16**, 135.
Énomé, S.: 1969, *Publ. Astron. Soc. Japan*, **21**, 367.
Énomé, S., Kakinuma, T., and Tanaka, H.: 1969, *Solar Phys.* **6**, 428.
Énomé, S. and Tanaka, H.: 1971, in R. Howard (ed.), *Solar Magnetic Fields*, IAU Symposium No. 43, D. Reidel Publ. Co., p. 413.
Epstein, R. I. and Feldman, P. A.: 1967, *Astrophys. J. Letters* **150**, L109.
Erickson, W. C. and Fisher, J. R.: 1974, *Radio Sci.* **9**, 387.
Erickson, W. C., Gergely, T. E., Kundu, M. R., and Mahoney, M. J.: 1977, *Solar Phys.* **54**, 57.
Eryushev, N. N.: 1967, *Izv. Krymsk. Astrofiz. Obs.* **36**, 121.
Eryushev, N. N., Tilin, M. V., and Tsvetkov, L. I.: 1971, *Izv. Krymsk. Astrofiz. Obs.* **43**, 3.
Eryushev, N. N. and Tsvetkov, L. I.: 1973, *Izv. Krymsk. Astrofiz. Obs.* **48**, 85.
Evans, L. G., Fainberg, J., and Stone, R. G.: 1971, *Solar Phys.* **21**, 198.
Evans, L. G., Fainberg, J., and Stone, R. G.: 1973, *Solar Phys.* **31**, 501.
Fainberg, J., Evans, L. G., and Stone, R. G.: 1972, *Science* **178**, 743.
Fainberg, J. and Stone, R. G.: 1970, *Solar Phys.* **15**, 222 and 433.
Fainberg, J. and Stone, R. G.: 1971, *Solar Phys.* **17**, 392.
Fainberg, J. and Stone, R. G.: 1974, *Space Sci. Rev.* **16**, 145.
Fajnshtejn, S. M.: 1968, *Izv. Vys. Ucheb. Zav. Radiofiz.* **11**, 217.
Falchi, A. D., Felli, M., and Tofani, G.: 1976, *Solar Phys.* **48**, 59.
Fan, C. Y., Pick, M., Ryle, R., Simpson, J. A., and Smith, D. R.: 1968, *J. Geophys. Res.* **73**, 1555.
Feix, G.: 9, *Solar Phys.* **9**, 265.
Feix, G.: 1970, *Astron. Astrophys.* **9**, 239.
Feix, G.: 1972, *Astron. Astrophys.* **16**, 268.
Felli, M. and Tofani, G.: 1970, *Solar Phys.* **13**, 194.
Felli, M., Pallavicini, R., and Tofani, G.: 1975a, *Solar Phys.* **44**, 135.
Felli, M., Pampaloni, P., and Tofani, G.: 1974, *Solar Phys.* **37**, 395.
Felli, M., Poletto, G., and Tofani, G.: 1977, *Solar Phys.* **51**, 65.
Felli, M., Tofani, G., Fürst, E., and Hirth, W.: 1975b, *Solar Phys.* **42**, 377.
Fermi, E.: 1949, *Phys. Rev.* **75**, 1169.
Fermi, E.: 1954, *Astrophys. J.* **119**, 1.
Fichtel, C. E. and McDonald, F. B.: 1967, *Ann. Rev. Astron. Astrophys.* **5**, 351.
Field, G. B.: 1961, *J. Geophys. Res.* **66**, 1395.
Findlay, J. W.: 1966, *Ann. Rev. Astron. Astrophys.* **4**, 77.
Firor, J.: 1956, *Astrophys. J.* **123**, 320.
Flamm, E. J. and Lingenfelter, R. E.: 1964, *Science* **144**, 1566.
Fokker, A. D.: 1960, Doctoral Thesis, Leiden.
Fokker, A. D.: 1963a, *Space Sci. Rev.* **2**, 70.
Fokker, A. D.: 1963b, *Bull. Astron. Inst. Neth.* **17**, 84.
Fokker, A. D.: 1965, *Bull. Astron. Inst. Neth.* **18**, 111.
Fokker, A. D.: 1967, *Solar Phys.* **2**, 316.
Fokker, A. D.: 1969, *Solar Phys.* **8**, 376.
Fokker, A. D.: 1970, *Solar Phys.* **11**, 92.
Fokker, A. D.: 1971, *Solar Phys.* **19**, 472.
Fokker, A. D.: 1972, in A. Abrami (ed.), *Proc. of the 2nd Meeting of CESRA*, Astron. Obs. Trieste, p. 13.
Fokker, A. D., Goh, T., Landre, E., and Roosen, J.: 1966, *Bull. Astron. Inst. Neth.* Suppl. **1**, 309.
Fokker, A. D. and Roosen, J.: 1961, *Bull. Astron. Inst. Neth.* **16**, 83.
Fokker, A. D. and Rutten, R. J.: 1967, *Bull. Astron. Inst. Neth.* **19**, 254.
Fomichev, V. V.: 1972, *Astron. Zh.* **49**, 348.
Fomichev, V. V. and Chernov, G. P.: 1973, *Astron. Zh.* **50**, 798.
Fomichev, V. V. and Chertok, I. M.: 1965, *Astron. Zh.* **42**, 1256; *Soviet Astron.* **9**, 976 (1966).
Fomichev, V. V. and Chertok, I. M.: 1967a, *Astron. Zh.* **44**, 495; *Soviet Astron.* **11**, 396.
Fomichev, V. V. and Chertok, I. M.: 1967b, *Geomagnetizm i Aeronomiya* **7**, 3.
Fomichev, V. V. and Chertok, I. M.: 1968a, *Astron. Zh.* **45**, 28.
Fomichev, V. V. and Chertok, I. M.: 1968b, *Astron. Zh.* **45**, 601.

Fomichev, V. V. and Chertok, I. M.: 1968c, *Astron. Zh.* **45**, 773.
Fomichev, V. V. and Chertok, I. M.: 1969a, *Astron. Zh.* **46**, 343.
Fomichev, V. V. and Chertok, I. M.: 1969a, *Astron. Zh.* **46**, 343.
Fomichev, V. V. and Chertok, I, M.: 1969b, *Astron. Zh.* **46**, 1319.
Fomichev, V. V. and Chertok, I. M.: 1970a, *Astron. Zh.* **47**, 226.
Fomichev, V. V. and Chertok, I. M.: 1970b, *Astron. Zh.* **47**, 322; *Soviet Astron.* **14**, 261.
Fomichev, V. V. and Chertok, I. M.: 1971, *Astron. Zh.* **48**, 1244.
Fomichev, V. V. and Chertok, I. M.: 1977, *Izv. Vys. Ucheb. Zav., Radiofizika* **20**, 1255.
Fourikis, N.: 1971, *Proc. Inst. Radio Electron. Engrs. Australia* **32**, 361.
Frank, L. A. and Gurnett, D. A.: 1972, *Solar Phys.* **27**, 446.
Franklin, K. L. and Burke, B. F.: 1958, *J. Geophys. Res.* **63**, 807.
Friedman, H.: 1963, *Ann. Rev. Astron. Astrophys.* **1**, 59.
Friedman, M. and Hamberger, S. M.: 1969, *Solar Phys.* **8**, 104 and 398.
Fritzová-Švestková, L., Chase, R. C., and Švestka, Z.: 1976, *Solar Phys.* **48**, 275.
Fritzová-Švestková, L. and Švestka, Z.: 1966, *Bull. Astron. Inst. Czech.* **17**, 249.
Fritzová-Švestková, L. and Švestka, Z.: 1967, *Solar Phys.* **2**, 87.
Frost, K. J.: 1969, *Astrophys. J. Letters* **158**, L159.
Frost, K. J. and Dennis, B. R.: 1971, *Astrophys. J.* **165**, 655.
Fung, P. C. W.: 1968a, *Can. J. Phys.* **46**, 1073.
Fung, P. C. W.: 1968b, *Plasma Phys.* **10**, 993.
Fung, P. C. W.: 1969a, *Plasma Phys.* **11**, 285.
Fung, P. C. W.: 1969b, *Astrophys. Space Sci.* **5**, 448.
Fung, P. C. W.: 1969c, *Can. J. Phys.* **47**, 161.
Fung, P. C. W.: 1969d, *Can. J. Phys.* **47**, 179.
Fung, P. C. W.: 1969e, *Can. J. Phys.* **47**, 757.
Fung, P. C. W. and Yip, W. K.: 1966, *Australian J. Phys.* **19**, 759.
Fürst, E.: 1972, *Solar Phys.* **25**, 178.
Fürst, E.: 1973, *Solar Phys.* **28**, 159.
Fürst, E., Hachenberg, O., and Hirth, W.: 1973a, *Solar Phys.* 28, 533.
Fürst, E., Hachenberg, O., and Hirth, W.: 1974, *Astron. Astrophys.* **36**, 123.
Fürst, E., Hachenberg, O., Zinz, W., and Hirth, W.: 1973b, *Solar Phys.* **32**, 445.
Fürst, E. and Hirth, W.: 1975, *Solar Phys.* **42**, 157.
Fürstenberg, F.: 1960, *Z. Astrophys.* **49**, 42.
Fürstenberg, F.: 1966, in L. Mollwo and W. Dittmar (ed.), *Solar Radio Observations during the IQSY*, HHI Berlin, p. 84.
Fürstenberg, F., Krüger, A., Olmr, J., and Tlamicha, A.: 1966, *Bull. Astron. Inst. Czech.* **17**, 213.
Furth, H. P.: 1969, in D. G. Wentzel and D. A. Tidman (ed.), *Plasma Instabilities in Astrophysics*, Gordon and Breach, p. 33.
Furth, H. P., Kileen, J., and Rosenbluth, M. N.: 1963, *Phys. Fluids* **6**, 459.
Gajlitis, A. K. and Tsytovich, V. N.: 1963, *Izv. Vys. Ucheb. Zav. Radiofiz.* **6**, 1104.
Galeev, A. A. and Zelenyj, L. M.: 1976, *Zh. Eksperim. Teor. Fiz.* **70**, 2133.
Gelfreikh, G. B.: 1962, *Solnechnye Dannye, Leningrad* No. 3, 50 and No. 5, 67.
Gelfreikh, G. B.: 1972, *Astron. Tsirkulyar AN SSSR* No. 699, p. 3.
Gelfreikh, G. B., Zhitnik, I. A., and Livshits, M. A.: 1970, *Astron. Zh.* **47**, 329; *Soviet Astron.* **14**, 266.
Gerasimova, N. N.: 1973, Preprint No. 26 Inst. Radiophys. Electr. Charkov.
Gergely, T. E.: 1974, Ph. D. Thesis, University of Maryland.
Gergely, T. E. and Erickson, W. C.: 1975, *Solar Phys.* **42**, 467.
Gergely, T. E. and Kundu, M. R.: 1974, *Solar Phys.* **34**, 433.
Gergely, T. E. and Kundu, M. R.: 1975, *Solar Phys.* **41**, 163.
Gerlakh, N. I. and Syrovatskij, S. I.: 1974, *Trudy Fiz. Inst. Lebedeva* **74**, 73.
Gershman, B. N.: 1960, *Zh. Eksperim. Teor. Fiz.* **38**, 912; *Soviet Phys. JETP.* **11**, 657.
Giachetti, R. and Pallavicini, R.: 1976, *Astron. Astrophys.* **53**, 347.
Gingerich, O., Noyes, R. W., Kalkofen, W., and Cuny, Y.: 1971, *Solar Phys.* **18**, 347.
Ginzburg, V. L.: 1951, *Doklady Akad. Nauk* **76**, 377.
Ginzburg, V. L.: 1953, *Uspekhi Fiz. Nauk* **51**, 343.
Ginzburg, V. L.: 1958, in J. G. Wilson and S. A. Wouthuysen (ed.), *Progress in Elementary Particles and Cosmic Ray Physics* Vol. IV, North-Holland Publ. Co., p. 339.

Ginzburg, V. L. and Frank, I. M.: 1946, *Zh. Experim. Teor. Fiz.* **16**, 25.
Ginzburg, V. L. and Ozernoj, S. M.: 1966, *Izv. Vys. Ucheb. Zav., Radifiz.* **9**, 221; *Astrophys. J.* **144**, 599.
Ginzburg, V. L., Sazonov, V. N., and Syrovatskij, S. I.: 1968, *Uspekhi Fiz. Nauk* **94**, 63.
Ginzburg, V. L. and Syrovatskij, S. I.: 1965, *Uspekhi Fiz. Nauk* **87**, 65; *Ann. Rev. Astron. Astrophys.* **3**, 297.
Ginzburg, V. L. and Syrovatskij, S. I.: 1969, *Ann. Rev. Astron. Astrophys.* **7**, 375.
Ginzburg, V. L. and Zheleznyakov, V. V.: 1958a, *Astron. Zh.* **35**, 694; *Soviet Astron.* **2**, 653.
Ginzburg, V. L. and Zheleznyakov, V. V.: 1958b, *Izv. Vys. Ucheb. Zav. Radiofiz.* **1**, 59.
Ginzburg, V. L. and Zheleznyakov, V. V.: 1959, *Astron. Zh.* **36**, 233; *Soviet Astron.* **3**, 235.
Ginzburg, V. L. and Zheleznyakov, V. V.: 1961, *Astron. Zh.* **38**, 3; *Soviet Astron.* **5**, 1.
Giovanelli, R. G.: 1946, *Nature* **158**, 81.
Giovanelli, R. G.: 1947, *Monthly Notices Roy. Astron. Soc.* **107**, 338.
Giovanelli, R. G.: 1948, *Monthly Notices Roy. Astron. Soc.* **108**, 163.
Gnevyshev, M. N.: 1963, *Astron. Zh.* **40**, 401.
Gnevyshev, M. N.: 1967, *Solar Phys.* **1**, 107.
Gold, T.: 1962, *Space Sci. Rev.* **1**, 100.
Gold, T. and Hoyle, F.: 1960, *Monthly Notices Roy. Astron. Soc.* **120**, 89.
Goldberg, L.: 1967, *Ann. Rev. Astron. Astrophys.* **5**, 279.
Goldstein, S. J., Jr. and Meisel, D. D.: 1969, *Nature* **224**, 349.
Golub, L., Krieger, A. S., Silk, J. K., Timothy, A. F., and Vaiana, G. S.: 1974, *Astrophys. J. Letters* **189**, L93.
Golub, L., Krieger, A. S., and Vaiana, G. S.: 1976, *Solar Phys.* **50**, 311.
Gopasyuk, S. I., Eryushev, N. N., and Neshpor, Y. I.: 1967, *Izv. Krymsk. Astrofiz. Obs.* **37**, 109.
Gopasyuk, S. I., Eryushev, N. N., and Neshpor, Y. I.: 1969, *Izv. Krymsk. Astrofiz. Obs.* **39**, 299.
Gordon, I. M.: 1968, *Astrophys. Letters* **2**, 49.
Gordon, I. M.: 1971, *Astrophys. Letters* **5**, 251.
Gorgolewski, S.: 1965, *Acta Astron.* **15**, 261.
Gopala Rao, U. V.: 1970, *Solar Phys.* **14**, 389.
Gosling, J. T., Asbridge, J. R., Bame, S. J., Hundhausen, A. J., and Strong, I. B.: 1967, *J. Geophys. Res.* **72**, 3357.
Gotwols, B. L.: 1972, *Solar Phys.* **25**, 232.
Gotwols, B. L.: 1973, *Solar Phys.* **33**, 475.
Gotwols, B. L. and Phipps, J.: 1972, *Solar Phys.* **26**, 386.
Graedel, T. E.: 1970, *Astrophys. J.* **160**, 301.
Graedel, T. E. and Lanzerotti, L.: 1971, *J. Geophys. Res.* **76**, 6972.
Graf, W. and Bracewell, R. N.: 1973, *Solar Phys.* **28**, 425.
Graf, W. and Bracewell, R. N.: 1973, *Solar Phys.* **33**, 75.
Graf. W. and Bracewell, R. N.: 1975, *Solar Phys.* **44**, 195.
Graham, M. H.: 1958, *Proc. Inst. Radio Engrs.* **46**, 1966.
Green, R. M. and Sweet, P. A.: 1967, *Astrophys. J.* **147**, 1153.
Greve, A.: 1974, *Solar Phys.* **36**, 85.
Greve, A.: 1975, *Solar Phys.* **44**, 371.
Grognard, R. J.-M. and McLean, D. J.: 1973, *Solar Phys.* **29**, 149.
Gribkov, V. A.: 1975, Private Communication.
Guidice, D. A. and Castelli, J. P.: 1975, *Solar Phys.* **44**, 155.
Gurevich, A. V.: 1967, *Zh. Eksperim. Teor. Fiz.* **53**, 953.
Gurevich, A. V.: 1968, *Zh. Eksperim. Teor. Fiz.* **54**, 522.
Gurnett, D. A. and Frank, L. A.: 1975, *Solar Phys.* **45**, 477.
Gusejnov, R. E.: 1961, *Astron. Zh.* **38**, 869.
Gurzadyan, G. A.: 1971, *Astron. Astrophys.* **13**, 348.
Gurzadyan, G. A.: 1972, *Astron. Astrophys.* **20**, 145.
Hachenberg, O.: 1959, in R. Righini (ed.), *Radioastronomia Solare*, Rendiconti S.I.F., XII, N. Zanichelli Edit., Bologna, p. 217.
Hachenberg, O.: 1965, in J. Aarons (ed.), *Solar Radio Astronomy*, Plenum Press, New York, p. 95.
Hachenberg, O., Fürstenberg, F., Helms, B., and Krüger, A.: 1963, *Radiostrahlungsausbrüche der Sonne im IGJ*, Teil I und II, HHI Berlin.

Hachenberg, O. and Krüger, A.: 1959, *J. Atmospheric Terrestr. Phys.* **17**, 20.
Hachenberg, O. and Krüger, A.: 1964, *Z. Astrophys.* **59**, 261.
Hachenberg, O. and Wallis, G.: 1961, *Z. Astrophys.* **52**, 42.
Haddock, F. T. and Alvarez, H.: 1973, *Solar Phys.* **29**, 183.
Haddock, F. T. and Graedel, T. E.: 1970, *Astrophys. J.* **160**, 293.
Hagen, J. P. and Barney, W. H.: 1968, *Astrophys. J.* **153**, 275.
Hagen, J. P. and Neidig, D. F., Jr.: 1971, *Solar Phys.* **18**, 305.
Hakura, Y.: 1976, *Space Sci. Rev.* **19**, 411.
Hakura, Y., Nishiraki, R., and Tao, K.: 1969, *J. Radio Res. Lab. Japan* **16**, 215.
Hall, D. E. and Sturrock, P. A.: 1967, *Phys. Fluids* **10**, 2620.
Hanasz, J.: 1966, *Australian Phys.* **19**, 635.
Hanasz. J., Schreiber, R., Welnowski, H., Wikierski, B., and Aksenov, V. I.: 1976, *Contr. Astron. Obs. Skalnate Pleso* **6**, 157.
Hanbury-Brown, R. and Twiss, R. Q.: 1954, *Phil. Mag.* **45**, 663.
Hansen, R. T., Garcia, C. J., Grognard, R. J.-M., and Sheridan, K. V.: 1971, *Proc. Astron. Soc. Australia* **2**, 57.
Harris, E. G.: 1959, *Phys. Rev. Letters* **2**, 34.
Harris, E. G.: 1961, *J. Nucl. Energy* **C2**, 138.
Harris, E. G.: 1969, in D. G. Wentzel, D. A. Tidman (ed.), *Plasma Instabilities in Astrophysics*, Gordon and Breach, p. 59.
Hartz, T. R.: 1964, *Ann. Astrophys.* **27**, 831.
Hartz, T. R.: 1969, *Planetary Space Sci.* **17**, 267.
Harvey, C. C.: 1968, *Nature* **217**, 50.
Harvey, C. C.: 1975, *Solar Phys.* **40**, 193.
Harvey, C. C.: 1976, *Astron. Astrophys.* **47**, 31.
Harvey, C. C. and Aubier, M. G.: 1973, *Astron. Astrophys.* **22**, 1.
Harvey, G. A. and McNarry, L. R.: 1970, *Solar Phys.* **11**, 467.
Harvey, K. L., Martin, S. F., and Riddle, A. C.: 1974, *Solar Phys.* **36**, 151.
Hatanaka, T.: 1956, *Publ. Astron. Soc. Japan* **8**, 73.
Haug, E.: 1972, *Solar Phys.* **25**, 425.
Haurwitz, M. W.: 1968, *Astrophys. J.* **151**, 351.
Helliwell, R. A.: 1967, *J. Geophys. Res.* **72**, 4773.
Hertweck, F. and Schlüter, A.: 1957, *Z. Naturforsch.* **12a**, 844.
Hewish, A., Dennison, P. A., and Pilkington, J. D. H.: 1966, *Nature* **209**, 1188.
Hewish, A., Scott, P. F., and Wills, D.: 1964, *Nature* **203**, 1214.
Hey, J. S.: 1946, *Nature* **157**, 47.
Heyvaerts, J.: 1974, *Solar Phys.* **38**, 419.
Heyvaerts, J.: 1975, *Astron. Astrophys.* **38**, 45.
Heyvaerts, J. and Verdier de Genduillac, G.: 1974, *Astron. Astrophys.* **30**, 211.
Hildebrandt, J.: 1976, Private Communication.
Hildner, E., Gosling, J. T., MacQueen, R. M., Munro, R. H., Poland, A. I., and Ross, C. L.: 1976, *Solar Phys.* **48**, 127.
Hinteregger, H. E. and Hall, L. A.: 1969, *Solar Phys.* **6**, 175.
Hirayama, T.: 1974, *Solar Phys.* **34**, 323.
Hirth, W.: 1967, *Z. Astrophys.* **65**, 48.
Hirth, W. and Wallis, G.: 1966, *Beitr. Plasmaphys.* **6**, 27.
Hoang, S., Poquérusse, M., and Steinberg, J. L.: 1977, *Astron. Astrophys.* **56**, 283.
Hoang, S. and Steinberg, J. L.: 1977, *Astron. Astrophys.* **58**, 287.
Hobbs, R. W., Jordan, S. D., Webster, W. J., Jr., Maran, S. P., and Caulk, H. W.: 1974, *Solar Phys.* **36**, 369.
Högbom, J. A.: 1960, *Monthly Notices Roy. Astron. Soc.* **120**, 530.
Hollweg, J. V.: 1972a, *Cosmic Electrodyn.* **2**, 423.
Hollweg, J. V.: 1972b, *Astrophys. J.* **177**, 255.
Hollweg, J. V.: 1974, *Publ. Astron. Soc. Pacific* **86**, 561.
Hollweg, J. V. and Harrington, J. V.: 1968, *J. Geophys. Res.* **73**, 7221.
Holt, S. S. and Cline, T. L.: 1968, *Astrophys. J.* **154**, 1027.

Holt, S. S. and Ramaty, R.: 1969, *Solar Phys.* **8**, 119.
Horan, D. M.: 1971, *Solar Phys.* **21**, 188.
House, L. L.: 1972, *Solar Phys.* **23**, 103.
Howard, R. and Švestka, Z.: 1977, *Solar Phys.* **54**, 65.
Hudson, H. S.: 1973, in R. Ramaty and R. G. Stone (ed.), *High Energy Phenomena on the Sun*, Symposium Proceedings, Goddard Space Flight Center, Greenbelt, Md., p. 207.
Hudson, H. S., Peterson, L. E., and Schwartz, D. A.: 1969, *Astrophys. J.* **157**, 389.
Hughes, M. P. and Harkness, L.: 1963, *Astrophys. J.* **138**, 239.
Hundhausen, A. J.: 1968, *Space Sci. Rev.* **8**, 690.
Hyder, C. L.: 1967, *Solar Phys.* **2**, 49 and 267.
Hyder, C. L., Epstein, G. L., and Hobbs, R. W.: 1973, *Astrophys. J.* **185**, 985.
Ikhsanova, V. N.: 1960, *Izv. Glav. Astron. Obs. Pulkovo* **21**, No. 164, 62.
Ikhsanova, V. N.: 1964, *Izv. Glav. Astron. Obs. Pulkovo* **23**, No. 172, 31.
Ikhsanova, V. N.: 1966, *Izv. Glav. Astron. Obs. Pulkovo* **24**, No. 180, 51.
Ishiguro, M.: 1974, *Astron. Astrophys. Suppl.* **15**, 431.
Ishiguro, M., Énomé, S., and Tanaka, H.: 1972, *Proc. Res. Inst. Atmosph. Nagoya Univ.* **19**, 105.
Ishiguro, M., Tanaka, H., Énomé, S., Torii, Ch., Tsukij, Y., Kobayashi, S., and Yoshimi, N.: 1975, *Proc. Res. Inst. Atmosph. Nagoya Univ.* **22**, 1.
Istomin, Ya. I. and Karpman, V. I.: 1972, *Zh. Eksperim. Teor. Fiz.* **63**, 131.
Ivanenko, D. D. and Pomeranchuk, I. Ya.: 1944, *Dokl. Akad. Nauk* **44**, 343.
Ivanov-Kholodnyj, G. S. and Nikolskij, G. M.: 1962, *Astron. Zh.* **39**, 777.
Jacobsen, C. and Carlquist, P.: 1964, *Icarus* **3**, 270.
Jaeger, J. C. and Westfold, K. C.: 1949, *Australian J. Sci. Res.* **A2**, 3.
Jaeger, J. C. and Westfold, K. C.: 1950, *Australian J. Sci. Res.* **A3**, 376.
Jäger, F. W.: 1966, in M. Cimino (ed.), *Atti del convegno sui campi magnetici solari*, G. Barbèra Edit., Firenze, p. 166.
Jäger, J. and Wallis, G.: 1961, *Beitr. Plasmaphys.* **1**, 44.
Jaggi, R. K.: 1963, *J. Geophys. Res.* **68**, 4429.
Jaggi, R. K.: 1965, in R. Lüst (ed.), *Stellar and Solar Magnetic Fields*, IAU Symposium No. 22, North-Holland Publ. Co., p. 384.
Jahn, H.: 1970, *Astrophys. Space Sci.* **9**, 153.
Jahn, H.: 1971, *Astrophys. Space Sci.* **12**, 34.
Jakimiec, J. and Jakimiec, M.: 1974, *Astron. Astrophys.* **34**, 415.
Jefferies, J. T. and Orrall, F. Q.: 1965, *Astrophys. J.* **141**, 509 and 519.
Joensen, P., McCutcheon, W. H., and Shuter, L. H.: 1974, *Solar Phys.* **39**, 309.
Johnson, D. J.: 1974, *J. Appl. Phys.* **45**, 1147.
Jokipii, J. R.: 1973, *Ann. Rev. Astron. Astrophys.* **11**, 1.
Jones, D.: 1976, *Nature*, **260**, 686.
Jones, D.: 1977a, *Geophys. Res. Letters* **4**, 121.
Jones, D.: 1977b, *Astron. Astrophys.* **55**, 245.
Kahler, S. W. and Kreplin, R. W.: 1970, *Solar Phys.* **14**, 372.
Kai, K.: 1964, *Bull. Tokyo Gakugei Univ.* **15**, 103.
Kai, K.: 1965a, *Ann. Tokyo Astron. Obs.* Ser. 2, **9**, 195.
Kai, K.: 1965b, *Publ. Astron. Soc. Japan* **17**, 294 and 309.
Kai, K.: 1967, *Proc. Astron. Soc. Australia* **1**, 49.
Kai, K.: 1968, *Publ. Astron. Soc. Japan* **20**, 140.
Kai, K.: 1969a, *Proc. Astron. Soc. Australia* **1**, 186.
Kai, K.: 1969b, *Proc. Astron. Soc. Australia* **1**, 189.
Kai, K.: 1969c, *Solar Phys.* **10**, 460.
Kai, K.: 1970a, *Solar Phys.* **11**, 310.
Kai, K.: 1970b, *Solar Phys.* **11**, 456.
Kai, K.: 1973, *Proc. Astron. Soc. Australia* **2**, 219.
Kai, K.: 1974, in G. Newkirk, Jr. (ed.), *Coronal Disturbances*, IAU Symposium No. 57, Reidel Publ. Co., p. 245.
Kai, K.: 1975, *Solar Phys.* **45**, 217.
Kai, K. and McLean, D. J.: 1968, *Proc. Astron. Soc. Australia* **1**, 141.
Kai, K. and Sekiguchi, H.: 1973, *Proc. Astron. Soc. Australia* **2**, 217.

Kai, K. and Sheridan, K. V.: 1974, *Solar Phys.* **35**, 181.
Kai, K. and Takayanagi, A.: 1973, *Solar Phys.* **29**, 461.
Kaiser, M. L.: 1975, *Solar Phys.* **45**, 181.
Kakinuma, T.: 1955, *Proc. Res. Inst. Atmosph. Nagoya Univ.* **3**, 96.
Kakinuma, T.: 1958, *Proc. Res. Inst. Atmosph. Nagoya Univ.* **5**, 71.
Kakinuma, T. and Swarup, G.: 1962, *Astrophys. J.* **136**, 975.
Kakinuma, T. and Tanaka, H.: 1961, *Proc. Res. Inst. Atmosph. Nagoya Univ.* **8**, 39.
Kakinuma, T. and Tanaka, H.: 1963, *Proc. Res. Inst. Atmosph. Nagoya Univ.* **10**, 25.
Kakinuma T. and Watanabe, T.: 1976, *Space Sci. Rev.* **19**, 611.
Kakinuma, T., Yamashita, T., and Énomé, S.: 1969, *Proc. Res. Inst. Atmosph. Nagoya Univ.* **16**, 127.
Kalaghan, P. M.: 1974, *Solar Phys.* **39**, 315.
Kane, S. R.: 1972, *Solar Phys.* **27**, 174.
Kane, S. R.: 1974, in G. Newkirk, Jr. (ed.), *Coronal Disturbances*, IAU Symp. No. 57, D. Reidel Publ. Co., p. 105.
Kane, S. R., Kreplin, R. W., Martres, M.-J., Pick, M., and Soru-Escaut, I.: 1974, *Solar Phys.* **38**, 483.
Kanno, M. and Tanaka, R.: 1975, *Solar Phys.* **43**, 63.
Kantrowitz, A. and Petschek, H. E.: 1966, in W. B. Kunkel (ed.), *Plasma Physics in Theory and Application*, McGraw-Hill, New York, p. 148.
Kaplan, S. A. and Tsytovich, V. N.: 1967a, *Astron. Zh.* **44**, 1036.
Kaplan, S. A. and Tsytovich, V. N.: 1967b, *Astron. Zh.* **44**, 1194; *Soviet Astron.* **11**, 956 (1968).
Kaplan, S. A. and Tsytovich, V. N.: 1969, *Astrophys. Space Sci.* **3**, 431.
Karlický, M.: 1977, *Bull. Astron. Inst. Czech.* **28**, 117.
Karlický, M. and Tlamicha, A.: 1976, *Bull. Astron. Inst. Czech.* **27**, 223.
Karpman, V. I., Alekhin, Yu. K., Borisov, N. D., and Ryabova, N. A. 1973, *Phys. Letters* **44A**, 205.
Karpman, V. I., Alekhin, Yu. K., Borisov, N. D., and Ryabova, N. A.: 1975, *Plasma Phys.* **17**, 361 and 937.
Kaufmann, P.: 1969, *Solar Phys.* **9**, 166.
Kaufmann, P.: 1972, *Solar Phys.* **23**, 178.
Kaufmann, P.: 1976, *Solar Phys.* **50**, 197.
Kaufmann, P., Matsuura, O. T., and Dos Santos, P. M.: 1970, *Solar Phys.* **14**, 190.
Kaufmann, P., Piazza, L. R., and Raffaelli, J. C.: 1977, *Solar Phys.* **54**, 179.
Kaufmann, P., Scalise, E., Jr. and Dos Santos, P. M.: 1970, *Solar Phys.* **15**, 195.
Kaverin, N. S., Kobrin, M. M., Korshunov, A. I., Panfilov, Yu. D., and Tikhomirov, V. A.: 1976, *Phys. Solariterr., Potsdam* **2**, 49.
Kawabata, K.: 1964, *Publ. Astron. Soc. Japan* **16**, 30.
Kawabata, K.: 1966a, *Rep. Ionosphere Space Res. Japan* **20**, 25 and 107.
Kawabata, K.: 1966b, *Rep. Ionosphere Space Res. Japan* **20**, 27 and 118.
Kawabata, K. and Sofue, Y.: 1972, *Publ. Astron. Soc. Japan* **24**, 469.
Kelogg, P. J., Lai, J. C., and Cartwright, D. G.: 1973, US Dept. Commerce, Spec. Report UAG-28, Boulder, Part II, p. 288.
Kerdraon, A.: 1973, *Astron. Astrophys.* **27**, 361.
Khaikin, S. E., Kaidanovskij, N. L., Esepkina, N. A., and Shivris, O. N.: 1960, *Izv. Glav. Astron. Obs. Pulkovo* **21**, No. 164, 3.
Kiepenheuer, K. O.: 1946, *Nature* **158**, 340.
Kiepenheuer, K. O.: 1950, *Phys. Rev.* **79**, 738.
Kippenhahn, R. and Schlüter, A.: 1957, *Z. Astrophys.* **43**, 36.
Kislyakov, A. G.: 1961, *Izv. Vys. Ucheb. zav. Radiofiz.* **4**, 433 and 760.
Kislyakov, A. G.: 1970, *Usp. Fiz. Nauk* **101**, 607; *Soviet Phys. Usp.* **13**, 495 (1971).
Kleczek, J., Krüger, A., and Wallis, G.: 1976, *Phys. Solariterr., Potsdam* **1**, 3.
Kleczek, J., Olmr, J., and Krüger, A.: 1968a, in K. O. Kiepenheuer (ed.), *Structure and Development of Solar Active Region*, IAU Symposium No. 35, D. Reidel Publ. Co., p. 594.
Kleczek, J., Olmr, J., and Krüger, A.: 1968b, *Bull. Astron. Inst. Czech.* **19**, 190.
Kleczek, J. and Olmr. J.: 1967, *Bull. Astron. Inst. Czech.* **18**, 68.
Ko, H. C.: 1973, in R. Ramaty and R. G. Stone (ed.), *High Energy Phenomena on the Sun*, Symposium Proceedings, Goddard Space Flight Center, Greenbelt, Md., p. 198.
Ko, H. C. and Chuang, C. W.: 1973, *Astrophys. Letters* **15**, 125.

Kobrin, M. M.: 1976, *Phys. Solariterr., Potsdam* **2**, 3.
Kobrin, M. M. and Korshunov, A. I.: 1972, *Solar Phys.* **25**, 339.
Kobrin, M. M., Korhunov, A. I., Arbusov, S. I., Pakhomov, V. V., and Fridman, V. M.: 1976, *Astron. Zh.* **53**, 789.
Kobrin, M. M., Pakhomov, V. V., and Prokofeva, N. A.: 1976, *Solar Phys.* **50**, 113.
Kolomenskij, A. A.: 1953, *Zh. Eksperim. Teor. Fiz.* **24**, 167.
Komelkov, V. S.: 1962, *Dokl. Akad. Nauk* **146**, 58; *Soviet Phys. Dokl.* **7**, 779 (1963).
Komesaroff, M.: 1958, *Australian J. Phys.* **11**, 201.
Kopecký, M.: 1956, *Bull. Astron. Inst. Czech.* **8**, 71.
Kopecký, M.: 1967, *Advan. Astron. Astrophys.* **5**, 189.
Kopecký, M.: 1971, *Bull. Astron. Inst. Czech.* **22**, 343.
Kopecký, M. and Křivský, L.: 1966, *Bull. Astron. Inst. Czech.* **17**, 360.
Kopp, R. A. and Orrall, F. Q.: 1976, *Astron. Astrophys.* **53**, 363.
Kopp, R. A. and Pneuman, G. W.: 1976, *Solar Phys.* **50**, 85.
Korchak, A. A.: 1963, *Astron. Zh.* **40**, 994.
Korchak, A. A.: 1965, *Geomagnetizm i Aeronomiya* **5**, 32 and 601.
Korchak, A. A.: 1967, *Astron. Zh.* **44**, 328; *Soviet Astron.* **11**, 258.
Korchak, A. A.: 1971, *Solar Phys.* **18**, 284.
Korchak, A. A. and Syrovatskij, S. I.: 1961, *Astron. Zh.* **38**, 885.
Korolkov, D. V., Soboleva, N. S., and Gelfreikh, G. B.: 1960, *Izv. Glav. Astron. Obs. Pulkovo* **21**, No. 164, 81.
Korolev, O. S.: 1973, *Astron. Zh.* **50**, 817.
Korolev, O. S., Markeev, A. K., Fomichev, V. V., and Chertok, I. M.: 1973, *Astron. Zh.* **50**, 1233.
Kovalev, V. A. and Korolev, O. S.: 1976, *Astron. Zh.* **53**, 130.
Kramers, H. A.: 1923, *Phil. Mag.* **46**, 836.
Krishnan, T. and Mullaly, R. F.: 1962, *Australian J. Phys.* **15**, 87.
Křivský, L.: 1968, in K. O. Kiepenheuer (ed.), *Structure and Development of Solar Active Regions*, IAU Symposium No. 35, D. Reidel Publ. Co., p. 465.
Křivský, L.: 1970, *Acta Phys. Acad. Sci. Hungar.* **29** Suppl. 2, 427.
Křivský, L.: 1976a, *Bull. Astron. Inst. Czech.* **27**, 276.
Křivský, L.: 1976b, *Bull. Astron. Inst. Czech.* **27**, 374.
Křivský, L.: 1977, Czech. Acad. Sci. Astron. Inst. Publ. No. 52.
Křivský, L. and Krüger, A.: 1966, *Bull. Astron. Inst. Czech.* **17**, 243.
Křivský, L. and Krüger, A.: 1973, *Bull. Astron. Inst. Czech.* **24**, 291.
Křivský, L., Valniček, B., Böhme, A., Fürstenberg, F., and Krüger, A.: 1976, *Contr. Astron. Obs. Skalnaté Pleso* **6**, 23.
Krüger, A.: 1961, Doctoral Thesis, Humboldt-Univ. Berlin.
Krüger, A.: 1962a, *Z. Astrophys.* **55**, 137.
Krüger, A.: 1962b, *Mber. DAW Berlin* **4**, 97.
Krüger, A.: 1963, *Astron. Nachr.* **287**, 119.
Krüger, A.: 1966, in L. Mollwo and W. Dittmar (ed.), *Solar Observations during the IQSY*, HHI, Berlin.
Krüger, A.: 1968, Dr. habil. Thesis, Univ. Rostock.
Krüger, A.: 1969a, in A. C. Stickland (ed.), *Annals of the IQSY*, MIT Press, Vol. 3, p. 70.
Krüger, A.: 1969b, *Geodät. Geophys. Veröff. Potsdam* II, **10**, 41.
Krüger, A.: 1969c, *Geodät. Geophys. Veröff. Potsdam* II, **13**, 27.
Krüger, A.: 1970, US Dept. Commerce, Spec. Report UAG–9, Boulder, p. 14.
Krüger, A.: 1972, *Solar Phys.* **27**, 217.
Krüger, A.: 1976a, *Contr. Astron. Obs. Skalnaté Pleso* **6**, 287.
Krüger, A.: 1976b, *Phys. Solariterr., Potsdam* **1**, 7.
Krüger, A., Bumba, V., Howard, R., and Kleczek, J.: 1968, *Bull. Astron. Inst. Czech.* **19**, 180.
Krüger, A., Fomichev, V. V., and Chertok, I. M.: 1972, *Astron. Zh.* **49**, 355.
Krüger, A. and Fürstenberg, F.: 1969, *Solnechnye Dannye, Leningrad* No. 3, 97 and *Solnechno-Zemnaya Fizika, Moscow* **1**, 70.
Krüger, A., Fürstenberg, F., and Fricke, K. H.: 1971, HHI Suppl. Ser. Solar Data Vol. II No. 5, 187, Berlin.
Krüger, A., Krüger, W., and Wallis, G.: 1964, *Z. Astrophys.* **59**, 37.

Krüger, A., Künzel, H., and Vyalshin, G. F.: 1975, *Solnechnye Dannye, Leningrad* No. 7, 87.
Krüger, A. and Olmr, J.: 1973, *Bull. Astron. Inst. Czech.* **24**, 202.
Krüger, A., Taubenheim, J., and Entzian, G.: 1969, in C. de Jager and Z. Švestka (ed.), *Solar Flares and Space Research*, North-Holland Publ. Co., p. 181.
Krüger, A. and Trinkkeller, B.: 1972, *Solar Phys.* **24**, 233.
Kruse, U. W., Marshall, L., and Platt, J. R.: 1956, *Astrophys. J.* **124**, 601.
Kuckes, A. F. and Sudan, R. N.: 1971, *Solar Phys.* **17**, 194.
Kuiper, T. B. H. and Pasachoff, J. M.: 1973, *Solar Phys.* **28**, 187.
Kuijpers, J.: 1974, *Solar Phys.* **36**, 157.
Kuijpers, J.: 1975a, *Solar Phys.* **44**, 173.
Kuijpers, J.: 1975b, *Astron. Astrophys.* **40**, 405.
Kuijpers, J. and Slottje, C.: 1976, *Solar Phys.* **46**, 247.
Kulsrud, R. M. and Ferrara, A.: 1971, *Astrophys. Space Sci.* **12**, 302.
Kundu, M. R.: 1959, *Ann. Astrophys.* **22**, 1.
Kundu, M. R.: 1960, *J. Geophys. Res.* **65**, 3903.
Kundu, M. R.: 1961a, *J. Geophys. Res.* **66**, 4308.
Kundu, M. R.: 1961b, *Astrophys. J.* **134**, 96.
Kundu, M. R.: 1962, *J. Geophys. Res.* **67**, 2695.
Kundu, M. R.: 1963, *Space Sci. Rev.* **2**, 438.
Kundu, M. R.: 1965, *Nature* **205**, 683.
Kundu, M. R.: 1970, *Solar Phys.* **13**, 348.
Kundu, M. R.: 1971a, *Solar Phys.* **21**, 130.
Kundu, M. R.: 1971b, in R. Howard, (ed.), *Solar Magnetic Fields*, IAU Symposium No. 43, D. Reidel Publ. Co., p. 642.
Kundu, M. R.: 1972, *Solar Phys.* **25**, 108.
Kundu, M. R.: 1974, in E. Schanda (ed.), *Proc. of the 4th Meeting of CESRA*, Inst. Appl. Phys., Univ., Berne, p. 21.
Kundu, M. R. and Alissandrakis, C. E.: 1975, *Monthly Notices Roy. Astron. Soc.* **173**, 65.
Kundu, M. R., Allissandrakis, C. E., and Kahler, S. W.: 1976, *Solar Phys.* **50**, 429.
Kundu, M. R., Becker, R. H., and Velusamy, T.: 1974, *Solar Phys.* **34**, 185.
Kundu, M. R. and Erickson, W. C.: 1974, *Solar Phys.* **36**, 179.
Kundu, M. R. and Erickson, W. C.: 1974, *Solar Phys.* **36**, 179.
Kundu, M. R., Erickson, W. C., Jackson, P. D., and Fainberg, J.: 1970, *Solar Phys.* **14**, 394.
Kundu, M. R. and Gergely, T. E.: 1973, *Solar Phys.* **31**, 461.
Kundu, M. R., Gergely, T. E., and Erickson, W. C.: 1977, *Solar Phys.* **53**, 489.
Kundu, M. R. and Haddock, F. T.: 1960, *Nature* **186**, 610.
Kundu, M. R. and Lantos, P.: 1977, *Solar Phys.* **52**, 393.
Kundu, M. R. and Liu, S.-Y.: 1973, *Solar Phys.* **29**, 409.
Kundu, M. R., Liu, S.-Y., and McCullough, T. P.: 1977, *Solar Phys.* **51**, 321.
Kundu, M. R. and McCullough, T. P.: 1972a, *Solar Phys.* **24**, 133.
Kundu, M. R. and McCullough, T. P.: 1972b, *Solar Phys.* **27**, 182.
Kundu, M. R., Roberts, J. A., Spencer, C. L., and Kuiper, J. W.: 1961, *Astrophys. J.* **133**, 255.
Kundu, M. R. and Spencer, C. L.: 1963, *Astrophys. J.* **137**, 572.
Kundu, M. R. and Velusamy, T.: 1974, *Solar Phys.* **34**, 125.
Kundu, M. R., Velusamy, T., and Becker, R. H.: 1974, *Solar Phys.* **34**, 217.
Künzel, H.: 1966, *Astron. Nachr.* **289**, 181.
Kuperus, M.: 1976a, *Solar Phys.* **47**, 79.
Kuperus, M.: 1976b, *Solar Phys.* **47**, 361.
Kuperus, M.: 1969, *Space Sci. Rev.* **9**, 713.
Kuseski, R. A. and Swanson, P. N.: 1976, *Solar Phys.* **48**, 41.
Labrum, N. R.: 1971, *Australian J. Phys.* **24**, 193.
Labrum, N. R. and Smerd, S. F.: 1968, *Proc. Astron. Soc. Australia* **1**, 140.
Labrum, N. R. and Stewart, R. T.: 1970, *Proc. Astron. Soc. Australia* **1**, 316.
Lacombe, C. and Mangeney, A.: 1969, *Astron. Astrophys.* **1**, 325.
Lacombe, C. and Møller-Pedersen, B.: 1971, *Astron. Astrophys.* **15**, 406.
Lamb, F. K.: 1970, *Solar Phys.* **12**, 186.

Landau, L. D.: 1946, *Zh. Eksperim. Teor. Fiz.* **16**, 574; *Soviet Phys. JETP* **10**, 25.
Lang, K. R.: 1974, *Solar Phys.* **36**, 351.
Lang, K. R.: 1977, *Solar Phys.* **52**, 63.
Lantos, P.: 1968, *Ann. Astrophys.* **31**, 105.
Lantos, P.: 1969, *Astron. Astrophys.* **1**, 492.
Lantos, P.: 1972, *Solar Phys.* **22**, 387.
Lantos, P. and Avignon, Y.: 1975, *Astron. Astrophys.* **41**, 137.
Lantos, P. and Kundu, M. R.: 1972, *Astron. Astrophys.* **21**, 119.
Lantos-Jarry, M. F.: 1970, *Solar Phys.* **15**, 40.
Leblanc, Y.: 1969, *Astron. Astrophys.* **1**, 467.
Leblanc, Y.: 1970, *Astron. Astrophys.* **4**, 315.
Leblanc, Y. and de la Noë, J.: 1977, *Solar Phys.* **52**, 133.
Leblanc, Y. and Lecacheux, A.: 1976, *Astron. Astrophys.* **46**, 257.
Leblanc, Y., Kuiper, T. B. H., and Hansen, S. F.: 1974, *Solar Phys.* **37**, 215.
Leblanc, Y. and le Squeren, A. M.: 1969, *Astron. Astrophys.* **1**, 239.
Lecacheux, A. and Rosolen, C.: 1975, *Astron. Astrophys.* **41**, 223.
Lehnert, B.: 1967, *Plasma Phys.* **9**, 301.
Lehnert, B.: 1972, Prepr. TRITA-EPP-72-23, Roy. Inst. of Technology Stockholm.
Leighton, R. B., Noyes, R. W., and Simon, G. W.: 1962, *Astrophys. J.* **135**, 474.
Le Squeren-Malinge, A. M.: 1964, *Ann. Astrophys.* **27**, 183.
Liemohn, H. B.: 1965, *Radio Sci.* **69D**, 741.
Lin, R. P.: 1970a, *Solar Phys.* **12**, 266.
Lin, R. P.: 1970b, *Solar Phys.* **15**, 453.
Lin, R. P.: 1974, *Space Sci. Rev.* **16**, 189.
Lin, R. P. (ed.): 1976, *Solar Phys.* **46**, 433.
Lin, R. P., Evans, L. G., and Fainberg, J.: 1973, *Astrophys. Letters* **14**, 191.
Lingenfelter, R. E., and Ramaty, R.: 1967, *Planet Space Sci.* **15**, 1303.
Linscott, I. R., Erkes, J. W., and Powell, N. R.: 1975, Dudley Obs. Rep. No. 10.
Linsky, J. L.: 1973, *Solar Phys.* **28**, 409.
Little, A. G., and Payne-Scott, R.: 1951, *Australian J. Sci. Res.* **A4**, 489.
Livshits, M. A., Obridko, V. N., and Pikelner, S. B.: 1966, *Astron. Zh.* **43**, 1135.
Livshits, M. A. and Tsytovich, V. N.: 1967, *Zh. Eksperim. Teor. Fiz.* **53**, 1610.
Locke, J. L.: 1976, *J. Roy. Astron. Soc. Can.* **71**, 9.
Loughhead, R. E., Roberts, J. A., and McCabe, M. K.: 1957, *Australian J. Phys.* **10**, 483.
Low, B. D.: 1973, *Astrophys. J.* **181**, 209.
Lybyshev, B. I.: 1977, *Astron. Zh.* **54**, 130.
Lynch, D. K.: 1975, Doctoral Thesis, University of Texas of Austin.
Machado, M. E. and Linsky, J. L.: 1975, *Solar Phys.* **42**, 395.
Machin, K. E. and Smith, F. G.: 1951, *Nature* **168**, 599.
Machin, K. E. and Smith, F. G.: 1952, *Nature* **170**, 319.
Machin, K. E., Ryle, M., and Vonberg, D. D.: 1952, *Proc. Inst. Electr. Engrs. London* **99**, 127.
Magun, A. and Mätzler, C.: 1973, *Solar Phys.* **30**, 489.
Magun, A., Stewart, R. T., and Robinson, R. D.: 1975, *Proc. Astron. Soc. Australia* **2**, 1.
Maisonnier C., Gourlan, C., Luzzi, G., Papagno, L., Pecorella, F., Rager, J. P., Robough, B. V., and Samuelli, M.: 1971, *Plasma Phys. and Contr. Nucl. Fusion Res.* 1971, Vol. 1, Int. Atomic Energy Agency, Vienna, p. 523.
Malinge, A. M.: 1963, *Ann. Astrophys.* **26**, 97.
Malitson, H. H. and Erickson, W. C.: 1966, *Astrophys. J.* **144**, 337.
Malitson, H. H., Fainberg, J., and Stone, R. G.: 1973a, *Astrophys. Letters* **14**, 111.
Malitson, H. H., Fainberg, J., and Stone, R. G.: 1973b, *Astrophys. J. Letters* **183**, L35.
Malville, J. M.: 1962, *Astrophys. J.* **136**, 266.
Malville, J. M.: 1967, *Solar Phys.* **2**, 484.
Malville, J. M. and Smith, S. F.: 1963, *J. Geophys. Res.* **68**, 3181.
Mandelshtam, S. L.: 1965a, *Space Sci. Rev.* **4**, 587.
Mandelshtam, S. L.: 1965b, *Ann. Astrophys.* **28**, 614.
Mangeney, A. and Veltri, P.: 1976, *Astron. Astrophys.* **47**, 181.

Mansfield, V. N.: 1967, *Astrophys. J.* **147**, 672.
Markeev, A. K. and Chernov, G. P.: 1970, *Astron. Zh.* **47**, 1044; *Soviet Astron.* **14**, 835 (1971).
Martres, M., Michard, R., and Soru-Iscovici, I.: 1966, *Ann. Astrophys.* **29**, 245.
Martres, M. J., Michard, R., Soru-Iscovici, I., and Tsap, T.: 1968, *Solar Phys.* **5**, 187.
Martres, M., Pick, M., and Parks, G. K.: 1970, *Solar Phys.* **15**, 58.
Martres, M. J., Pick, M., Soru-Escaut, I., and Axisa, F.: 1972, *Nature Phys. Sci.* **236**, 25.
Martyn, D. F.: 1946, *Nature* **158**, 308 and 632.
Martyn, D. F.: 1947, *Nature* **159**, 26.
Matsuura, O. T.: 1969, *Solar Phys.* **9**, 173.
Matsuura, O. T. and Nave, M. F. F.: 1971, *Solar Phys.* **16**, 417.
Matsuura, O. T. and Nave, M. F. F.: 1971, *Solar Phys.* **16**, 417.
Mattoo, S. K. and Bhonsle, R. V.: 1974, *Solar Phys.* **38**, 223.
Mätzler, C.: 1973, *Solar Phys.* **32**, 241.
Mätzler, Ch.: 1976, *Solar Phys.* **49**, 117.
Maxwell, A.: 1963, *Planet. Space Sci.* **11**, 897.
Maxwell, A.: 1965, in D de Jager (ed.), *The Solar Spectrum*, D. Reidel Publ. Co., p. 342.
Maxwell, A. and Fitzwilliam, J.: 1973, *Astrophys. Letters* **13**, 237.
Maxwell, A., Defouw, R. J., and Cummings, P.: 1964, *Planet Space Sci.* **12**, 435.
Maxwell, A., Hughes, M. P., and Thomson, A. R.: 1963, *J. Geophys. Res.* **68**, 1347.
Maxwell, A. and Swarup, G.: 1958, *Nature* **181**, 36.
Maxwell, A. and Thompson, A. R.: 1962, *Astrophys. J.* **135**, 138.
Mayer, C. H., McCullough, T. P., and Sloanaker, R. M.: 1958, *Astrophys. J.* **127**, 1 and 11.
Mayfield, E. B., Higman, J., and Samson, C.: 1970, *Solar Phys.* **13**, 372.
McCracken, K. G. and Rao, U. R.: 1970, *Space Sci. Rev.* **11**, 155.
McCready, L. L., Pawsey, J. L., and Payne-Scott, R.: 1947, *Proc. Roy. Soc. London* **A190**, 357.
McDonald, F. B., Cline, T. L., and Simnett, G. M.: 1972, *J. Geophys. Res.* **77**, 2213.
McKenzie, J. F.: 1964, *Proc. Phys. Soc. London* **84**, 269.
McKenzie, J. F.: 1966, *Proc. Phys. Soc. London* **87**, 349.
McKenzie, J. F.: 1967, *Proc. Phys. Soc. London* **91**, 532 and 537.
McLean, D. J.: 1959, *Australian J. Phys.* **12**, 404.
McLean, D. J.: 1967, *Proc. Astron. Soc. Australia* **1**, 47.
McLean, D. J.: 1969, *Proc. Astron. Soc. Australia* **1**, 188.
McLean, D. J.: 1970, *Proc. Astron. Soc. Australia* **1**, 315.
McLean, D. J.: 1971, *Australian J. Phys.* **24**, 201.
McLean, D. J.: 1973, *Proc. Astron. Soc. Australia* **2**, 222.
McLean, D. J.: 1974, in G. Newkirk, Jr. (ed.), *Coronal Disturbances*, IAU Symposium No. 57, D. Reidel Publ. Co., p. 301.
McLean, D. J. and Sheridan, K. V.: 1972, *Solar Phys.* **26**, 176.
McLean, D. J. and Sheridan, K. V.: 1973, *Solar Phys.* **32**, 485.
McLean, D. J., Sheridan, K. V., Stewart, R. T., and Wild, J. P.: 1971, *Nature* **234**, 140.
Melrose, D. B.: 1968, *Astrophys. Space Sci.* **2**, 171.
Melrose, D. B.: 1970, *Australian J. Phys.* **23**, 871.
Melrose, D. B.: 1971a, *Astrophys. Space Sci.* **10**, 186.
Melrose, D. B.: 1971b, *Astrophys. Space Sci.* **10**, 197.
Melrose, D. B.: 1973, *Proc. Astron. Soc. Australia* **2**, 208.
Melrose, D. B.: 1974a, *Solar Phys.* **34**, 421.
Melrose, D. B.: 1974b, *Solar Phys.* **35**, 441.
Melrose, D. B.: 1974c, *Solar Phys.* **37**, 353.
Melrose, D. B.: 1974d, *Solar Phys.* **38**, 205.
Melrose, D. B.: 1974e, *Australian J. Phys.* **27**, 31 and 43.
Melrose, D. B.: 1975a, *Solar Phys.* **43**, 79.
Melrose, D. B.: 1975b, *Solar Phys.* **43**, 211.
Melrose, D. B.: 1975c, *Australian J. Phys.* **28**, 101.
Melrose, D. B. and Sy, W.: 1971, *Proc. Astron. Soc. Australia* **2**, 56.
Mercier, C.: 1973, *Solar Phys.* **33**, 177.
Mercier, C.: 1975, *Solar Phys.* **45**, 169.

Mercier, C. and Rosenberg, H.: 1974, *Solar Phys.* **39**, 193.
Mewe, R.: 1972, *Solar Phys.* **22**, 459.
Meyer, F.: 1968, in K. O. Kiepenheuer (ed.), *Structure and Development of Solar Active Regions*, IAU Symposium No. 35, D. Reidel Publ. Co., p. 485.
Meyer, P. and Vogt, R. E.: 1961, *Phys. Rev. Letters* **6**, 193.
Michard, R. and Ribes, E.: 1968, in K. O. Kiepenheur (ed.), *Structure and Development of Solar Active Regions*, IAU Symposium No. 35, D. Reidel Publ. Co., p. 420.
Michel, H.-St.: 1965, *Mber DAW Berlin* **7**, 763.
Milkey, R. W.: 1971, *Solar Phys.* **16**, 465.
Mills, B. Y. and Little, A. G.: 1953, *Australian J. Phys.* **6**, 272.
Milne, E. A.: 1921, *Monthly Notices Roy. Astron. Soc.* **81**, 361.
Minnaert, M.: 1959, in R. Righini (ed.), *Radioastronomia Solare*, Rendiconti S.I.F., XII, Bologna, p. 1.
Minnaert, M.: 1965, in C. de Jager (ed.), *The Solar Spectrum*, D. Reidel Publ. Co., p. 3.
Mogilevskij, E. I.: 1976, Preprint No. 5a IZMIRAN, Moscow.
Mogilevskij, E. I. and Akinyan, S. T.: 1961, *Geomagnetizm i Aeronomiya* **1**, 921.
Moiseev, I. G.: 1977, *Izv. Krymsk. Astrofiz. Obs.* **56**, 100.
Molchanov, A. P.: 1961, *Astron. Zh.* **38**, 849.
Møller-Pedersen, B.: 1974, *Astron. Astrophys.* **37**, 163.
Mollwo, L.: 1957, *Archiv elektr. Übertr.* **11**, 295.
Mollwo, L.: 1965, *Beitr. Plasmaphys.* **5**, 123.
Mollwo, L.: 1969, Private Communication.
Mollwo, L.: 1970, *Solar Phys.* **12**, 125.
Mollwo, L.: 1971, *Solar Phys.* **19**, 128.
Mollwo, L.: 1973, *Solar Phys.* **30**, 497.
Mollwo, L. and Sauer, K.: 1977, *Solar Phys.* **51**, 435.
Molodenskij, M. M. and Korchak, A. A.: 1966, *Geomagnetizm i Aeronomiya* **6**, 512.
Moreton, G. E.: 1964, *Astron. J.* **69**, 145.
Moreton, G. E. and Severnij, A. B.: 1968, *Solar Phys.* **3**, 282.
Morimoto, M. and Kai, K.: 1962, *J. Phys. Soc. Japan* **17**, Suppl. A2, 220.
Moriyama, F.: 1961, *Ann. Tokyo Astron. Obs.* **7**, 132.
Mosier, S. R. and Fainberg, J.: 1975, *Solar Phys.* **40**, 501.
Mullaly, R. F. and Watkinson, A.: 1968, *Nature* **218**, 539.
Mullan, D. J.: 1976, *Astron. Astrophys.* **52**, 305.
Mullan, D. J.: 1977, *Solar Phys.* **54**, 183.
McMullin, J. N. and Helfer, H. L.: 1977, *Solar Phys.* **53**, 471.
Mustel, E. R.: 1966, *Astron. Zh.* **43**, 1121.
Nagata, T.: 1975, *Space Sci. Rev.* **17**, 205.
Nakada, M. P., Neupert, W. M., and Thomas, R. J.: 1974, *Solar Phys.* **37**, 429.
Nakagawa, Y.: 1973, *Solar Phys.* **33**, 87.
Nakagawa, Y. and Raadu, M. A.: 1972, *Solar Phys.* **25**, 127.
Nakagawa, Y., Raadu, M. A., Billings, D. E., and McNamara, D.: 1971, *Solar Phys.* **19**, 72.
Nakagawa, Y., Wu, S. T., and Han, S. M.: 1973, *Solar Phys.* **30**, 111.
Nakagawa, Y., Wu, S. T., and Tandberg-Hanssen, E.: 1975, *Solar Phys.* **41**, 387.
Nelson, G. J. and Robinson, R. D.: 1975, *Proc. Astron. Soc. Australia* **2**, 6.
Neupert, W. M.: 1967, Preprint NASA X-614-67-90 Greenbelt, Md., *Solar Phys.* **2**, 294.
Neupert, W. M.: 1969, *Ann. Rev. Astron. Astrophys.* **7**, 121.
Neupert, W. M.: 1971, *Phil. Trans. Roy. Soc. London* **A270**, 143.
Newkirk, G., Jr.: 1961, *Astrophys. J.* **133**, 983.
Newkirk, G., Jr.: 1967, *Ann. Rev. Astron. Astrophys.* **5**, 213.
Newkirk, G., Jr.: 1971, in R. Howard (ed.), *Solar Magnetic Fields*, IAU Symposium No. 43, D. Reidel Publ. Co., p. 547.
Newkirk, G., Jr., Trotter, D. E., Altschuler, M. D., and Howard, R.: 1972, *Solar Phys.* **24**, 370.
Newman, C. E.: 1973, *J. Math. Phys.* **14**, 502.
Newman, C. E. and Petrosian, V.: 1975, *Phys. Fluids* **18**, 547.
Neylan, A. A.: 1959, *Australian J. Phys.* **12**, 399.
Nishida, A. and Nagayama, N.: 1973, *J. Geophys. Res.* **78**, 3782.

Noci, G.: 1971, *Solar Phys.* **17**, 369.
Noyes, R. W.: 1967, in R. N. Thomas (ed.), *Aerodynamic Phenomena in Stellar Atmospheres*, IAU Symposium No. 28, Academic Press, New York, San Francisco, London, p. 293.
Noyes, R. W. and Hall, D. N. B.: 1972, *Astrophys. J. Letters* **176**, L89.
Obayashi, T.: 1975, *Space Sci. Rev.* **17**, 195.
O'Brien, P. A. and Bell, C. J.: 1954, *Nature* **173**, 219.
Olmr, J., Tlamicha, A., Fürstenberg, F., and Krüger, A.: 1967, *Bull. Astron. Inst. Czech.* **18**, 70.
Oort, J. H. and Walraven, T.: 1956, *Bull. Astron. Inst. Neth.* **12**, 285.
Oster, L.: 1959, *Z. Astrophys.* **47**, 169.
Oster, L.: 1961a, *Rev. Mod. Phys.* **33**, 525.
Oster, L.: 1961b, *Astrophys. J.* **134**, 1010.
Oster, L.: 1964, *Phys. Fluids* **7**, 263.
Oster, L. and Sofia, S.: 1966, *Astrophys. J.* **143**, 944.
Osterbrock, D. E.: 1961, *Astrophys. J.* **134**, 347.
Pallavicini, R. and Vaiana, G. S.: 1976, *Solar Phys.* **49**, 297.
Palmer, I. D.: 1974, *Solar Phys.* **37**, 443.
Palmer, I. D. and Lin, R. P.: 1972, *Proc. Astron. Soc. Australia* **2**, 101.
Palmer, I. D. and Smerd, S. F.: 1972, *Solar Phys.* **26**, 460.
Palmer, I. D., Smerd, S. F., and Riddle, A. C.: 1972, *Proc. Astron. Soc. Australia* **2**, 103.
Papagiannis, M. D., Zerefos, C. S., and Repais, C. C.: 1972, *Solar Phys.* **27**, 208.
Parijskij, Yu. N., Korolkov, L. V., Shivris, O. N., Kajdanovskij, N. L., Esepkina, N. A., Zverev, Yu. K., Stotskij, A. A., Akhmedov, Sh. B., Bogod, V. M., Boldyrev, S. I., Gelfreikh, G. B., Ipatova, I. A., Korzhavin, A. N., and Romantsov, V. V.: 1976, *Astron. Zh.* **53**, 1017.
Parker, E. N.: 1957, *Phys. Rev.* **107**, 830.
Parker, E. N.: 1958, *Phys. Rev.* **109**, 1328.
Parker, E. N.: 1960, *Astrophys. J.* **132**, 1445.
Parker, E. N.: 1963, *Astrophys. J. Suppl.* **8**, 177.
Parker, E. N.: 1964a, *Astrophys. J.* **140**, 1170.
Parker, E. N.: 1964b, in H. Odishaw (ed.), *Research in Geophysics*, Vol. 1, M.I.T. Press Cambridge, Mass., p. 99.
Parker, E. N.: 1965, *Space Sci. Rev.* **4**, 666.
Parker, E. N.: 1969, *Space Sci. Rev.* **9**, 325.
Parker, E. N.: 1970, *Astrophys. J.* **160**, 383.
Parker, E. N.: 1973a, *J. Plasma Phys.* **9**, 49.
Parker, E. N.: 1973b, *Astrophys. J.* **180**, 247.
Parks, G. K. and Winckler, J. R.: 1969, *Astrophys. J. Letters* **155**, L117.
Parks, G. K. and Winckler, J. R.: 1971, *Solar Phys.* **16**, 186.
Pawsey, J. L.: 1946, *Nature* **158**, 633.
Pawsey, J. L. and Smerd, S. F.: 1953, in G. P. Kuiper (ed.), *The Sun*, Univ. Chicago Press, p. 466.
Payne-Scott, R. and Little, A. G.: 1951, *Australian J. Sci. Res.* **A4**, 508.
Payne-Scott, R., Yabsley, D. E., and Bolton, J. G.: 1947, *Nature* **160**, 256.
Perry, R. M. and Altschuler, M. D.: 1973, *Solar Phys.* **28**, 435.
Peterova, N. G. and Akhmedov, S. B.: 1973, *Astron. Zh.* **50**, 1220.
Peterson, A. M. and Hower, G. L.: 1963, *J. Geophys. Res.* **68**, 723.
Peterson, A. M. and Hower, G. L.: 1966, in B. M. McCormac (ed.), *Radiation Trapped in the Earth Magnetic Field*, D. Reidel Publ. Co., p. 714.
Peterson, L. E. and Winckler, J. R.: 1959, *J. Geophys. Res.* **64**, 697.
Petschek, H. E.: 1964, in N. Hess (ed.), *AAS-NASA Symposium on the Physics of Solar Flares*, NASA, Washington, p. 425.
Philipp, W. G. and Benz, A. O.: 1977, *Astron. Astrophys.* **56**, 39.
Pick, M.: 1961, *Ann. Astrophys.* **24**, 183.
Pick, M.: 1965, in J. Aarons (ed.), *Solar System Radio Astronomy*, Plenum Press, New York, p. 81.
Pick, M.: 1966, *Ann. Géophys.* **22**, 310.
Pick, M., Martres, M. J., Axisa, F., and Mercier, C.: 1975, *Solar Phys.* **42**, 461.
Piddington, J. H.: 1954, *Astrophys. J.* **119**, 531.
Piddington, J. H.: 1967, *Planetary Space Sci.* **15**, 733.

Piddington, J. H.: 1973, *Solar Phys.* **31**, 229.
Piddington, J. H.: 1974, *Solar Phys.* **38**, 465.
Piddington, J. H. and Minnett, H. C.: 1951, *Australian J. Sci. Res.* **A4**, 131.
Pikelner, S. B.: 1965, in A. Beer (ed.), *Vistas in Astronomy*, Vol. 6, Pergamon Press, p. 131.
Pikelner, S. B. and Ginzburg, M. A.: 1963, *Astron. Zh.* **40**, 842; *Soviet Astron.* **7**, 639 (1964).
Pines, D.: 1956, *Rev. Mod. Phys.* **28**, 184.
Pintér, Š.: 1969, *Solar Phys.* **8**, 149.
Pintér, Š.: 1972a, *Bull. Astron. Inst. Czech.* **23**, 69.
Pintér, Š.: 1972, in G. E. Kocharov and V. A. Dergacheva (ed.), *Proc. of the IVth International Seminar on Particle Acceleration in Different Scales of the Cosmos*. Academy of Sciences of the USSR, Phys.-Techn. A. F. Ioffe-Institute and Inst. of Nuclear Phys., Moscow State University, Leningrad, p. 63.
Pintér, Š. and Olmr, J.: 1970, *Bull. Astron. Inst. Czech.* **21**, 373.
Pneuman, G. W.: 1966, *Astrophys. J.* **145**, 242.
Pneuman, G. W.: 1967, *Solar Phys.* **2**, 462.
Pneuman, G. W.: 1968, *Solar Phys.* **3**, 578.
Pneuman, G. W.: 1969, *Solar Phys.* **6**, 255.
Pneuman, G. W.: 1973, *Solar Phys.* **28**, 247.
Pneuman, G. W. and Kopp, R. A.: 1970, *Solar Phys.* **13**, 176.
Pneuman, G. W. and Kopp, R. A.: 1971, *Solar Phys.* **18**, 258.
Poquérusse, M.: 1977, *Astron. Astrophys.* **56**, 251.
Post, R. F. and Rosenbluth, M. N.: 1965, *Phys. Fluids* **9**, 730.
Pottasch, S. R.: 1964, *Space Sci. Rev.* **3**, 816.
Pounds, K. A.: 1965, *Ann. Astrophys.* **28**, 132.
Priese, J.: 1969, *Solar Phys.* **9**, 235.
Priese, J.: 1972, Dr. sc. Thesis, Akademie der Wissenschaften der D.D.R., Berlin.
Priest, E. R.: 1973, *Astrophys. J.* **181**, 227.
Priest, E. R. and Cowley, S. W. H.: 1975, *J. Plasma Phys.* **14**, 271.
Priest, E. R. and Heyvaerts, J.: 1974, *Solar Phys.* **36**, 433.
Priest, E. R. and Raadu, M. A.: 1975, *Solar Phys.* **43**, 177.
Priest, E. R. and Smith, D. F.: 1972, *Astrophys. Letters* **12**, 25.
Priest, E. R. and Soward, A. M.: 1975, Paper presented at the IAU Symposium No. 71 on 'Basic Mechanisms of Solar Activity', Prague, August 1975.
Priester, W. and Rosenberg, J.: 1965, NASA Techn. Note D-2888, Washington, D.C.
Pustilnik, L. A.: 1973, *Astron. Zh.* **50**, 1211.
Raadu, M. A. and Nakagawa, Y.: 1971, *Solar Phys.* **20**, 64.
Rabben, M.: 1962, *Z. Astrophys.* **55**, 73.
Rädler, K.-H.: 1968, *Z. Naturforsch.* **23a**, 1841.
Ramaty, R.: 1968a, *J. Geophys. Res.* **72**, 879.
Ramaty, R.: 1968b, *J. Geophys. Res.* **73**, 3573.
Ramaty, R.: 1969a, *Astrophys. J.* **158**, 753.
Ramaty, R.: 1969b, *Astrophys. Letters* **4**, 43.
Ramaty, R.: 1973, in R. Ramaty and R. G. Stone (ed.), *High Energy Phenomena on the Sun*, Symposium Proceedings, Goddard Space Flight Center, Greenbelt, Md., p. 188.
Ramaty, R. and Holt, S. S.: 1970, *Nature* **226**, 68.
Ramaty, R. and Lingenfelter, R. E.: 1966, *J. Geophys. Res.* **71**, 3687.
Ramaty, R. and Lingenfelter, R. E.: 1967, *J. Geophys. Res.* **72**, 879.
Ramaty, R. and Lingenfelter, R. E.: 1968, *Solar Phys.* **5**, 531.
Ramaty, R. and Lingenfelter, R. E.: 1975, in S. R. Kane (ed.), *Solar Gamma-, X-, and EUV Radiation*, IAU Symposium No. 68, D. Reidel Publ. Co., p. 363.
Ramaty, R., Kozlovsky, B., and Lingenfelter, R. E.: 1975, *Space Sci. Rev.* **18**, 341.
Ramaty, R. and Petrosian, V.: 1972, *Astrophys. J.* **178**, 241.
Rao, U. R.: 1971, *Space Sci. Rev.* **12**, 719.
Rao, U. V. G.: 1965, *Australian J. Phys.* **18**, 283.
Rao, U. V. G.: 1970, *Solar Phys.* **14**, 389.
Rawer, K. and Suchy, K.: 1967, in S. Flügge (ed.), *Handbuch der Physik*, Vol. 49, Geophysics III.2, Springer-Verlag Berlin, Heidelberg, New York, p. 1.

Razin, V. A.: 1957, Thesis, Gorky State Univ.
Razin, V. A.: 1960a, *Izv. Vys. Ucheb. Zav. Radiofiz.* **3**, 584.
Razin, V. A.: 1960b, *Izv. Vys. Ucheb. Zav. Radiofiz.* **3**, 921.
Reber, G.: 1944, *Astrophys. J.* **100**, 279.
Riddle, A. C.: 1969, *Solar Phys.* **7**, 434.
Riddle, A. C.: 1970a, *Solar Phys.* **13**, 448.
Riddle, A. C.: 1970b, *Proc. Astron. Soc. Australia* **1**, 310.
Riddle, A. C.: 1970c, *Proc. Astron. Soc. Australia* **1**, 375.
Riddle, A. C.: 1972a, *Proc. Astron. Soc. Australia* **2**, 98.
Riddle, A. C.: 1972b, *Proc. Astron. Soc. Australia* **2**, 148.
Riddle, A. C.: 1974a, *Solar Phys.* **34**, 181.
Riddle, A. C.: 1974b, *Solar Phys.* **35**, 153.
Riddle, A. C.: 1974c, *Solar Phys.* **36**, 375.
Riddle, A. C.: 1975, *Astron. Astrophys.* **38**, 153.
Riddle, A. C. and Sheridan, K. V.: 1971, *Proc. Astron. Soc. Australia* **2**, 62.
Riddle, A. C., Tandberg-Hanssen, E., and Hansen, R.: 1974, *Solar Phys.* **35**, 171.
Roberts, J. A.: 1958, *Australian J. Phys.* **11**, 215.
Roberts, J. A.: 1959a, in R. N. Bracewell (ed.), *Paris Symposium on Radio Astronomy*, Stanford Univ. Press, p. 194.
Roberts, J. A.: 1959b, *Australian J. Phys.* **12**, 327.
Robinson, R. D.: 1974, *Proc. Astron. Soc. Australia* **2**, 258.
Robinson, R. D. and Smerd, S. F.: 1975, *Proc. Astron. Soc. Australia* **2**, 1.
Roederer, J. G.: 1964, *Space Sci. Rev.* **3**, 487.
Romanchuk, P. R. and Krivodubskij, V. N.: 1974, Kiev State University, Astron. Obs., Preprint No. 2 (in russ.).
Roosen, J.: 1969, *Solar Phys.* **8**, 204.
Roosen, J. and Goh, T.: 1967, *Solar Phys.* **1**, 242.
Rosenberg, H.: 1970, *Astron. Astrophys.* **9**, 159.
Rosenberg, H.: 1971a, in A. Abrami (ed.), *Proc. of the 2nd Meeting of CESRA*, Astron. Obs., Trieste, p. 59.
Rosenberg, H.: 1971b, in R. Howard (ed.), *Solar Magnetic Fields*, IAU Symposium No. 43, D. Reidel Publ. Co., p. 652.
Rosenberg, H.: 1972, *Solar Phys.* **25**, 188.
Rosenberg, H.: 1975, *Solar Phys.* **42**, 247.
Rosenberg, H.: 1976, *Phil. Trans. Roy. Soc. London* **281**, 461.
Rugge, H. R. and Walker, A. B. C.: 1971, *Solar Phys.* **18**, 244.
Rumyantsev, A. A. and Zotov, A. A.: 1976, *Astron. Zh.* **53**, 560.
Rust, D. M.: 1972, *Solar Phys.* **25**, 141.
Ryle, M.: 1952, *Proc. Roy. Soc. London* **A211**, 351.
Ryle, M. and Hewish, A.: 1960, *Monthly Notices Roy. Astron. Soc.* **120**, 220.
Ryle, M. and Vonberg, D. D.: 1946, *Nature* **158**, 339.
Ryle, M. and Vonberg, D. D.: 1948, *Proc. Roy. Soc. London* **A193**, 98.
Sagdeev, R. Z.: 1962, *Soviet Phys. Techn. Phys.* **6**, 867.
Sagdeev, R. Z. and Shafranov, V. D.: 1960, *Zh. Eksperim. Teor. Fiz.* **39**, 181.
Saito, K.: 1972, Ann. Tokyo Astron. Obs., 2nd Ser., **13**, 93.
Sakurai, K.: 1970, *J. Geophys. Res.* **75**, 225.
Sakurai, K.: 1971a, *Solar Phys.* **16**, 125.
Sakurai, K.: 1971b, *Solar Phys.* **16**, 198.
Sakurai, K.: 1972a, Preprint NASA X-693-72-45, Greenbelt, Md.
Sakurai, K.: 1972b, Preprint NASA X-693-72-82, Greenbelt, Md.
Sakurai, K.: 1972c, *Astrophys. J.* **174**, 135.
Sakurai, K.: 1973, *Solar Phys.* **31**, 483.
Sakurai, K.: 1974, *Solar Phys.* **36**, 171.
Sakurai, K. and Stone, R.: 1971, *Solar Phys.* **19**, 247.
Sakurai, T. and Uchida, Y.: 1977, *Solar Phys.* **52**, 97.
Salomonovich, A. E.: 1960, *Astron. Zh.* **37**, 969; *Soviet Astron.* **4**, 904 (1961).
Santin, P.: 1971, *Solar Phys.* **18**, 87.

Santin, P.: 1973, *Solar Phys.* **30**, 159.
Santin, P.: 1976, *Astron. Astrophys.* **49**, 193.
Sastry, C. V.: 1972, *Astrophys. Letters* **11**, 47.
Sastry, C. V.: 1973, *Solar Phys.* **28**, 197.
Sauer, K.: 1972, in R. Rompe and M. Steenbeck (ed.), *Ergebnisse der Plasmaphysik und der Gaselektronik*, Akademie-Verlag Berlin, Vol. 3, p. 253.
Sawant, H. S., Alurkar, S. K., and Bhonsle, R. V.: 1975, *Nature* **253**, 339.
Sawant, H. S., Bhonsle, R. V., and Alurkar, S. K.: 1976, *Solar Phys.* **50**, 481.
Sawyer, C.: 1977, *Solar Phys.* **51**, 203.
Sazonov, V. N.: 1969a, *Astron. Zh.* **46**, 502.
Sazonov, V. N.: 1969b, *Zh. Eksperim. Teor. Fiz.* **56**, 1075.
Sazonov, V. N.: 1969c, *Astron. Zh.* **46**, 1016.
Sazonov, V. N. and Tsytovich, V. N.: 1968, *Izv. Vys. Ucheb. Zav. Radiofiz.* **11**, 1287.
Scalise, E., Jr., Basu, D., and Marques Dos Santos, P.: 1971, *Astron. Astrophys.* **13**, 471.
Scarf, F. L., Bernstein, W., and Fredricks, R. W.: 1965, *J. Geophys. Res.* **70**, 9.
Schatten, K. H.: 1970, *Solar Phys.* **11**, 236.
Schatten, K. H., Wilcox, J. M., and Ness, N. F.: 1969, *Solar Phys.* **6**, 442.
Schatzman, E.: 1966, *Compt. Rend. Acad. Sci. Paris* **262**, 93.
Schatzman, E.: 1967, *Solar Phys.* **1**, 411.
Schatzman, E. and Souffrin, P.: 1967, *Ann. Rev. Astron. Astrophys.* **5**, 67.
Scheffler, H.: 1958, *Z. Astrophys.* **45**, 113.
Scheffler, H.: 1959, *Astron. Nachr.* **285**, 153.
Scherrer, P. H. and El-Raey, M.: 1974, *Solar Phys.* **35**, 361.
Scheuer, P. A. G.: 1960, *Monthly Notices Roy. Astron. Soc.* **120**, 231.
Scheuer, P. A. G.: 1968, *Astrophys. J. Letters* **151**, L139.
Schmahl, E. J.: 1973, *Australian J. Phys. Ap. Suppl.* **29**, 1.
Schmidt, H.U.: 1964, in W. N. Hess (ed.), *AAS-NASA Symposium on the Physics of Solar Flares*, NASA, Washington, p. 107.
Schmidt, H. U.: 1966, in M. Cimino (ed.), *Atti del convegno sui campi magnetici solari*, G. Barbèra Edit., Firenze, p. 233.
Schott, E.: 1947, *Phys. Blätter* **3**, 159.
Schwarzschild, K.: 1914, *Sitz. Ber. Preuß. Akad. Wiss.* 1183.
Schwinger, J.: 1949, *Phys. Rev.* **75**, 1912.
Seehafer, N.: 1975, *Astron. Nachr.* **296**, 177.
Sen, H. K. and White, M. L.: 1972, *Solar Phys.* **23**, 146.
Sengupta, P. R.: 1971, *Solar Phys.* **17**, 160.
Sentman, D. D. and Shawhan, S. D.: 1974, *Solar Phys.* **35**, 83.
Seshadri, S. R. and Tuan, H. S.: 1965, *Radio Sci.* **69D**, 767.
Severnij, A. B.: 1958a, *Izv. Krymsk. Astrofiz. Obs.* **19**, 72.
Severnij, A. B.: 1958b, *Izv. Krymsk. Astrofiz. Obs.* **20**, 22.
Severnij, A. B.: 1964, *Ann. Rev. Astron. Astrophys.* **2**, 363 and *Space Sci. Rev.* **3**, 451.
Severnij, A. B.: 1966, *Astron. Zh.* **43**, 465.
Severnij, A. B. and Shabanskij, V. P.: 1960, *Astron. Zh.* **37**, 609.
Severnij, A. B. and Shabanskij, V. P.: 1961, *Izv. Krymsk. Astrofiz. Obs.* **25**, 88.
Sheridan, K. V.: 1967, *Proc. Astron. Soc. Australia* **1**, 58.
Sheridan, K. V.: 1968, *Proc. Astron. Soc. Australia* **1**, 138.
Sheridan, K. V.: 1970a, *Proc. Astron. Soc. Australia* **1**, 304.
Sheridan, K. V.: 1970b, *Proc. Astron. Soc. Australia* **1**, 376.
Sheridan, K. V.: 1971, *Proc. Astron. Soc. Australia* **2**, 57.
Sheridan, K. V., Labrum, N. R., and Payten, W. J.: 1972, *Nature Phys. Sci.* **238**, 115.
Sheridan, K. V. and McLean, D. J.: 1971, *Applied Optics* **10**, 2427.
Sheridan, K. V., McLean, D. J., and Smerd, S. F.: 1973, *Astrophys. Letters* **15**, 139.
Shimabukuro, F. I.: 1970, *Solar Phys.* **12**, 438.
Shimabukuro, F. I.: 1970, *Solar Phys.* **15**, 424.
Shimabukuro, F. I.: 1971, *Solar Phys.* **18**, 247.
Shimabukuro, F. I., Chapman, G. A., and Mayfield, E. B.: 1973, *Solar Phys.* **30**, 163.

Shiomi, Y.: 1969, *Solar Phys.* **6**, 276.
Shmeleva, O. P. and Syrovatskij, S. I.: 1973, *Solar Phys.* **33**, 341.
Shuter, W. L. H.: 1976, *Solar Phys.* **48**, 85.
Shuter, W. L. H. and McCutcheon, W. H.: 1973, *Nature Phys. Sci.* **241**, 140.
Simnett, G. M.: 1971, *Solar Phys.* **20**, 448.
Simnett, G. M.: 1974, *Space Sci. Rev.* **16**, 257.
Simnett, G. M. and Holt, S. S.: 1971, *Solar Phys.* **16**, 208.
Simon, A.: 1969, in D. G. Wentzel and D. A. Tidman (ed.), *Plasma Instabilities in Astrophysics*, Gordon and Breach, New York, London, Paris, p. 49.
Simon, M.: 1969a, *Astrophys. J.* **156**, 341.
Simon, M.: 1969b, *Astrophys. Letters* **3**, 23.
Simon, M.: 1971, *Solar Phys.* **21**, 297.
Simon, M. and Shimabukuro, F. I.: 1971, *Astrophys. J.* **168**, 525.
Simon, P.: 1960, *Ann. Astrophys.* **23**, 102.
Sitenko, A. G. and Stepanov, K. N.: 1956, *Zh. Eksperim. Teor. Fiz.* **31**, 642.
Slee, O. B.: 1961, *Monthly Notices Roy. Astr. Soc.* **123**, 223.
Slee, O. B.: 1966, *Planet. Space Sci.* **14**, 255.
Slonim, Yu. M.: 1968, *Astron. Zh.* **45**, 286 and 726.
Slonim, Yu. M.: 1969, *Astron. Zh.* **46**, 570 and 697.
Slonim, Yu. M.: 1973, *Solar Processes and their Observation* (in russ.) Izdat. 'FAN' Usbek. SSR, Tashkent.
Slottje, C.: 1971, in A. Abrami (ed.), *Proc. of the 2nd Meeting of CESRA*, Astron. Obs., Trieste, p. 88.
Slottje, C.: 1972, *Solar Phys.* **25**, 210.
Slottje, C.: 1974, *Astron. Astrophys.* **32**, 107.
Slysh, V. I.: 1967a, *Astron. Zh.* **44**, 94; *Soviet Astron.* **11**, 72.
Slysh, V. I.: 1967b, *Astron. Zh.* **44**, 487; *Soviet Astron.* **11**, 389.
Slysh, V. I.: 1967c, *Kosm. Issled.* **5**, 897, *Cosmic Res.* **5**, 759.
Smerd, S. F.: 1950, *Australian J. Sci. Res.* **A3**, 34.
Smerd, S. F.: 1965, in C. de Jager (ed.), *The Solar Spectrum*, D. Reidel Publ. Co., p. 398.
Smerd, S. F.: 1968, *Proc. Astron. Soc. Australia* **1**, 124.
Smerd, S. F.: 1970, *Proc. Astron. Soc. Australia* **1**, 305.
Smerd, S. F.: 1971, *Australian J. Phys.* **24**, 229.
Smerd, S. F.: 1976, *Solar Phys.* **46**, 493.
Smerd, S. F. and Dulk, G. A.: 1971, in R. Howard (ed.), *Solar Magnetic Fields*, IAU Symposium No. 43, D. Reidel Publ. Co., p. 616.
Smerd, S. F. and Westfold, K. C.: 1949, *Phil. Mag.* **40**, 831.
Smerd, S. F., Wild, J. P., and Sheridan, K. V.: 1962, *Australian J. Phys.* **15**, 180.
Smith, D. F.: 1970a, *Adv. Astron. Astrophys.* **7**, 147.
Smith, D. F.: 1970b, *Solar Phys.* **13**, 444.
Smith, D. F.: 1970c, *Solar Phys.* **15**, 202.
Smith, D. F.: 1971, *Astrophys. J.* **170**, 559.
Smith, D. F.: 1972, *Solar Phys.* **23**, 191.
Smith, D. F.: 1973, *Solar Phys.* **33**, 213.
Smith, D. F.: 1974a, *Space Sci. Rev.* **16**, 91.
Smith, D. F.: 1974b, *Solar Phys.* **34**, 393.
Smith, D. F.: 1976, Preprint, Max-Planck-Inst. f. Physik und Astrophysik, Garching b. München.
Smith, D. F. and Davies, W. D.: 1975, *Solar Phys.* **41**, 439.
Smith, D. F. and Fung, P. C. W.: 1971, *J. Plasma Phys.* **5**, 1.
Smith, D. F. and Pneuman, G. W.: 1972, *Solar Phys.* **25**, 461.
Smith, D. F. and Priest, E. R.: 1972, *Astrophys. J.* **176**, 487.
Smith, D. F. and Riddle, A. C.: 1975, *Solar Phys.* **44**, 471.
Smith, D. F. and Sturrock, P. A.: 1971, *Astrophys. Space Sci.* **12**, 411.
Smith, E. v. P.: 1968, in Y. Öhman (ed.), *Mass Motions in Solar Flares and Related Phenomena*, Nobel Symp. No. 9, Almquist and Wiksell, Stockholm/John Wiley, New York, London, Sidney, p. 137.
Smith, E. v. P. and Gottlieb, D. M.: 1974, *Space Sci. Rev.* **16**, 771.
Smith, S. F. and Harvey, K. L.: 1971, in C. Macris (ed.), *Physics of the Solar Corona*, D. Reidel Publ. Co., p. 156.

Smith, S. F. and Howard, R.: 1968, in K. O. Kiepenheuer (ed.), *Structure and Development of Solar Active Regions*, IAU Symposium No. 35, D. Reidel Publ. Co., p. 33.
Smith, S. F. and Ramsey, H. E.: 1966, *Astron. J.* **71**, 197.
Smith, R. A. and de la Noë, J.: 1976, *Astrophys. J.* **207**, 605.
Snyder, W. and Helliwell, R. A.: 1952, *J. Geophys. Res.* **57**, 73.
Sofue, Y., Kawabata, K., Takahashi, F., and Kawajiri, N.: 1976, *Solar Phys.* **50**, 465.
Sokolov, A. A., Ternov, I. M., Bagrov, V. G., Galtsov, D. V., and Shukovskij, V. T.: 1968, *Z. Phys.* **211**, 1.
Solar Radio Group Utrecht: 1974, *Space Sci. Rev.* **16**, 45.
Sollfrey, W. and Yura, H. T.: 1965, *Phys. Rev.* **139**, A48.
Somov, B. V.: 1974, *Solar Phys.* **42**, 235.
Somov, B. V.: 1976, *Trudy Fiz. Inst. Lebedeva* **88**, 127.
Somov, B. V., Spektor, A. R., and Syrovatskij, S. I.: 1977, *Izv. Akad. Nauk Ser. Fiz.* **41**, 273.
Somov, B. V. and Syrovatskij, S. I.: 1974, *Trudy Fiz. Inst. Lebedeva* **74**, 14.
Somov, B. V. and Syrovatskij, S. I.: 1976, *Usp. Fiz. Nauk* **120**, 217.
Somov, B. V. and Syrovatskij, S. I.: 1977, Preprint No. 27, P.N. Lebedev Physical Inst., Moscow.
Sonnenrup, B. U. O.: 1970, *J. Plasma Phys.* **4**, 161.
Soper, G. K. and Harris, E. G.: 1965, *Phys. Fluids* **8**, 984.
Southworth, G. C.: 1945, *J. Franklin Inst.* **239**, 285.
Southworth, G. C.: 1956, *Sci. Monthly* **82**, 55.
Soward, A. M. and Priest, E. R.: 1977, *Phil. Trans. Roy. Soc.*, Ser. A, **284**, 370.
Spangler, S. R. and Shawhan, S. D.: 1974, *Solar Phys.* **37**, 189.
Speiser, T. W.: 1970, *Planet. Space Sci.* **18**, 613.
Spicer, D. S.: 1977a, *Solar Phys.* **51**, 431.
Spicer, D. S.: 1977b, *Solar Phys.* **53**, 305.
Spitzer, L., Jr. and Härm, R.: 1953, *Phys. Rev.* **89**, 977.
Steenbeck, M., Krause, K., and Rädler, K.-H.: 1966, *Z. Naturforsch.* **21a**, 369 and 1285.
Stein, R. F. and Leibacher, J.: 1974, *Ann. Rev. Astron. Astrophys.* **12**, 407.
Stein, W. A. and Ney, E. P.: 1963, *J. Geophys. Res.* **68**, 65.
Steinberg, J. L.: 1972, *Astron. Astrophys.* **18**, 382.
Steinberg, J. L., Aubier-Giraud, M., Leblanc, Y., and Boischot, A.: 1971, *Astron. Astrophys.* **10**, 362.
Steinberg, J. L. and Caroubalos, C.: 1970, *Astron. Astrophys.* **9**, 329.
Steinberg, J. L., Caroubalos, C., and Bougeret, J. L.: 1974, *Astron. Astrophys.* **37**, 109.
Stenflo, J. O.: 1969, *Solar Phys.* **8**, 115.
Stenflo, J. O.: 1972, *Cosmic Electrodyn.* **2**, 309.
Stepanov, A. V.: 1973, *Astron. Zh.* **50**, 1243.
Stepanov, K. N.: 1958, *Zh. Eksperim. Teor. Fiz.* **35**, 1155.
Stepanov, K. N.: 1959, *Zh. Eksperim. Teor. Fiz.* **36**, 1457.
Stepanov, K. N. and Pakhomov, V. I.: 1960, *Zh. Eksperim. Teor. Fiz.* **38**, 1564.
Stepanov, A. V.: 1970, *Izv. Vys. Ucheb. Zav. Radiofiz.* **13**, 1342.
Stepnayan, A. A. and Vladimirskij, B. M.: 1961, *Astron. Zh.* **38**, 439.
Stern, D. P.: 1966, *Space Sci. Rev.* **6**, 147.
Stewart, R. T.: 1965, *Australian J. Phys.* **18**, 67.
Stewart, R. T.: 1967, *Proc. Astron. Soc. Australia* **1**, 59.
Stewart, R. T.: 1971, *Australian J. Phys.* **24**, 209.
Stewart, R. T.: 1972, *Proc. Astron. Soc. Australia* **2**, 100.
Stewart, R. T.: 1974, *Solar Phys.* **39**, 451.
Stewart, R. T.: 1975, *Solar Phys.* **40**, 417.
Stewart, R. T.: 1976, *Solar Phys.* **50**, 437.
Stewart, R. T. and Hardwick, B.: 1969, *Proc. Astron. Soc. Australia* **1**, 185.
Stewart, R. T., Howard, R. A., Hansen, F., Gergely, T., and Kundu, M.: 1974, *Solar Phys.* **36**, 219.
Stewart, R. T. and Labrum, N. R.: 1972, *Solar Phys.* **27**, 192.
Stewart, R. T., McCabe, M. K., Koomen, M. J., Hansen, R. T., and Dulk, G. A.: 1974, *Solar Phys.* **36**, 203.
Stewart, R. T. and Sheridan, K. V.: 1970, *Solar Phys.* **12**, 229.
Stewart, R. T. and Sheridan, K. V.: 1971, *Proc. Astron. Soc. Australia* **2**, 60.

Stewart, R. T., Sheridan, K. V., and Kai, K.: 1970, *Proc. Astron. Soc. Australia* **1**, 313.
Stix, T. H.: 1976, *Phys. Rev. Letters* **36**, 521.
Stone, R. G. and Fainberg, J.: 1971, *Solar Phys.* **20**, 106.
Stone, R. G. and Fainberg, J.: 1975, *Solar Phys.* **42**, 179.
Straka, R. M.: 1971, *Solar Phys.* **21**, 469.
Straka, R. M., Papagiannis, M. D., and Kogut, J.: 1975, *Solar Phys.* **45**, 131.
Strauss, F. M. and Papagiannis, M. D.: 1971, *Astrophys. J.* **164**, 369.
Sturrock, P. A.: 1961a, *Nature* **192**, 58.
Sturrock, P. A.: 1961b, in J. A. Drummond (ed.), *Plasma Physics*, McGraw Hill Book Co., p. 124.
Sturrock, P. A.: 1964, in W. N. Hess (ed.), *AAS-NASA Symposium on the Physics of Solar Flares*, NASA, Washington, p. 357.
Sturrock, P. A.: 1966, *Nature* **211**, 695.
Sturrock, P. A.: 1968, in K. O. Kiepenheuer (ed.), *Structure and Development of Solar Active Regions*, IAU Symposium No. 35, D. Reidel Publ. Co., p. 471.
Sturrock, P. A.: 1972, *Solar Phys.* **23**, 438.
Sturrock, P. A.: 1973, in R. Ramaty and R. G. Stone (ed.), *High Energy Phenomena on the Sun*, Symposium Proceedings, Goddard Space Flight Center, Greenbelt, Md., p. 3.
Sturrock, P. A.: 1974, in G. Newkirk, Jr. (ed.), *Coronal Disturbances*, IAU Symposium No. 57, D. Reidel Publ. Co., p. 437.
Sturrock, P. A. and Antiochos, S. K.: 1976, *Solar Phys.* **49**, 359.
Sturrock, P. A., Baum, P. J., Beckers, J. M., Newman, C. E., Priest, E. R., Rosenberg, H., Smith, D. F., and Wentzel, D. G. (eds.): 1976, *Solar Phys.* **46**, 411.
Sturrock, P. A. and Coppi, B.: 1964, *Nature* **204**, 4953.
Sturrock, P. A. and Coppi, B.: 1966, *Astrophys. J.* **143**, 3.
Suemoto, Z. and Hiei, E.: 1959, *Publ. Astron. Soc. Japan* **11**, 152.
Suemoto, Z., Hiei, E., and Hirayama, T.: 1962, *J. Phys. Soc. Japan* **17**, Suppl. A-2, 231.
Suzuki, I., Kawabata, K.-A., and Ogawa, H.: 1976, *Solar Phys.* **46**, 205.
Suzuki, S.: 1974, *Solar Phys.* **38**, 3.
Suzuki, S., Attwood, C. F., and Sheridan, K. V.: 1966, *IEEE Transact. Antennas Propag.* **AP-14**, 91.
Suzuki, S. and Tsuchiya, A.: 1958, *Proc. Inst. Radio Engrs.* **46**, 190.
Svalgaard, L. and Wilcox, J. M.: 1975, *Solar Phys.* **41**, 461.
Svalgaard, L. and Wilcox, J. M.: 1976, *Solar Phys.* **49**, 177.
Švestka, Z.: 1966, *Space Sci Rev.* **5**, 388.
Švestka, Z.: 1967, *Bull. Astron. Inst. Czech.* **18**, 55.
Švestka, Z.: 1968, *Solar Phys.* **4**, 18.
Švestka, Z.: 1972, *Ann. Rev. Astron. Astrophys.* **10**, 1.
Švestka, Z. (ed.): 1977, *Proc. of the Meeting 'How can Flares be Understood?', Solar Phys.* **53**, 213.
Švestka, Z. and Fritzová-Švestková, L.: 1974, *Solar Phys.* **36**, 417.
Švestka, Z. and Olmr, J.: 1966, *Bull. Astron. Inst. Czech.* **17**, 4.
Švestka, Z. and Simon, P.: 1969, *Solar Phys.* **10**, 3.
Swann, W. F. G.: 1933, *Phys. Rev.* **43**, 217 and **44**, 224.
Swanson, P. N.: 1973, *Solar Phys.* **32**, 77.
Swarup, G.: 1964, in N. Hess (ed.), *AAS-NASA Symposium on the Physics of Solar Flares*, NASA, Washington, p. 179.
Swarup, G. and Kapahi, V. K.: 1970, *Solar Phys.* **14**, 404.
Swarup, G., Stone, P. H., and Maxwell, A.: 1960, *Astrophys. J.* **131**, 725.
Sweet, P. A.: 1958, in B. Lehnert (ed.), *Electromagnetic Phenomena in Cosmical Physics*, IAU Symposium No. 6, University Press, Cambridge, p. 123.
Sweet, P. A.: 1965, in R. Lüst (ed.), *Stellar and Solar Magnetic Fields*, IAU Symposium No. 22, North-Holland Publ. Co., p. 377.
Sweet, P. A.: 1969, *Ann. Rev. Astron. Astrophys.* **7**, 140.
Swenson, G. W., Jr. and Kellermann, K. I.: 1975, *Science* **188**, 1263.
Sy, W. N.: 1973, *Proc. Astron. Soc. Australia* **2**, 215.
Sy, W. N.: 1974, *Solar Phys.* **34**, 427.
Syrovatskij, S. I.: 1962, *Astron. Zh.* **39**, 987.
Syrovatskij, S. I.: 1966a, *Astron. Zh.* **43**, 340.

Syrovatskij, S. I.: 1966b, *Zh. Eksperim. Teor. Fiz.* **50**, 1133.
Syrovatskij, S. I.: 1967, in H. van Woerden (ed.), *Radio Astronomy and the Galactic System*, IAU Symposium No. 31, North-Holland Publ. Co., p. 133.
Syrovatskij, S. I.: 1968, *Zh. Eksperim. Teor. Fiz.* **54**, 1422.
Syrovatskij, S. I.: 1969, in C. de Jager and Z. Švestka (ed.), *Solar Flares and Space Research*, North-Holland Publ. Co., p. 346.
Syrovatskij, S. I.: 1972, *Comments Astrophys. Space Phys.* **4**, 65.
Syrovatskij, S. I.: 1974, *Trudy Fiz. Inst. Lebedeva* **74**, 3.
Syrovatskij, S. I.: 1975, *Izv. Akad. Nauk SSSR, Ser. Fiz.* **39**, 359.
Syrovatskij, S. I., Frank, A. G., and Khodzaev, A. Z.: 1972, Preprint No. 142, FIAN, Moscow.
Syrovatskij, S. I. and Shmeleva, O. P.: 1972, *Astron. Zh.* **49**, 334.
Takakura, T.: 1956, *Publ. Astron. Soc. Japan* **8**, 182.
Takakura, T.: 1959, *Publ. Astron. Soc. Japan* **11**, 55 and 71.
Takakura, T.: 1960a, *Publ. Astron. Soc. Japan* **12**, 55.
Takakura, T.: 1960b, *Publ. Astron. Soc. Japan* **12**, 325 and 352.
Takakura, T.: 1961a, *Publ. Astron. Soc. Japan* **13**, 166.
Takakura, T.: 1961b, *Publ. Astron. Soc. Japan* **13**, 312.
Takakura, T.: 1962, *J. Phys. Soc. Japan* **17**, Suppl. A-2, 243.
Takakura, T.: 1963a, *Publ. Astron. Soc. Japan* **15**, 327.
Takakura, T.: 1963b, *Publ. Astron. Soc. Japan* **15**, 462.
Takakura, T.: 1964, *Publ. Astron. Soc. Japan* **16**, 230.
Takakura, T.: 1966, *Space Sci. Rev.* **5**, 80.
Takakura, T.: 1967, *Solar Phys.* **1**, 304.
Takakura, T.: 1969a, *Solar Phys.* **6**, 133.
Takakura, T.: 1969b, in C. de Jager and Z. Švestka (ed.), *Solar Flares and Space Research*, North-Holland, p. 144 and 165.
Takakura, T.: 1971a, *Solar Phys.* **19**, 186.
Takakura, T.: 1971b, in R. Howard (ed.), *Solar Magnetic Fields*, IAU Symposium No. 43, D. Reidel Publ. Co., p. 390.
Takakura, T.: 1972, *Solar Phys.* **26**, 151.
Takakura, T.: 1973, in R. Ramaty and R. G. Stone (ed.), *High Energy Phenomena on the Sun*, Symposium Proceedings, Goddard Space Flight Center, Greenbelt, Md., p. 179.
Takakura, T.: 1975, in S. R. Kane (ed.), *Solar Gamma-, X-, and EUV-Radiation*, IAU Symposium No. 68, D. Reidel Publ. Co., p. 299.
Takakura, T.: 1977, *Solar Phys.* **52**, 429.
Takakura, T. and Kai, K.: 1961, *Publ. Astron. Soc. Japan* **13**, 94.
Takakura, T. and Kai, K.: 1966, *Publ. Astron. Soc. Japan* **18**, 57.
Takakura, T., Naito, Y., and Ohki, K.: 1975, *Solar Phys.* **41**, 153.
Takakura, T. and Scalise, E., Jr.: 1970, *Solar Phys.* **11**, 434.
Takakura, T. and Shibahashi, H.: 1976, *Solar Phys.* **46**, 323.
Takakura, T., Tsuchiya, A., Morimoto, M., and Kai, K.: 1967, *Proc. Astron. Soc. Australia* **1**, 56.
Takakura, T. and Uchida, Y.: 1968a, *Astrophys. Letters* **1**, 147.
Takakura, T. and Uchida, Y.: 1968b, *Astrophys. Letters* **2**, 87.
Takakura, T., Uchida, Y., and Kai, K.: 1968, *Solar Phys.* **4**, 45.
Takakura, T. and Yousef, S.: 1974, *Solar Phys.* **36**, 451.
Takakura, T. and Yousef, S.: 1975, *Solar Phys.* **40**, 421.
Tanaka, H.: 1964, *Proc. Res. Inst. Atmosph. Nagoya Univ.* **11**, 41.
Tanaka, H.: 1975, *Solar Radio Emission*, Instruction Manual for Monthly Report, ICSU-STP-IAU WDC-C 2, Toyokawa, Japan.
Tanaka, H., Castelli, J. P., Covington, A. E., Krüger, A., Landecker, T. L., and Tlamicha, A.: 1973, *Solar Phys.* **29**, 243.
Tanaka, H. and Énomé, S.: 1970, *Nature* **225**, 435.
Tanaka, H. and Énomé, S.: 1971, *Solar Phys.* **17**, 408.
Tanaka, H. and Énomé, S.: 1975, *Solar Phys.* **40**, 123.
Tanaka, H., Énomé, S., Torii, C., Tsukiji, Y., Kobayashi, S., Ishiguro, M., and Arisawa, M.: 1970, *Proc. Res. Inst. Atmosph. Nagoya Univ.* **17**, 57.

Tanaka, H. and Kakinuma, T.: 1960, *Proc. Res. Inst. Atmosph. Nagoya Univ.* **7**, 72 and 79.
Tanaka, H. and Kakinuma, T.: 1961, *Proc. Res. Inst. Atmosph. Nagoya Univ.* **8**, 39.
Tanaka, H. and Kakinuma, T.: 1962, *J. Phys. Soc. Japan* **17**, *Suppl.* A–2, 211.
Tanaka, H. and Kakinuma, T.: 1964, *Rep. Ionosphere Space Res. Japan* **18**, 32.
Tanaka, H., Kakinuma, T., Énomé, S., Torii, C., Tsukiji, Y., and Kobayashi, S.: 1969, *Proc. Res. Inst. Atmosph. Nagoya Univ.* **16**, 113.
Tanaka, H. and Torii, C.: 1972, *Proc. Res. Inst. Atmosph. Nagoya Univ.* **19**, 101.
Tanaka, K. and Nakagawa, Y.: 1973, *Solar Phys.* **33**, 187.
Tarnstrom, G. L.: 1974a, in E. Schanda (ed.), *Proc. of the 4th Meeting of CESRA*, Inst. Appl. Phys., Univ., Berne, p. 159.
Tarnstrom, G. L.: 1974b, Report 19, AARNE Karjalainen Obs. Univ. of Oulu, Finland.
Tarnstrom, G. L. and Philip, K. W.: 1972a, *Astron. Astrophys.* **16**, 21.
Tarnstrom, G. L. and Philip, K. W.: 1972b, *Astron. Astrophys.* **17**, 267.
Tataronis, J. A. and Crawford, F. W.: 1970, *J. Plasma Phys.* **4**, 231 and 249.
Teske, R. G.: 1971, *Solar Phys.* **19**, 356.
Teske, R. G., Soyumer, T., and Hudson, H. S.: 1971, *Astrophys. J.* **165**, 615.
Thomas, R. J. and Teske, R. G.: 1971, *Solar Phys.* **16**, 431.
Thompson, A. R.: 1961, *Astrophys. J.* **133**, 643.
Thompson, A. R. and Maxwell, A.: 1962, *Astrophys. J.* **136**, 546.
Tidman, D. A.: 1965, *J. Planet. Space Sci.* **13**, 781.
Tidman, D. A., Birmingham, T. T., and Stainer, H. M.: 1966, *Astrophys. J.* **146**, 207.
Tidman, D. A. and Dupree, T. H.: 1965, *Phys. Fluids* **8**, 1860.
Timothy, A. F., Krieger, A. S., and Vaiana, G. S.: 1975, *Solar Phys.* **42**, 135.
Tindo, I. P., Ivanov, V. D., Mandelshtam, S. L., and Shurygin, A. I.: 1970, *Solar Phys.* **14**, 204.
Tindo, I. P., Ivanov, V. D., Mandelshtam, S. L., and Shurygin, A. I.: 1972a, *Solar Phys.* **24**, 429.
Tindo, I. P., Ivanov, V. D. Valniček, B., and Livshits, M. A.: 1972b, *Solar Phys.* **27**, 426.
Tindo, I. P., Mandelshtam, S. L., and Shurygin, A. I.: 1973, *Solar Phys.* **32**, 469.
Titulaer, C.: 1967, *Bull. Astron. Inst. Neth.* **19**, 82.
Tlamicha, A.: 1969, *Solar Phys.* **10**, 150.
Tlamicha, A. and Karlický, M.: 1976, *Bull. Astron. Inst. Czech.* **27**, 6.
Tlamicha, A., Olmr, J., Fürstenberg, F., and Krüger, A.: 1967, *Bull. Astron. Inst. Czech.* **18**, 70.
Tlamicha, T. and Takakura, T.: 1963, *Nature* **200**, 999.
Tolbert, C. W. and Straiton, A. W.: 1962, *J. Geophys. Res.* **67**, 1741.
Tousey, R.: 1963, *Space Sci. Rev.* **2**, 3.
Trakhtengerts, V. Yu.: 1966, *Astron. Zh.* **43**, 356; *Soviet Astron.* **10**, 281.
Trubnikov, V. A.: 1958, *Dokl. Akad. Nauk* **118**, 913.
Tsuchiya, A.: 1969, *Solar Phys.* **7**, 268.
Tsuchiya, A. and Takahashi, K.: 1968, *Solar Phys.* **3**, 346.
Tsytovich, V. N.: 1951, *Vestnik Mos. Gos. Univ.* **11**, 27.
Tsytovich, V. N.: 1966, *Usp. Fiz. Nauk* **90**, 435; *Soviet Uspekhi* **9**, 805.
Tsytovich, V. N.: 1973, *Ann. Rev. Astron. Astrophys.* **11**, 363.
Tur, T. J. and Priest, E. R.: 1976, *Solar Phys.* **48**, 89.
Twiss, R. Q.: 1954, *Phil. Mag.* **45**, 249.
Twiss, R. Q.: 1958, *Australian J. Phys.* **11**, 564.
Twiss, R. Q.: 1962, *Astrophys. J.* **136**, 438.
Twiss, R. Q. and Roberts, J. A.: 1958, *Australian J. Phys.* **11**, 424.
Uchida, Y.: 1960, *Publ. Astron. Soc. Japan* **12**, 376.
Uchida, Y.: 1968, *Solar Phys.* **4**, 30.
Uchida, Y.: 1974, *Solar Phys.* **39**, 431.
Uchida, Y., Altschuler, M. D., and Newkirk, G., Jr.: 1973, *Solar Phys.* **28**, 495.
Uchida, Y. and Kaburaki, O.: 1974, *Solar Phys.* **35**, 451.
Uchida, Y. and Sakurai, T.: 1977, *Solar Phys.* **51**, 413.
Ulich, B. L. and Haas, R. W.: 1976, *Astrophys. J., Suppl. Ser.*, **30**, 247.
Unsöld, A.: 1960, *Z. Astrophys.* **50**, 48 and 57.
Unsöld, A.: 1968, *Roy. Astron. Soc. Quart. J.* **9**, 294.
Uralov, A. M. and Nefedev, V. P.: 1976, *Astron. Zh.* **53**, 1041.

Urbarz, H.: 1967, *Z. Astrophys.* **66**, 321.
Urbarz, H. W., Fomichev, V. V., and Chertok, I. M.: 1977, *Astron. Zh.* **54**, 137.
Uscinski, B. J.: 1968, *Phil. Trans. Roy. Soc. London*, **A 262**, 609.
Vaiana, G. S., Krieger, A. S., and Timothy, A. F.: 1973, *Solar Phys.* **32**, 81.
Vainshtejn, S. I.: 1965, *Zh. Eksperim. Teor. Fiz.* **65**, 550.
Van Allen, J. A. and Krimigis, S. M.: 1965, *J. Geophys. Res.* **70**, 5737.
Van Beek, H. F., De Feiter, L. D., and De Jager, C.: 1974a, in D. E. Page (ed.), *Correlated Interplanetary and Magnetospheric Observations*, D. Reidel Publ. Co., p. 533.
Van Beek, H. F., De Feiter, L. D., and De Jager, C.: 1974b, in M. J. Rycroft and R. D. Reasenberg, (ed.), *Space Research XIV*, Akademie-Verlag, Berlin, p. 447.
Van Bueren, H. G. and Kuperus, M.: 1970, *Solar Phys.* **14**, 208.
Vandakurov, Yu. V.: 1974, *Astron. Zh.* **51**, 672; *Solnechnye Dannye, Leningrad* **5**, 72.
Van de Hulst, H. C.: 1953, in G. P. Kuiper (ed.), *The Sun*, The University of Chicago Press, p. 207.
Van Hoven, G.: 1976, *Solar Phys.* **49**, 95.
Van Hoven, G., Chiuderi, C., and Giachetti, R.: 1977, *Astrophys. J.* **213**, 869.
Van Nieuwkoop, J.: 1971, Doctoral Thesis, Univ. Utrecht.
Vasyliunas, V. M.: 1975, *Rev. Geophys. Space Phys.* **13**, 303.
Vernazza, J. E., Avrett, E. H., and Loeser, R.: 1973, *Astrophys. J.* **184**, 605.
Vernazza, J. E., Avrett, E. H., and Loeser, R.: 1976, *Astrophys. J. Suppl. Ser.* **30**, 1, 1–A2 and *Astrophys. J.* **202**, 844.
Vesecky, J. F. and Meadows, A. J.: 1969, *Solar Phys.* **6**, 80.
Veselovskij, I. S.: 1974, Isled. Kosm. Postranst., Vol. 4, VINTI Moscow, p. 7.
Vilkovskij, E. Ya.: 1973, *Astron. Zh.* **50**, 1001.
Vitkevich, V. V.: 1951, *Dokl. Akad. Nauk* **77**, 585.
Vitkevich, V. V.: 1953, *Dokl. Akad. Nauk* **91**, 1301.
Vitkevich, V. V. and Gorelova, M. V.: 1960, *Astron. Zh.* **37**, 622.
Vjalshin, G. F., Karaev, A., Krat, V. A., Křivský, L., Krüger, A., Künzel, H., and Mamedov, S.: 1967, *Solnechnye Dannye, Leningrad No. 3*, 90.
Vjalshin, G. F. and Krüger, A.: 1973, *Solnechnye Dannye, Leningrad No. 8*, 68.
Vladimirskij, B. M. and Levitskij, L. S.: 1974, *Izv. Krymsk. Astrofiz. Obs.* **51**, 74.
Vladimirskij, V. V.: 1948, *Zh. Eksperim. Teor. Fiz.* **18**, 393.
Vlasov, A. A.: 1938, *Eksperim. Teor. Fiz.* **8**, 291.
Vlasov, A. A.: 1945, *J. Phys. USSR* **9**, 25.
Vorpahl, J. A.: 1972, *Solar Phys.* **26**, 397.
Vorpahl, J. A.: 1973a, *Solar Phys.* **28**, 115.
Vorpahl, J. A.: 1973b, *Solar Phys.* **29**, 447.
Vorpahl, J. A. and Zirin, H.: 1970, *Solar Phys.* **11**, 285.
Waldmeier, M. and Müller, H.: 1950, *Z. Astrophys.* **27**, 58.
Wagner, W. J.: 1976, *Astrophys. J.* **206**, 583.
Walker, A. B. C.: 1972, *Space Sci. Rev.* **13**, 672.
Wallis, G.: 1959, in R. N. Bracewell (ed.), *Paris Symposium on Radio Astronomy*, Stanford Univ. Press, p. 595.
Warwick, C. S.: 1962, *J. Geophys. Res.* **67**, 1333.
Warwick, C. S.: 1966, *Astrophys. J.* **145**, 215.
Warwick, C. S. and Wood-Haurwitz, M.: 1962, *J. Geophys. Res.* **67**, 1317.
Warwick, J. W.: 1962, *Publ. Astron. Soc. Pacific* **74**, 302.
Warwick, J. W.: 1965a, in J. Aarons (ed.), *Solar System Radio Astronomy*, Plenum Press, New York, p. 131.
Warwick, J. W.: 1965b, *Australian J. Phys.* **18**, 167.
Warwick, J. W.: 1967, *Astrophys. J.* **150**, 1081.
Warwick, J. W. and Dulk, G. A.: 1969, *Astrophys. J. Letters* **158**, L123.
Wassenberg, W.: 1971, *Solar Phys.* **20**, 130.
Watanabe, T., Washimi, H., Kakinuma, T., Kojima, M., Maruyama, K., and Ishida, Y.: 1971, *Proc. Res. Inst. Atmosph. Nagoya Univ.*, **18**, 59.
Weber, R. R., Fitzenreiter, R. J., Novaco, J. C., and Fainberg, J.: 1977, *Solar Phys.* **54**, 431.
Wefer, F. L. and Bleiweiss, M. P.: 1976, *Solar Phys.* **48**, 77.

Weiss, A. A.: 1963a, *Australian J. Phys.* **16**, 240.
Weiss, A. A.: 1963b, *Australian J. Phys.* **16**, 526.
Weiss, A. A.: 1965, *Australian J. Phys.* **18**, 167.
Weiss, A. A. and Stewart, R. T.: 1965, *Australian J. Phys.* **18**, 143.
Welly, J. D.: 1963, *Z. Naturforsch.* **18a**, 1157.
Wende, C. D.: 1969, *J. Geophys. Res.* **74**, 4649 and 6471.
Wentzel, D. G.: 1963, *Astrophys. J.* **137**, 135.
Wentzel, D. G.: 1964, *Astrophys. J.* **140**, 1563.
Wentzel, D. G.: 1965, *J. Geophys. Res.* **70**, 2716.
Wentzel, D. G.: 1974, *Solar Phys.* **39**, 129.
Wentzel, D. G.: 1977, *Solar Phys.* **52**, 163.
Westfold, K. C.: 1959, *Astrophys. J.* **130**, 241.
White, W. A.: 1963, Preprint NASA X-614-63-193, Greenbelt, Md.
White, W. A.: 1964, Preprint NASA X-610-64-62, Greenbelt, Md.
Wibberenz, G.: 1974, *J. Geophys.* **40**, 667.
Wilcox, J. M.: 1967, *Solar Phys.* **1**, 437.
Wilcox, J. M.: 1968, *Space Sci.* **8**, 258.
Wilcox, J. M.: 1973, Stanford University Institute for Plasma Research, Report No. 544, 26.
Wilcox, J. M.: 1975, *J. Atmospheric Terrest. Phys.* **37**, 237.
Wilcox, J. M. and Howard, R.: 1968, *Solar Phys.* **5**, 564.
Wilcox, J. M. and Ness, N. F.: 1965, *J. Geophys. Res.* **70**, 5793.
Wilcox, J. M. and Ness, N. F.: 1967, *Solar Phys.* **1**, 437.
Wild, J. P.: 1950a, *Australian J. Sci. Res.* **A3**, 399.
Wild, J. P.: 1950b, *Australian J. Sci. Res.* **A3**, 541.
Wild, J. P.: 1957, in H. C. Van der Hulst (ed.), *Radio Astronomy*, IAU Symposium No. 4, University Press, Cambridge, p. 321.
Wild, J. P.: 1962, *J. Phys. Soc. Japan* **17**, Suppl. A–2, 249.
Wild, J. P.: 1963, in J. W. Evans (ed.), *The Solar Corona*, IAU Symposium No. 16, Academic Press, p. 115.
Wild, J. P.: 1964, *Nature* **203**, 1128.
Wild, J. P.: 1965, *Proc. Roy. Soc.* **A286**, 499.
Wild, J. P.: 1967, *Proc. Astron. Soc. Australia* **1**, 38.
Wild, J. P.: 1968a, Proc. of Conference on *'Plasma Instabilities in Astrophysics'* held at Asilomar, Calif., p. 119.
Wild, J. P.: 1968b, *Proc. Astron. Soc. Australia* **1**, 137.
Wild, J. P.: 1968c, *The Australian Physicist* **5**, 117.
Wild, J. P.: 1969a, *Proc. Astron. Soc. Australia* **1**, 181.
Wild, J. P.: 1969b, *Solar Phys.* **9**, 260.
Wild, J. P.: 1970a, *Australian J. Phys.* **23**, 113.
Wild, J. P.: 1970b, *Proc. Astron. Soc. Australia* **1**, 348.
Wild, J. P.: 1970c, *Proc. Astron. Soc. Australia* **1**, 365.
Wild, J. P.: 1971, Invited and Rapporteur Papers, *12th International Conference on Cosmic Rays*, Hobart, p. 29.
Wild, J. P.: 1974, *Highlights of Astronomy*, 3–19.
Wild, J. P. and Hill, E. R.: 1971, *Australian J. Phys.* **24**, 43.
Wild, J. P. and McCready, L. L.: 1950, *Australian J. Sci. Res.* **A3**, 387.
Wild, J. P., Murray, J. D., and Rowe, W. C.: 1954a, *Australian J. Phys.* **7**, 439.
Wild, J. P., Roberts, J. A., and Murray, J. D.: 1954b, *Nature* **173**, 532.
Wild, J. P. and Sheridan, K. V.: 1958, *Proc. Inst. Radio Engrs.* **46**, 160.
Wild, J. P. and Sheridan, K. V.: 1968, *Nature* **218**, 536.
Wild, J. P., Sheridan, K. V., and Kai, K.: 1968, *Nature* **218**, 536.
Wild, J. P., Sheridan, K. V., and Neylan, A. A.: 1959a, *Australian J. Phys.* **12**, 369.
Wild, J. P., Sheridan, K. V., and Trent, G. H.: 1959b, in R. N. Bracewell (ed.), *Paris Symposium on Radio Astronomy*, Stanford Univ. Press, p. 176.
Wild, J. P. and Smerd, S. F.: 1972, *Ann. Rev. Astron. Astrophys.* **10**, 159.
Wild, J. P., Smerd, S. F., and Weiss, A. A.: 1963, *Ann. Rev. Astron. Astrophys.* **1**, 291.

Wild, J. P. and Tlamicha, A.: 1965, *Nature* **203**, 1128.
Withbroe, G. L. and Vernazza, J. E.: 1976, *Solar Phys.* **50**, 127.
Wrixon, G. T. and Hogg, D. C.: 1971, *Astron. Astrophys.* **10**, 193.
Yeh, T. and Axford, W. I.: 1970, *J. Plasma Phys.* **4**, 207.
Yeh, T. and Dryer, M.: 1973, *Astrophys. J.* **182**, 301.
Yeh, T. and Pneuman, G. W.: 1977, *Solar Phys.* **54**, 419.
Yip, W. K.: 1967, *Australian J. Phys.* **20**, 421.
Yip, W. K.: 1970a, *Planetary Space Sci.* **18**, 867.
Yip, W. K.: 1970b, *Australian J. Phys.* **23**, 161.
Yip, W. K.: 1970c, *Planet. Space Sci.* **18**, 479.
Yip, W. K.: 1972, *Solar Phys.* **24**, 197.
Yip, W. K.: 1973, *Solar Phys.* **30**, 513.
Young, C. W., Spencer, C. L., Moreton, G. E., and Roberts, J. A.: 1961, *Astrophys. J.* **133**, 243.
Yudin, O. I.: 1968, *Dokl. Akad. Nauk* **180**, 821.
Yurovskij, Yu. F.: 1969a, *Izv. Krymsk. Astrofiz. Obs.* **40**, 147.
Yurovskij, Yu. F.: 1969b, *Solnechno-Zemnaya Fizika, Moscow* **1,** 15.
Yurovskij, Yu. F. and Nikolaev, N. Ya.: 1976, *Izv. Krymsk. Astrofiz. Obs.* **55**, 74.
Zaitsev, V. V.: 1965, *Astron. Zh.* **42**, 740; *Soviet Astron.* **9**, 572 (1966).
Zaitsev, V. V.: 1966, *Astron. Zh.* **43**, 1148.
Zaitsev, V. V.: 1967, *Astron. Zh.* **44**, 490.
Zaitsev, V. V.: 1968, *Astron. Zh.* **45**, 766; *Soviet Astron.* **12**, 610 (1969).
Zaitsev, V. V.: 1970, *Izv. Vys. Ucheb. Zav. Radiofiz.* **13**, 837.
Zaitsev, V. V.: 1971, *Solar Phys.* **20**, 95.
Zaitsev, V. V., Mityakov, N. A., and Rapoport, V. O.: 1972, *Solar Phys.* **24**, 444.
Zaitsev, V. V. and Stepanov, A. V.: 1974, *Solnechnye Dannye, Leningrad* No. 11, p. 71.
Zaitsev, V. V. and Stepanov, A. V.: 1975, *Astron. Astrophys.* **45**, 135.
Zanelli, C. and Zlobec, P.: 1977, *Solar Phys.* **53**, 497.
Zaumen, W. T. and Acton, L. W.: 1974, *Solar Phys.* **36**, 139.
Zelenyj, L. M.: 1977, *Fiz. Plasmy* **3**, 66.
Zheleznyakov, V. V.: 1959, *Izv. Vys. Ucheb. Zav. Radiofiz.* **2**, 858.
Zheleznyakov, V. V.: 1962, *Astron. Zh.* **39**, 5.
Zheleznyakov, V. V.: 1963, *Astron. Zh.* **40**, 829.
Zheleznyakov, V. V.: 1964, *Izv. Vys. Ucheb. Zav. Radiofiz.* **7**, 67.
Zheleznyakov, V. V.: 1965, *Astron. Zh.* **42**, 244; *Soviet Astron.* **9**, 191.
Zheleznyakov, V. V.: 1966a, *Zh. Eksperim. Teor. Fiz.* **51**, 570; *Soviet Phys. JETP* **24**, 381 (1967).
Zheleznyakov, V. V.: 1966b, *Izv. Vys. Ucheb. Zav. Radiofiz.* **9**, 1057.
Zheleznyakov, V. V.: 1967a, *Astron. Zh.* **44**, 42; *Soviet Astron.* **11**, 33.
Zheleznyakov, V. V.: 1967b, *Astrophys. J.* **148**, 849.
Zheleznyakov, V. V.: 1968, *Astrophys. Space Sci.* **2**, 417.
Zheleznyakov, V. V.: 1972, *Astrophys. Space Sci.* **15**, 24.
Zheleznyakov, V. V. and Suvorov, E. V.: 1968, *Zh. Eksperim. Teor. Fiz.* **54**, 627.
Zheleznyakov, V. V. and Suvorov, E. V.: 1972, *Astrophys. Space Sci.* **15**, 24.
Zheleznyakov, V. V., Suvorov, E. V., and Saposhnikov, V. E.: 1974, *Astron. Zh.* **51**, 243.
Zheleznyakov, V. V. and Trakhtengerts, V. Yu.: 1965, *Astron. Zh.* **42**, 1005.
Zheleznyakov, V. V. and Zaitsev, V. V.: 1970, *Astron. Zh.* **47**, 60 and 308.
Zheleznyakov, V. V. and Zlotnik, E. Y.: 1971, *Solar Phys.* **20**, 85.
Zheleznyakov, V. V. and Zlotnik, E. Ya.: 1974, *Solar Phys.* **36**, 443.
Zheleznyakov, V. V. and Zlotnik, E. Ya.: 1975, *Solar Phys.* **43**, 431 and **44**, 447 and 461.
Zimmermann, G.: 1971, *Astron. Astrophys.* **15**, 433.
Zirin, H.: 1972, *Solar Phys.* **22**, 34.
Zirin, H. and Tanaka, K.: 1973, *Solar Phys.* **32**, 173.
Zlobec, P.: 1972, in J. Delannoy and F. Pomeyrol (ed.), *Proc. of the 3rd Meeting of CESRA*, Obs. Univ. Bordeaux, Floriac, p. 151.
Zlobec, P.: 1975, *Solar Phys.* **43**, 453.
Zlotnik, E. Ya.: 1968, *Astron. Zh.* **45**, 310 and 585.

LIST OF SYMBOLS

$\mathbf{A}$	vector potential
$A(\alpha)$	aerial pattern
A_e	effective area
A_p	physical aperture
$\mathbf{B}, \mathbf{H}$	magnetic field vector
$B_\nu(T)$	Kirchhoff–Planck function
B_0	heliographic latitude
c	velocity of light in vacuum
D	directivity
$\mathbf{D}$	electric induction
e	elementary charge (electron)
e_α	charge of a particle of sort α
E	energy (particles)
$\mathbf{E}$	electric field vector
$f(v)$	(velocity) distribution function
$F(E)$	isotropic energy distribution
$g_{r,s}$	statistical weight
g	Gaunt factor
$g(\alpha, \beta)$	aerial gain
G	receiver gain
G_j	(longitudinal) wave polarization coefficient
h	height, scale height
$h = 2\pi h$	Planck's constant
i	inclination
i	imaginary unit
I_ν, I_ω	intensity (brightness) of radiation
I, Q, U, V	Stokes parameters
I_{iq}	polarization tensor
j (subscript)	number of wave mode
$\mathbf{j}$	current density
$\mathbf{k}, \mathbf{k}_j$	wave vector ($k = \|\mathbf{k}\|$ – wave number)
k_D	Debye wave number
K	Boltzmann's constant
l_D	Debye length
L	characteristic length
L_0	heliographic longitude

m, m_e	electron mass
m_0	rest mass of electron
m_α	mass of a particle of sort α
M	Mach number
$(M) = M_{mn}$	instrumental matrix
n, n_j	refractive index
N	number density, e.g.:
	N_e, N_i electron, ion density
	N_k plasmon density
N_R	noise factor
p	axial ratio of the polarization ellipse
$\mathbf{p}$	momentum
P_g	gas pressure
P_k	kinetic pressure
P_m	magnetic pressure
P	beam pattern
P	ray parameter
P, p	power
	P_R receiver noise power
$\mathbf{P}$	dielectric polarization vector
Q_ν, Q_ω	electron emissivity
R, r, ϑ, φ	spherical coordinates
r	degree of ionization
r_H	gyroradius
R	radius ($R_\odot$ – solar radius)
R	sunspot number
R	directivity factor
R_j	(transverse) wave polarization coefficient
s	ray path length
s	harmonic number
S_ν, S_ω	flux density
t	time
	t_c self-collision time
	t_d duration, dissipation time
	t_D deflection time
	t_{eq} equipartition time
	t_E energy exchange time
	t_s slowing-down time
T	temperature
	T_a effective aerial temperature
	T_b brightness temperature
	T_e, T_{eff} apparent disk temperature, effective temperature
	T_e electron temperature
	T_i ion temperature
	T_R noise temperature
	T_u temperature of undisturbed Plasma

T	period, e.g.
	T_{syn} synodic rotation period
$(T) = T_{mn}$	transmission matrix
u_r	partition function of r-times ionized atoms
u_ν, u_ω	spectral energy of waves
u_σ, u_k	transition (emission) probabilities
$v = \|\mathbf{v}\|, w = \|\mathbf{w}\|$	velocity
	v_A, v_a Alfvén wave velocity
	v_e electron speed, exciter speed
	v_i ion speed
	v_g group velocity
	v_k phase velocity
	v_T, v_{th} thermal velocity
	v_s sound velocity
	v_{II} type II burst exciter velocity
	$v_{\|\|}$ $v \cos\theta$
	$v_\perp$ $v \sin\theta$
V	volume
W_j	energy density of a wave mode j
$W_{k,j}, W_{\omega,j}$	spectral densities of a wave mode j
x, y, z	Cartesian coordinates
$X = (\omega_p/\omega)^2$, $Y = \omega_H/\omega$, $Z = \nu_{\mathrm{coll}}/\omega$	magneto-ionic parameters
$Y_T = Y \sin\theta$	
$Y_L = Y \cos\theta$	
Y	emission measure
z	charge number
α	force-free field parameter
α, β	position angles
$\alpha_{\nu,j}, \alpha_{\omega,j}$	absorption coefficient
$\alpha_{\sigma\to j}$	wave mode conversion coefficient
α(subscript)	sort of particles
β	v/c $(\beta_T = v_T/c)$
β_σ	wave mode transformation coefficient
γ	exponent of power-law distributions
γ	polytropic exponent
γ	growth rate of instabilities
γ^*	Lorentz factor
γ^+	Euler's constant
$\delta(x)$	delta function
ε	photon energy
ε	electric permittivity, ε_{il} – dielectric tensor
ε_A	aperture efficiency
ε_B	beam efficiency
ε_L	loss efficiency

ζ	angular velocity
η	electric resistivity
η_ν, η_ω	(volume) emissivity, η_k – emissivity in the wave vector space
ϑ	$\sphericalangle(\mathbf{k}_j, \mathbf{v}_{g,j})$.
θ	$\sphericalangle(\mathbf{B}, \mathbf{k}_j)$
θ	stereo angle
θ	step function
κ	thermal conductivity
λ	wavelength
λ	heliographic longitude
λ_f	mean free path
Λ	Coulomb logarithm for electron-ion collisions
μ	magnetic moment
μ_j	real part of the refractive index
ν	wave frequency
ν_{coll}	collision frequency
ν_H	gyrofrequency
ν_p	plasma frequency
ν_c	critical frequency of synchrotron radiation
ρ	degree of polarization
	ρ_l degree of linear polarization
	ρ_c degree of circular polarization
ρ	mass density
ρ	charge density
σ	electric conductivity
$\sigma = \cot p$	polarization parameter
τ_ν, τ_ω	optical depth
φ	angle of incidence $= \sphericalangle(\mathbf{k}_j, \mathbf{r})$
ϕ	$\sphericalangle(\mathbf{v}_e, \mathbf{B})$ (pitch angle)
ϕ	heliographic latitude
ϕ	phase angle
Φ	(scalar) potential
χ	orientation angle
$\chi_{r,s}$	ionization energy
χ_j	absorption index
ψ	$\sphericalangle(\mathbf{v}_e, \mathbf{k}_j)$
$\omega = 2\pi\nu$	angular wave frequency
	ω_{uh} upper hybrid frequency
	ω_{pe} electron plasma frequency
	ω_{He} electron gyrofrequency
Ω	longitude of the ascending mode
Ω	solid angle
	Ω_A beam solid angle
	Ω_M main beam solid angle
	$\Omega_\odot$ solid angle of the optical Sun seen from Earth.

SUBJECT INDEX

GEOPHYSICS AND ASTROPHYSICS MONOGRAPHS

AN INTERNATIONAL SERIES OF FUNDAMENTAL TEXTBOOKS

1. R. Grant Athay, *Radiation Transport in Spectral Lines.*
2. J. Coulomb, *Sea Floor Spreading and Continental Drift.*
3. G. T. Csanady, *Turbulent Diffusion in the Environment.*
4. F. E. Roach and Janet L. Gordon, *The Light of the Night Sky.*
5. H. Alfvén and G. Arrhenius, *Structure and Evolutionary History of the Solar System.*
6. J. Iribarne and W. Godson, *Atmospheric Thermodynamics.*
7. Z. Kopal, *The Moon in the Post-Apollo Era.*
8. Z. Švestka, *Solar Flares.*
9. A. Vallance Jones, *Aurora.*
10. C.-J. Allègre and G. Michard, *Introduction to Geochemistry.*
11. J. Kleczek, *The Universe.*
12. E. Tandberg-Hanssen, *Solar Prominences.*
13. A. Giraud and M. Petit, *Ionospheric Techniques and Phenomena.*
14. E. L. Chupp, *Gamma Ray Astronomy.*
15. W. D. Heintz, *Double Stars.*
16. A. Krüger, *Introduction to Solar Radio Astronomy and Radio Physics.*
17. V. Kourganoff, *Introduction to Advanced Astrophysics.*
18. J. Audouze and S. Vauclaire, *An Introduction to Nuclear Astrophysics.*